YANMAR

INBOARD DIESEL
1975-98 REPAIR MANUAL
GM, GM/HM, JH AND JH2 SERIES

SELOC™

President	Dean F. Morgantini, S.A.E.
Vice President–Finance	Barry L. Beck
Vice President–Sales	Glenn D. Potere
Executive Editor	Kevin M. G. Maher, A.S.E.
Manager–Marine/Recreation	James R. Marotta, A.S.E., S.T.S.
Manager–Professional	George B. Heinrich III, A.S.E., S.A.E.
Manager–Consumer	Richard Schwartz, A.S.E.
Manager–Production	Ben Greisler, S.A.E.
Production Assistant	Melinda Possinger
Project Managers	Will Kessler, A.S.E., S.A.E., Todd W. Stidham, A.S.E., Ron Webb
Schematics Editor	Christopher G. Ritchie, A.S.E.
Editor	Christopher Bishop, A.S.E.

Brought to you by

CHILTON™ MARINE

Manufactured in USA
© 1999 W. G. Nichols
1020 Andrew Drive
West Chester, PA 19380
ISBN 0-89330-049-7
1234567890 8765432109

www.Chiltononline.com

Contents

B3-27-00

Contents

**ENGINE AND
ENGINE OVERHAUL**   **7**

TRANSMISSIONS   **8**

**POWERTRAIN
INSTALLATION AND
ALIGMENT**   **9**

REMOTE CONTROLS   **10**

GLOSSARY

MASTER INDEX

Other titles
Brought to you by

CHILTON™
MARINE

Title	Part #
Chrysler Outboards, All Engines, 1962-84	018-7
Force Outboards, All Engines, 1984-96	024-1
Honda Outboards, All Engines 1988-98	1200
Johnson/Evinrude Outboards, 1-2 Cyl, 1956-70	007-1
Johnson/Evinrude Outboards, 1-2 Cyl, 1971-89	008-X
Johnson/Evinrude Outboards, 1-2 Cyl, 1990-95	026-8
Johnson/Evinrude Outboards, 3, 4 & 6 Cyl, 1973-91	010-1
Johnson/Evinrude Outboards, 3-4 Cyl, 1958-72	009-8
Johnson/Evinrude Outboards, 4, 6 & 8 Cyl, 1992-96	040-3
Kawasaki Personal Watercraft, 1973-91	032-2
Kawasaki Personal Watercraft, 1992-97	042-X
Marine Jet Drive, 1961-96	029-2
Mariner Outboards, 1-2 Cyl, 1977-89	015-2
Mariner Outboards, 3, 4 & 6 Cyl, 1977-89	016-0
Mercruiser Stern Drive, Type I, Alpha/MR, Bravo I & II, 1964-92	005-5
Mercruiser Stern Drive, Alpha I (Generation II), 1992-96	039-X
Mercruiser Stern Drive, Bravo I, II & III, 1992-96	046-2
Mercury Outboards, 1-2 Cyl, 1965-91	012-8
Mercury Outboards, 3-4 Cyl, 1965-92	013-6
Mercury Outboards, 6 Cyl, 1965-91	014-4
Mercury/Mariner Outboards, 1-2 Cyl, 1990-94	035-7
Mercury/Mariner Outboards, 3-4 Cyl, 1990-94	036-5
Mercury/Mariner Outboards, 6 Cyl, 1990-94	037-3
Mercury/Mariner Outboards, All Engines, 1995-99	1416
OMC Cobra Stern Drive, Cobra, King Cobra, Cobra SX, 1985-95	025-X
OMC Stern Drive, 1964-86	004-7
Polaris Personal Watercraft, 1992-97	045-4
Sea Doo/Bombardier Personal Watercraft, 1988-91	033-0
Sea Doo/Bombardier Personal Watercraft, 1992-97	043-8
Suzuki Outboards, All Engines, 1985-99	1600
Volvo/Penta Stern Drives 1968-91	011-X
Volvo/Penta Stern Drives, Volvo Engines, 1992-93	038-1
Volvo/Penta Stern Drives, GM and Ford Engines, 1992-95	041-1
Yamaha Outboards, 1-2 Cyl, 1984-91	021-7
Yamaha Outboards, 3 Cyl, 1984-91	022-5
Yamaha Outboards, 4 & 6 Cyl, 1984-91	023-3
Yamaha Outboards, All Engines, 1992-98	1706
Yamaha Personal Watercraft, 1987-91	034-9
Yamaha Personal Watercraft, 1992-97	044-6
Yanmar Inboard Diesels, 1975-98	7400

SAFETY NOTICE

Proper service and repair procedures are vital to the safe, reliable operation of all marine engines, as well as the personal safety of those performing repairs. This manual outlines procedures for servicing and repairing inboard diesel engines using safe, effective methods. The procedures contain many NOTES, CAUTIONS and WARNINGS which should be followed, along with standard procedures, to eliminate the possibility of personal injury or improper service which could damage the vessel or compromise its safety.

It is important to note that repair procedures and techniques, tools and parts for servicing marine engines, as well as the skill and experience of the individual performing the work, vary widely. It is not possible to anticipate all of the conceivable ways or conditions under which these engines may be serviced, or to provide cautions as to all possible hazards that may result. Standard and accepted safety precautions and equipment should be used during cutting, grinding, chiseling, prying, or any other process that can cause material removal or projectiles.

Some procedures require the use of tools specially designed for a specific purpose. Before substituting another tool or procedure, you must be completely satisfied that neither your personal safety, nor the performance of the marine engine, will be compromised.

Although information in this manual is based on industry sources and is complete as possible at the time of publication, the possibility exists that some vehicle manufacturers made later changes which could not be included here. While striving for total accuracy, Chilton Marine cannot assume responsibility for any errors, changes or omissions that may occur in the compilation of this data.

PART NUMBERS

Part numbers listed in this reference are not recommendations by Chilton Marine for any product brand name. They are references that can be used with interchange manuals and aftermarket supplier catalogs to locate each brand supplier's discrete part number.

SPECIAL TOOLS

Special tools are recommended by the marine manufacturer to perform a specific task. Use has been kept to a minimum, but, where absolutely necessary, they are referred to in the text by the part number of the tool manufacturer. These tools can be purchased, under the appropriate part number, from your local dealer or regional distributor, or an equivalent tool can be purchased locally from a tool supplier or parts outlet. Before substituting any tool for the one recommended, read the SAFETY NOTICE at the top of this page.

ALL RIGHTS RESERVED

ACKNOWLEDGMENTS

Chilton Marine expresses sincere appreciation to the following companies who supported the production of this manual by providing information, products and general assistance:

- Yanmar Diesel America Corporation—Buffalo Grove, IL

- Mack Boring and Parts Company—Union, NJ

- Engine City Technical Institute—Union, NJ

Special thanks to Tom Watson and David Fox, for providing the literature to get this book started; to Steve McGovern, for inviting us into your fine facility and providing us with teardown engines; to Larry Berlin, for teaching us everything you know about marine diesels (Yes, Larry, you can leave your lab coat on for the photograph); and to Ed Reed for your overall guidance in the finer points of marine engine repair.

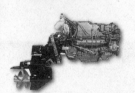

1

GENERAL INFORMATION AND BOATING SAFETY

HOW TO USE THIS MANUAL

This manual is designed to be a handy reference guide to maintaining and repairing your Yanmar diesel marine engine. We strongly believe that regardless of how many or how few years experience you may have, there is something new waiting here for you.

This manual covers the topics that a factory service manual (designed for factory trained technicians) and a manufacturer owner's manual (designed more by lawyers these days) covers. This manual will take you through the basics of maintaining and repairing your marine engine, step-by-step, to help you understand what the factory trained technicians already know by heart. By using the information in this manual, any boat owner should be able to make better informed decisions about what he or she needs to do to maintain and enjoy his/her boat.

Even if you never plan on touching a wrench (and if so, we hope that you will change your mind), this manual will still help you understand what a technician needs to do in order to maintain your engine.

Can You Do It?

If you are not the type who is prone to taking a wrench to something, NEVER FEAR. The procedures in this manual cover basic topics at a level virtually anyone will be able to handle. And just the fact that you purchased this manual shows your interest in better understanding your marine engine.

You may find that maintaining your marine engine yourself is preferable in most cases. From a monetary standpoint, it could also be beneficial. The money spent on hauling your boat to a marina and paying a tech to service the engine could buy you fuel for a whole weekend's sailing. If you are unsure of your own mechanical abilities, at the very least you should fully understand what a marine mechanic does to your boat. You may decide that anything other than maintenance and adjustments should be performed by a technician (and that's your call), but know that every time you board your vessel, you are placing faith in the mechanic's work and trusting him or her with your well-being, and maybe your life.

It should also be noted that in most areas a factory trained technician will charge an hourly rate from the time he leaves his shop to the time he returns home. The cost savings in doing the job yourself should be readily apparent at this point.

Where to Begin

Before spending any money on parts, and before removing any nuts or bolts, read through the entire procedure or topic. This will give you the overall view of what tools and supplies will be required to perform the procedure or what questions need to be answered before purchasing parts. So read ahead and plan ahead. Each operation should be approached logically and all procedures thoroughly understood before attempting any work.

Avoiding Trouble

Some procedures in this manual may require you to "label and disconnect . . ." a group of lines, hoses or wires. Don't be lulled into thinking you can remember where everything goes — you won't. If you reconnect or install a part incorrectly, things may operate poorly, if at all. If you hook up electrical wiring incorrectly, you may instantly learn a very, very expensive lesson.

A piece of masking tape, for example, placed on a hose and another on its fitting will allow you to assign your own label such as the letter "A", or a short name. As long as you remember your own code, the lines can be reconnected by matching letters or names. Do remember that tape will dissolve when saturated in fluids. If a component is to be washed or cleaned, use another method of identification. A permanent felt-tipped marker can be very handy for marking metal parts; but remember that fluids will remove permanent marker.

SAFETY is the most important thing to remember when performing maintenance or repairs. Be sure to read the information on safety in this manual.

Maintenance or Repair?

▶ See Figure 1

Proper maintenance is the key to long and trouble-free engine life, and the work can yield its own rewards. A properly maintained engine performs better than one that is neglected. As a conscientious boat owner, set aside a Saturday morning, at least once a month, to perform a thorough check of items which could cause problems. Keep your own personal log to jot down which services you performed, how much the parts cost you, the date, and the amount of hours on the engine at the time. Keep all receipts for parts purchased, so that they may be referred to in case of related problems or to determine operating expenses. As a do-it-yourselfer, these receipts are the only proof you have that the required maintenance was performed. In the event of a warranty problem, these receipts will be invaluable.

It's necessary to mention the difference between maintenance and repair. Maintenance includes routine inspections, adjustments, and

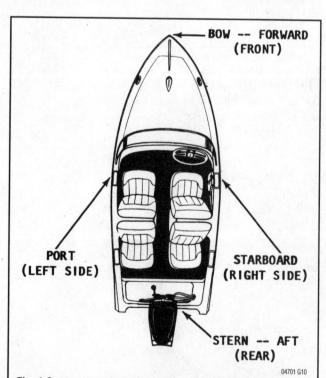

Fig. 1 Common terminology used for reference designation on boats of all size. These terms are used though out the manual

04701 G10

replacement of parts that show signs of normal wear. Maintenance compensates for wear or deterioration. Repair implies that something has broken or is not working. A need for repair is often caused by lack of maintenance.

For example: draining and refilling the engine oil is maintenance recommended by all manufacturers at specific intervals. Failure to do this can allow internal corrosion or damage and impair the operation of the engine, requiring expensive repairs. While no maintenance program can prevent items from breaking or wearing out, a general rule can be stated: MAINTENANCE IS CHEAPER THAN REPAIR.

Two basic rules should be mentioned here. First, whenever the Port side of the engine is referred to, it is meant to specify the left side of the engine when you are sitting at the helm. Conversely, the Starboard means your right side. Most screws and bolts are removed by turning counterclockwise, and tightened by turning clockwise. An easy way to remember this is: righty-tighty; lefty-loosey. Corny, but effective. And if you are really dense (and we have all been so at one time or another), buy a ratchet that is marked ON and OFF, or mark your own.

Professional Help

Occasionally, there are some things when working on a marine engine that are beyond the capabilities or tools of the average Do-It-Yourselfer (DIYer). This shouldn't include most of the topics of this manual, but you will have to be the judge. Some engines require special tools or a selection of special parts, even for basic maintenance.

Talk to other boaters who use the same model of engine and speak with a trusted marina to find if there is a particular system or component on your engine that is difficult to maintain. For example, although the technique of valve adjustment on some engines may be easily understood and even performed by a DIYer, it might require a handy assortment of shims in various sizes and a few hours of disassembly to get to that point. Not having the assortment of shims handy might mean multiple trips back and forth to the parts store, and this might not be worth your time.

You will have to decide for yourself where basic maintenance ends and where professional service should begin. Take your time and do your research first (starting with the information in this manual) and then make your own decision. If you really don't feel comfortable with attempting a procedure, DON'T DO IT. If you've gotten into something that may be over your head, don't panic. Tuck your tail between your legs and call a marine mechanic. Marinas and independent shops will be able to finish a job for you. Your ego may be damaged, but your boat will be properly restored to its full running order. So, as long as you approach jobs slowly and carefully, you really have nothing to lose and everything to gain by doing it yourself.

Purchasing Parts

▶ See Figures 2, 3 and 4

When purchasing parts there are two things to consider. The first is quality and the second is to be sure to get the correct part for your engine. To get quality parts, always deal directly with a reputable retailer. To get the proper parts always refer to the information tag on your engine prior to calling the parts counter. An incorrect part can adversely affect your engine performance and fuel economy, and will cost you more money and aggravation in the end.

Fig. 2 By far the most important asset in purchasing parts is a knowledgeable and enthusiastic parts person

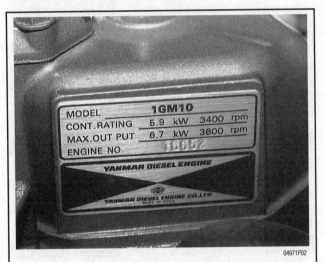
Fig. 3 Always refer to the numbers on your engine identification tag when ordering parts

Fig. 4 Parts catalogs, giving application and part number information, are provided by manufacturers for most replacement parts

Just remember, a tow back to shore will cost you an hourly rate. That is an hourly rate from the time the towboat leaves his home port, to the time he returns to his home port. Get the picture?

So who should you call for parts? Well, there are many sources for the parts you will need. Where you shop for parts will be determined by what kind of parts you need, how much you want to pay, and the types of stores in your neighborhood.

Your marina can supply you with many of the common parts you require. Using a marina for as your parts supplier may be hand because of location (just walk right down the dock) or because the marina specializes in your particular brand of engine. In addition, it is always a good idea to get to know the marina staff (especially the marine technician).

The marine parts jobber, who is usually listed in the yellow pages or whose name can be obtained from the marina, is another excellent source for parts. In addition to supplying local marinas, he also does a sizeable business in over-the-counter parts sales for the do-it-yourselfer.

Almost every community has one or more convenient marine chain stores. These stores often offer the best retail prices and the convenience of one-stop shopping for all your needs. Since they cater to the do-it-yourselfer, these stores are almost always open weeknights, Saturdays, and Sundays, when the jobbers are usually closed.

The lowest prices for parts are most often found in discount stores or the auto department of mass merchandisers. Parts sold here are name and private brand parts bought in huge quantities, so they can offer a competitive price. Private brand parts are made by major manufacturers and sold to large chains under a store label.

Avoiding the Most Common Mistakes

There are 3 common mistakes in mechanical work:

1. Incorrect order of assembly, disassembly or adjustment. When taking something apart or putting it together, performing steps in the wrong order usually just costs you extra time; however, it CAN break something. Read the entire procedure before beginning disassembly. Perform everything in the order in which the instructions say you should, even if you can't immediately see a reason for it. When you're taking apart something that is very intricate, you might want to draw a picture of how it looks when assembled at one point in order to make sure you get everything back in its proper position. When making adjustments, perform them in the proper order; often, one adjustment affects another, and you cannot expect satisfactory results unless each adjustment is made only when it cannot be changed by any other.

2. Overtorquing (or undertorquing). While it is more common for overtorquing to cause damage, undertorquing may allow a fastener to vibrate loose causing serious damage. Especially when dealing with aluminum parts, pay attention to torque specifications and utilize a torque wrench in assembly. If a torque figure is not available, remember that if you are using the right tool to perform the job, you will probably not have to strain yourself to get a fastener tight enough. The pitch of most threads is so slight that the tension you put on the wrench will be multiplied many times in actual force on what you are tightening.

3. Crossthreading. This occurs when a part such as a bolt is screwed into a nut or casting at the wrong angle and forced. Crossthreading is more likely to occur if access is difficult. It helps to clean and lubricate fasteners, then to start threading with the part to be installed positioned straight in. Always start a fastener, etc. with your fingers. If you encounter resistance, unscrew the part and start over again at a different angle until it can be inserted and turned several times without much effort. Keep in mind that some parts may have tapered threads, so that gentle turning will automatically bring the part you're threading to the proper angle, but only if you don't force it or resist a change in angle. Don't put a wrench on the part until it has been tightened a couple of turns by hand. If you suddenly encounter resistance, and the part has not seated fully, don't force it. Pull it back out to make sure it's clean and threading properly.

BOATING SAFETY

In 1971 Congress ordered the U.S. Coast Guard to improve recreational boating safety. In response, the Coast Guard drew up a set of regulations.

Beside these federal regulations, there are state and local laws you must follow. These sometimes exceed the Coast Guard requirements. This section discusses only the federal laws. State and local laws are available from your local Coast Guard. As with other laws, "Ignorance of the boating laws is no excuse." The rules fall into two groups: regulations for your boat and required safety equipment on your boat.

Regulations For Your Boat

Most boats on waters within Federal jurisdiction must be registered or documented. These waters are those that provide a means of transportation between two or more states or to the sea. They also include the territorial waters of the United States.

DOCUMENTING OF VESSELS

A vessel of five or more net tons may be documented as a yacht. In this process, papers are issued by the U.S. Coast Guard as they are for large ships. Documentation is a form of national registration. The boat must be used solely for pleasure. Its owner must be a U.S. citizen, a partnership of U.S. citizens, or a corporation controlled by U.S. citizens. The captain and other officers must also be U.S. citizens. The crew need not be.

If you document your yacht, you have the legal authority to fly the yacht ensign. You also may record bills of sale, mortgages, and other papers of title with federal authorities. Doing so gives legal notice that such instruments exist. Documentation also permits preferred status for mortgages. This gives you additional security and aids financing and transfer of title. You must carry the original documentation papers aboard your vessel. Copies will not suffice.

REGISTRATION OF BOATS

If your boat is not documented, registration in the state of its principal use is probably required. If you use it mainly on an ocean, a gulf, or other similar water, register it in the state where you moor it.

If you use your boat solely for racing, it may be exempt from the requirement in your state. States may also exclude dinghies. Some require registration of documented vessels and non-power driven boats.

All states, except Alaska, register boats. In Alaska, the U.S. Coast Guard issues the registration numbers. If you move your vessel to a new state of principal use, a valid registration certificate is good for 60 days. You must have the registration certificate (certificate of number) aboard your vessel when it is in use. A copy will not suffice. You may be cited if you do not have the original on board.

NUMBERING OF VESSELS

A registration number is on your registration certificate. You must paint or permanently attach this number to both sides of the forward half of your boat. Do not display any other number there.

The registration number must be clearly visible. It must not be placed on the obscured underside of a flared bow. If you can't place the number on the bow, place it on the forward half of the hull. If that doesn't work, put it on the superstructure. Put the number for an inflatable boat on a bracket or fixture. Then, firmly attach it to the forward half of the boat. The letters and numbers must be plain block characters and must read from left to right. Use a space or a hyphen to separate the prefix and suffix letters from the numerals. The color of the characters must contrast with that of the background, and they must be at least three inches high.

In some states your registration is good for only one year. In others, it is good for as long as three years. Renew your registration before it expires. At that time you will receive a new decal or decals. Place them as required by state law. You should remove old decals before putting on the new ones. Some states require that you show only the current decal or decals. If your vessel is moored, it must have a current decal even if it is not in use.

If your vessel is lost, destroyed, abandoned, stolen, or transferred, you must inform the issuing authority. If you lose your certificate of number or your address changes, notify the issuing authority as soon as possible.

SALES & TRANSFERS

Your registration number is not transferable to another boat. The number stays with the boat unless its state of principal use is changed.

HULL IDENTIFICATION NUMBER

A Hull Identification Number (HIN) is like the Vehicle Identification Number (VIN) on your car. Boats built between November 1, 1972 and July 31, 1984 have old format HINs. Since August 1, 1984 a new format has been used. Your boat's HIN must appear in two places. If it has a transom, the primary number is on its starboard side within two inches of its top. If it does not have a transom or if it was not practical to use the transom, the number is on the starboard side. In this case, it must be within one foot of the stern and within two inches of the top of the hull side. On pontoon boats, it is on the aft crossbeam within one foot of the starboard hull attachment. Your boat also has a duplicate number in an unexposed location. This is on the boat's interior or under a fitting or item of hardware.

LENGTH OF BOATS

▶ **See Figure 5**

For some purposes, boats are classed by length. Required equipment, for example, differs with boat size. Manufacturers may measure a boat's length in several ways. Officially, though, your boat is measured along a straight line from its bow to its stern. This line is parallel to its keel.

The length does not include bowsprits, boomkins, or pulpits. Nor does it include rudders, brackets, outboard motors, outdrives, diving platforms, or other attachments.

CAPACITY INFORMATION

▶ **See Figure 5**

Manufacturers must put capacity plates on most recreational boats less than 20 feet long. Sailboats, canoes, kayaks, and inflatable boats are usually exempt. Outboard boats must display the maximum permitted horsepower of their engines. The plates must also show the allowable maximum weights of the people on board. And they must show the allowable maximum combined weights of people, engines, and gear. Inboards and stern drives need not show the weight of their engines on their capacity plates. The capacity plate must appear where it is clearly visible to the operator when underway. This information serves to remind you of the capacity of your boat under normal circumstances. You should ask yourself, "Is my boat loaded above its recommended capacity" and, "Is my boat overloaded for the present sea and wind conditions?" If you are stopped by a legal authority, you may be cited if you are overloaded.

Fig. 5 A U.S. Coast Guard certification plate indicates the amount of occupants and gear appropriate for safe operation of the vessel

CERTIFICATE OF COMPLIANCE

Manufacturers are required to put compliance plates on motorboats greater than 20 feet in length. The plates must say, "This boat," or "This equipment complies with the U. S. Coast Guard Safety Standards in effect on the date of certification." Letters and numbers can be no less than one-eighth of an inch high. At the manufacturer's option, the capacity and compliance plates may be combined.

VENTILATION

A cup of gasoline spilled in the bilge has the potential explosive power of 15 sticks of dynamite. This statement, commonly quoted over 20 years ago, may be an exaggeration. However, it illustrates a fact. Gasoline fumes in the bilge of a boat are highly explosive and a serious danger. They are heavier than air and will stay in the bilge until they are vented out.

Because of this danger, Coast Guard regulations require ventilation on many power boats. There are several ways to supply fresh air to engine and gasoline tank compartments and to remove dangerous vapors. Whatever the choice, it must meet Coast Guard standards. The following discussion does not deal with all of the regulations nor does it cover all recreational boats. It deals only with boats built after July 31, 1980, vessels made or used non-commercially, vessels leased, rented, or chartered for noncommercial use, and boats carrying six or fewer passengers for hire.

The following is not intended to be a complete discussion of the regulations. It is limited to the majority of recreational vessels. Contact your local Coast Guard office for further information.

General Precautions

Ventilation systems will not remove raw gasoline that leaks from tanks or fuel lines. If you smell gasoline fumes, you need immediate repairs. The best device for sensing gasoline fumes is your nose. Use it! If you smell gasoline in an engine compartment or elsewhere, don't start your engine. The smaller the compartment, the less gasoline it takes to make an explosive mixture.

Ventilation for Open Boats

In open boats, gasoline vapors are dispersed by the air that moves through them. So they are exempt from ventilation requirements.

To be "open," a boat must meet certain conditions. Engine and fuel tank compartments and long narrow compartments that join them must be open to the atmosphere." This means they must have at least 15 square inches of open area for each cubic foot of net compartment volume. The open area must be in direct contact with the atmosphere. There must also be no long, unventilated spaces open to engine and fuel tank compartments into which flames could extend.

Ventilation for All Other Boats

Powered and natural ventilation are required in an enclosed compartment with a permanently installed gasoline engine that has a cranking motor. A compartment is exempt if its engine is open to the atmosphere. Diesel powered boats are also exempt.

VENTILATION SYSTEMS

There are two types of ventilation systems. One is "natural ventilation." In it, air circulates through closed spaces due to the boat's motion. The other type is "powered ventilation." In it, air is circulated by a motor driven fan or fans.

Natural Ventilation System Requirements

A natural ventilation system has an air supply from outside the boat. The air supply may also be from a ventilated compartment or a compartment open to the atmosphere. Intake openings are required. In addition, intake ducts may be required to direct the air to appropriate compartments.

The system must also have an exhaust duct that starts in the lower third of the compartment. The exhaust opening must be into another ventilated compartment or into the atmosphere. Each supply opening and supply duct, if there is one, must be above the usual level of water in the bilge. Exhaust openings and ducts must also be above the bilge water. Openings and ducts must be at least three square inches in area or two inches in diameter. Openings should be placed so exhaust gasses do not enter the fresh air intake. Exhaust fumes must not enter cabins or other enclosed, non-ventilated spaces. The carbon monoxide gas in them is deadly.

Intake and exhaust openings must be covered by cowls or similar devices. These registers keep out rain water and water from breaking seas. Most often, intake registers face forward and exhaust openings aft. This aids the flow of air when the boat is moving or at anchor since most boats face into the wind when anchored.

Power Ventilation System Requirements

Powered ventilation systems must meet the standards of a natural system. They must also have one or more exhaust blowers. The

blower duct can serve as the exhaust duct for natural ventilation if fan blades do not obstruct the air flow when not powered. Openings in engine compartment, for carburetion are in addition to ventilation system requirements.

Required Safety Equipment

Coast Guard regulations require that your boat have certain equipment aboard. These requirements are minimums. Exceed them whenever you can.

TYPES OF FIRES

Fire extinguishers have labels that tell the types of fires for which they are designed. There are four common classes of fires:
* Class A—fires are in ordinary combustible materials such as paper or wood.
* Class B—fires involve gasoline, oil and grease.
* Class C—fires are electrical.
* Class D—fires involve ferrous metals

The extinguishers on motorboats must be for Class B fires. Never use water on Class B or Class C fires. Water spreads Class B fires. Water may cause you to be electrocuted in a Class C fire.

FIRE EXTINGUISHERS

▶ See Figure 6

If your motorboat meets one or more of the following conditions, you must have at least one fire extinguisher aboard. The conditions are:
* Inboard or stern drive engines
* Closed compartments under seats where portable fuel tanks can be stored

04701P29

Fig. 6 An approved fire extinguisher should be mounted close to the helmsman for emergency use

- Double bottoms not sealed together or not completely filled with flotation materials
- Closed living spaces
- Closed stowage compartments in which combustible or flammable materials are stored
- Permanently installed fuel tanks
- It is 26 feet or more in length.

Contents of Extinguishers

Fire extinguishers use a variety of materials. Those used on boats usually contain dry chemicals, Halon, or carbon dioxide (CO3). Dry chemical extinguishers contain chemical powders such as sodium bicarbonate—baking soda.

Carbon dioxide is a colorless and odorless gas when released from an extinguisher. It is not poisonous but caution must be used in entering compartments filled with it. It will not support life and keeps oxygen from reaching your lungs. A fire-killing concentration of carbon dioxide is lethal.

If you are in a compartment with a high concentration of CO3, you will have no difficulty breathing. But the air does not contain enough oxygen to support life. Unconsciousness or death can result.

HALON EXTINGUISHERS

Some fire extinguishers and `built-in' or `fixed' automatic fire extinguishing systems contain a gas called Halon. Like carbon dioxide it is colorless and odorless and will not support life. Some Halons may be toxic if inhaled.

To be acceptable to the Coast Guard, a fixed Halon system must have an indicator light at the vessel's helm. A green light shows the system is ready. Red means it is being discharged or has been discharged. Warning horns are available to let you know the system has been activated. If your fixed Halon system discharges, ventilate the space thoroughly before you enter it. There are no residues from Halon but it will not support life.

Although Halon has excellent fire fighting properties, it is thought to deplete the earth's ozone layer and has not been manufactured since January 1, 1994. Halon extinguishers can be refilled from existing stocks of the gas until they are used up, but high federal excise taxes are being charged for the service. If you discontinue using your Halon extinguisher, take it to a recovery station rather than releasing the gas into the atmosphere. Compounds such as FE 241, designed to replace Halon, are now available.

Fire Extinguisher Approval

Fire extinguishers must be Coast Guard approved. Look for the approval number on the nameplate. Approved extinguishers have the following on their labels: "Marine Type USCG Approved, Size . . . , Type . . . , 162.208/," etc. In addition, to be acceptable by the Coast Guard, an extinguisher must be in serviceable condition and mounted in its bracket. An extinguisher not properly mounted in its bracket will not be considered serviceable during a Coast Guard inspection.

Care and Treatment

Make certain your extinguishers are in their stowage brackets and are not damaged. Replace cracked or broken hoses. Nozzles should be free of obstructions. Sometimes, wasps and other insects nest inside nozzles and make them inoperable. Check your extinguishers frequently. If they have pressure gauges, is the pressure within acceptable limits? Do the locking pins and sealing wires show they have not been used since recharging?

Don't try an extinguisher to test it. Its valves will not reseat properly and the remaining gas will leak out. When this happens, the extinguisher is useless.

Weigh and tag carbon dioxide, Halon or compound FE 241 extinguishers twice a year. If their weight loss exceeds 10 percent of the weight of the charge, recharge them. Check to see that they have not been used. They should have been inspected by a qualified person within the past six months, and they should have tags showing all inspection and service dates. The problem is that they can be partially discharged while appearing to be fully charged.

Some Halon extinguishers have pressure gauges the same as dry chemical extinguishers. Don't rely too heavily on the gauge. The extinguisher can be partially discharged and still show a good gauge reading. Weighing a Halon extinguisher is the only accurate way to assess its contents.

If your dry chemical extinguisher has a pressure indicator, check it frequently. Check the nozzle to see if there is powder in it. If there is, recharge it. Occasionally invert your dry chemical extinguisher and hit the base with the palm of your hand. The chemical in these extinguishers packs and cakes due to the boat's vibration and pounding. There is a difference of opinion about whether hitting the base helps, but it can't hurt. It is known that caking of the chemical powder is a major cause of failure of dry chemical extinguishers. Carry spares in excess of the minimum requirement. If you have guests aboard, make certain they know where the extinguishers are and how to use them.

Using a Fire Extinguisher

A fire extinguisher usually has a device to keep it from being discharged accidentally. This is a metal or plastic pin or loop. If you need to use your extinguisher, take it from its bracket. Remove the pin or the loop and point the nozzle at the base of the flames. Now, squeeze the handle, and discharge the extinguisher's contents while sweeping from side to side. Recharge a used extinguisher as soon as possible.

If you are using a Halon or carbon dioxide extinguisher, keep your hands away from the discharge. The rapidly expanding gas will freeze them. If your fire extinguisher has a horn, hold it by its handle.

Legal Requirements for Extinguishers

You must carry fire extinguishers as defined by Coast Guard regulations. They must be firmly mounted in their brackets and immediately accessible.

A motorboat less than 26 feet long must have at least one approved hand-portable, Type B-1 extinguisher. If the boat has an approved fixed fire extinguishing system, you are not required to have the Type B-1 extinguisher. Also, if your boat is less than 26 feet long, is propelled by an outboard motor, or motors, and does not have any of the first six conditions described at the beginning of this section, it is not required to have an extinguisher. Even so, it's a good idea to have one, especially if a nearby boat catches fire, or if a fire occurs at a fuel dock.

A motorboat 26 feet to under 40 feet long, must have at least two Type B-1 approved hand-portable extinguishers. It can, instead, have at least one Coast Guard approved Type B-2. If you have an approved fire extinguishing system, only one Type B-1 is required.

A motorboat 40 to 65 feet long must have at least three Type B-1 approved portable extinguishers . It may have, instead, at least one Type B-1 plus a Type B-2. If there is an approved fixed fire extinguishing system, two Type B-1 or one Type B-2 is required.

WARNING SYSTEM

Various devices are available to alert you to danger. These include fire, smoke, gasoline fumes, and carbon monoxide detectors. If your boat has a galley, it should have a smoke detector. Where possible, use wired detectors. Household batteries often corrode rapidly on a boat.

You can't see, smell, nor taste carbon monoxide gas, but it is lethal. As little as one part in 10,000 parts of air can bring on a headache. The symptoms of carbon monoxide poisoning—headaches, dizziness, and nausea—are like sea sickness. By the time you realize what is happening to you, it may be too late to take action. If you have enclosed living spaces on your boat, protect yourself with a detector. There are many ways in which carbon monoxide can enter your boat.

PERSONAL FLOTATION DEVICES

Personal Flotation Devices (PFDs) are commonly called life preservers or life jackets. You can get them in a variety of types and sizes. They vary with their intended uses. To be acceptable, they must be Coast Guard approved.

Type I PFDs

A Type I life jacket is also called an offshore life jacket. Type I life jackets will turn most unconscious people from facedown to a vertical or slightly backward position. The adult size gives a minimum of 22 pounds of buoyancy. The child size has at least 11 pounds. Type I jackets provide more protection to their wearers than any other type of life jacket. Type I life jackets are bulkier and less comfortable than other types. Furthermore, there are only two sizes, one for children and one for adults.

Type I life jackets will keep their wearers afloat for extended periods in rough water. They are recommended for offshore cruising where a delayed rescue is probable.

Type II PFDs

◆ See Figure 7

A Type II life jacket is also called a near-shore buoyant vest. It is an approved, wearable device. Type II life jackets will turn some unconscious people from facedown to vertical or slightly backward positions. The adult size gives at least 15.5 pounds of buoyancy. The medium child size has a minimum of 11 pounds. And the small child and infant sizes give seven pounds. A Type II life jacket is more comfortable than a Type I but it does not have as much buoyancy. It is not recommended for long hours in rough water. Because of this, Type IIs are recommended for inshore and inland cruising on calm water. Use them where there is a good chance of fast rescue.

Type III PFDs

Type III life jackets or marine buoyant devices are also known as flotation aids. Like Type IIs, they are designed for calm inland or close offshore water where there is a good chance of fast rescue. Their minimum buoyancy is 15.5 pounds. They will not turn their wearers face up.

Type III devices are usually worn where freedom of movement is necessary. Thus, they are used for water skiing, small boat sailing, and fishing among other activities. They are available as vests and flotation coats. Flotation coats are useful in cold weather. Type IIIs come in many sizes from small child through large adult.

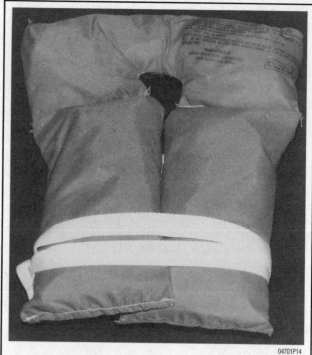

04701P14

Fig. 7 Type II approved flotation devices are recommended for inshore and inland cruising on calm water. Use them where there is a good chance of fast rescue

Life jackets come in a variety of colors and patterns—red, blue, green, camouflage, and cartoon characters. From a safety standpoint, the best color is bright orange. It is easier to see in the water, especially if the water is rough.

Type IV PFDs

◆ See Figures 8 and 9

Type IV ring life buoys, buoyant cushions and horseshoe buoys are Coast Guard approved devices called throwables. They are made to be thrown to people in the water, and should not be worn. Type IV cushions are often used as seat cushions. Cushions are hard to hold onto in the water. Thus, they do not afford as much protection as wearable life jackets.

The straps on buoyant cushions are for you to hold onto either in the water or when throwing them. A cushion should never be worn on your back. It will turn you face down in the water.

Type IV throwables are not designed as personal flotation devices for unconscious people, non-swimmers, or children. Use them only in emergencies. They should not be used for, long periods in rough water.

Ring life buoys come in 18, 20, 24, and 30 inch diameter sizes. They have grab lines. You should attach about 60 feet of polypropylene line to the grab rope to aid in retrieving someone in the water. If you throw a ring, be careful not to hit the person. Ring buoys can knock people unconscious

Type V PFDs

Type V PFDs are of two kinds, special use devices and hybrids. Special use devices include boardsailing vests, deck suits, work vests, and others. They are approved only for the special uses or conditions indicated on their labels. Each is designed and intended for the particular application shown on its label. They do not meet legal requirements for general use aboard recreational boats.

Fig. 8 Type II buoyant cushions are made to be thrown to people in the water. If you can squeeze air out of the cushion, it is faulty and should be replaced

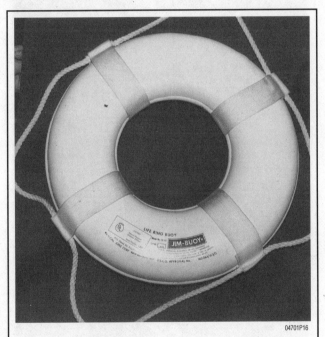

Fig. 9 Type IV throwables, such as this ring life buoy, are not designed as personal flotation devices for unconscious people, non-swimmers, or children

Hybrid life jackets are inflatable devices with some built-in buoyancy provided by plastic foam or kapok. They can be inflated orally or by cylinders of compressed gas to give additional buoyancy. In some hybrids the gas is released manually. In others it is released automatically when the life jacket is immersed in water.

The inherent buoyancy of a hybrid may be insufficient to float a person unless it is inflated. The only way to find this out is for the user to try it in the water. Because of its limited buoyancy when deflated, a hybrid is recommended for use by anon-swimmer only if it is worn with enough inflation to float the wearer.

If they are to count against the legal requirement for the number of life jackets you must carry on your vessel, hybrids manufactured before February 8, 1995 must be worn whenever a boat is underway and the wearer is not below decks or in an enclosed space. To find out if your Type V hybrid must be worn to satisfy the legal requirement, read its label. If its use is restricted it will say, "REQUIRED TO BE WORN" in capital letters.

Hybrids cost more than other life jackets, but this factor must be weighed against the fact that they are more comfortable than Type I, II, or III life jackets. Because of their greater comfort, their owners are more likely to wear them than are the owners of Type I, II, or III life jackets.

The Coast Guard has determined that improved, less costly hybrids can save lives since they will be bought and used more frequently. For these reasons a new federal regulation was adopted effective February 8, 1995. The regulation increases both the deflated and inflated buoyancys of hybrids, makes them available in a greater variety of sizes and types, and reduces their costs by reducing production costs.

Even though it may not be required, the wearing of a hybrid or a life jacket is encouraged whenever a vessel is underway. Like life jackets, hybrids are now available in three types. To meet legal requirements, a Type I hybrid can be substituted for a Type I life jacket. Similarly Type II and III hybrids can be substituted for Type II and Type III life jackets. A Type I hybrid, when inflated, will turn most unconscious people from facedown to vertical or slightly backward positions just like a Type I life jacket. Type I and III hybrids function like Type II and III life jackets. If you purchase a new hybrid, it should have an owner's manual attached which describes its life jacket type and its deflated and inflated buoyancys. It warns you that it may have to be inflated to float you. The manual also tells you how to don the life jacket and how to inflate it. It also tells you how to change its inflation mechanism, recommended testing exercises, and inspection and maintenance procedures. The manual also tells you why you need a life jacket and why you should wear it. A new hybrid must be packaged with at least three gas cartridges. One of these may already be loaded into the inflation mechanism. Likewise, if it has an automatic inflation mechanism, it must be packaged with at least three of these water sensitive elements. One of these elements may be installed.

Legal Requirements

A Coast Guard approved life jacket must show the manufacturer's name and approval number. Most are marked as Type I, II, III, IV, or V. All of the newer hybrids are marked for type.

You are required to carry at least one wearable life jacket or hybrid for each person on board your recreational vessel. If your vessel is 16 feet or more in length and is not a canoe or a kayak, you must also have at least one Type IV on board. These requirements apply to all recreational vessels that are propelled or controlled by machinery, sails, oars, paddles, poles, or another vessel. Sailboards are not required to carry life jackets.

You can substitute an older Type V hybrid for any required Type I, II, or III life jacket provided that its approval label shows it is approved for the activity the vessel is engaged in, approved as a substitute for a life jacket of the type required on the vessel, used as required on the labels, and used in accordance with any requirements in its owner's manual, if the approval label makes reference to such a manual.

A water skier being towed is considered to be on board the vessel when judging compliance with legal requirements.

You are required to keep your Type I, II, or III life jackets or equivalent hybrids readily accessible, which means you must be able to get to them quickly if they are needed. Accessible means you must be able to reach out and get them when needed. All life jackets must be in good, serviceable condition. All Type V hybrids must be worn whenever the boat is underway and the wearer is not below decks or in an enclosed space.

General Considerations

The proper use of a life jacket requires the wearer to know how it will perform. You can gain this knowledge only through experience. Each person on your boat should be assigned a life jacket. Next, it should be fitted to the person who will wear it. Only then can you be sure that it will be ready for use in an emergency. Boats can sink fast. There may be no time to look around for a life jacket. Fitting one on you in the water is almost impossible. This advice is good even if the water is calm, and you intend to boat near shore. Most drownings occur in inland waters within a few feet of safety. Most victims had life jackets, but they weren't wearing them.

Keeping life jackets in the plastic covers they came wrapped in and in a cabin assures that they will stay clean and unfaded. But this is no way to keep them when you are on the water. When you need a life jacket it must be readily accessible and adjusted to fit you. You can't spend time hunting for it or learning how to fit it.

There is no substitute for the experience of entering the water while wearing a life jacket. Children, especially, need practice. If possible, give, your guests this experience. Tell them they should keep their arms to their sides when jumping in to keep the life jacket from riding up. Let them jump in and see how the life jacket responds. Is it adjusted so it does not ride up? Is it the proper size? Are all straps snug? Are children's life jackets the right sizes for them? Are they adjusted properly? If a child's life jacket fits correctly, you can lift the child by the jacket's shoulder straps and the child's chin and ears will not slip through. Non-swimmers, children, handicapped persons, elderly persons and even pets should always wear life jackets when they are aboard. Many states require that everyone aboard wear them in hazardous waters. Inspect your life-saving equipment from time to time. Leave any questionable or unsatisfactory equipment on shore. An emergency is no time for you to conduct an inspection.

Indelibly mark your life jackets with your vessel's name, number, and calling port. This can be important in a search and rescue effort. It could help concentrate effort where it will do the most good.

Care of Life Jackets

Given reasonable care, life jackets last many years. Thoroughly dry them before putting them away. Stow them in dry, well ventilated places. Avoid the bottoms of lockers and deck storage boxes where moisture may collect. Air and dry them frequently.

Life jackets should not be tossed about or used as fenders or cushions. Many contain kapok or fibrous glass material enclosed in plastic bags. The bags can rupture and are then unserviceable. Squeeze your life jacket gently. Does air leak out? If so, water can leak in and it will no longer be safe to use. Cut it up so no one will use it, and throw it away. The covers of some life jackets are made of nylon or polyester. These materials are plastics. Like many plastics, they break down after extended exposure to the ultraviolet light in sunlight. This process may be more rapid when the materials are dyed with bright dyes such as "neon" shades.

Ripped and badly faded fabric are clues that the covering of your life jacket is deteriorating. A simple test is to pinch the fabric between your thumbs and forefingers. Now try to tear the fabric. If it can be torn, it should definitely be destroyed and discarded. Compare the colors in protected places to those exposed to the sun. If the colors have faded, the materials have been weakened. A fabric covered life jacket should ordinarily last several boating seasons with normal use. A life jacket used every day in direct sunlight should probably be replaced more often.

SOUND PRODUCING DEVICES

▶ See Figure 10

All boats are required to carry some means of making an efficient sound signal. Devices for making the whistle or horn noises required by the Navigation Rules must be capable of a four second blast. The blast should be audible for at least one-half mile. Athletic whistles are not acceptable on boats 12 meters or longer. Use caution with athletic whistles. When wet, some of them come apart and loose their "pea." When this happens, they are useless.

If your vessel is 12 meters long and less than 20 meters, you must have a power whistle (or power horn) and a bell on board. The bell must be in operating condition and have a minimum diameter of at least 200 mm (7.9 inches) at its mouth.

04701P19

Fig. 10 All boats are required to carry some means of making an efficient sound signal, such as this horn

VISUAL DISTRESS SIGNALS

▶ See Figure 11

Visual Distress Signals (VDS) attract attention to your vessel if you need help. They also help to guide searchers in search and rescue situations. Be sure you have the right kinds, and use them properly.

It is illegal to fire flares improperly. In addition, they cost the Coast Guard and its Auxiliary many wasted hours in fruitless searches. If you signal a distress with flares and then someone helps you, please let the Coast Guard or the appropriate Search And Rescue Agency (SAR) know so the distress report will be canceled.

Recreational boats less than 16 feet long must carry visual distress signals on coastal waters at night. Coastal waters are:

• The ocean (territorial sea)
• The Great Lakes
• Bays or sounds
• Rivers over two miles across at their mouths upstream to where they narrow to two miles.

Recreational boats 16 feet or longer must carry VDS at all times on coastal waters. The same requirement applies to boats carrying six or fewer passengers for hire. Open sailboats less than 26 feet long without engines are exempt in the daytime as are manually propelled boats. Also exempt are boats in organized races, regattas, parades, etc. Boats owned in the United States and operating on the high seas must be equipped with VDS.

A wide variety of signaling devices meet Coast Guard regulations. For pyrotechnic devices, a minimum of three must be carried. Any combination can be carried as long as it adds up to at least three signals for day use and at least three signals for night use. Three day/night signals meet both requirements. If possible, carry more than the legal requirement. These devices are listed in table 2-5. The American flag flying upside down is a commonly recognized distress signal. It is not recognized in the Coast Guard regulations, though. In an emergency, your efforts would probably be better used in more effective signaling methods.

Types of VDS

VDS are divided into two groups; daytime and nighttime use. Each of these groups is subdivided into pyrotechnic and non-pyrotechnic devices.

DAYTIME NON-PYROTECHNIC SIGNALS

A bright orange flag with a black square over a black circle is the simplest VDS. It is usable, of course, only in daylight. It has the advantage of being a continuous signal. A mirror can be used to good advantage on sunny days. It can attract the attention of other boaters and of aircraft from great distances. Mirrors are available with holes in their centers to aid in "aiming." In the absence of a mirror, any shiny object can be used. When another boat is in sight, an effective VDS is to extend your arms from your sides and move them up and down. Do it slowly. If you do it too fast the other people may think you are just being friendly. This simple gesture is seldom misunderstood, and requires no equipment.

DAYTIME PYROTECHNIC DEVICES

Orange smoke is a useful daytime signal. Hand-held or floating smoke flares are very effective in attracting attention from aircraft. Smoke flares don't last long, and are not very effective in high wind or poor visibility. As with other pyrotechnic devices, use them only when you know there is a possibility that someone will see the display.

To be usable, smoke flares must be kept dry. Keep them in airtight containers and store them in dry places. If the "striker" is damp, dry it out before trying to ignite the device. Some pyrotechnic devices require a forceful "strike" to ignite them.

All hand-held pyrotechnic devices may produce hot ashes or slag when burning. Hold them over the side of your boat in such a way that they do not burn your hand or drip into your boat.

Nighttime Non-Pyrotechnic Signals

An electric distress light is available. This light automatically flashes the international morse code SOS distress signal (··· ···). Flashed four to six times a minute, it is an unmistakable distress signal. It must show that it is approved by the Coast Guard. Be sure the batteries are fresh. Dated batteries give assurance that they are current.

Under the Inland Navigation Rules, a high intensity white light flashing 50-70 times per minute is a distress signal. Therefore, use strobe lights on inland waters only for distress signals.

Nighttime Pyrotechnic Devices

▶ See Figure 12

Aerial and hand-held flares can be used at night or in the daytime. Obviously, they are more effective at night.

Currently, the serviceable life of a pyrotechnic device is rated at 42 months from its date of manufacture. Pyrotechnic devices are expensive. Look at their dates before you buy them. Buy them with as much time remaining as possible.

Like smoke flares, aerial and hand-held flares may fail to work if they have been damaged or abused. They will not function if they are or have been wet. Store them in dry, airtight containers in dry places. But store them where they are readily accessible.

Aerial VDSs, depending on their type and the conditions they are used in, may not go very high. Again, use them only when there is a good chance they will be seen.

04701G09

Fig. 11 Internationally accepted distress signals

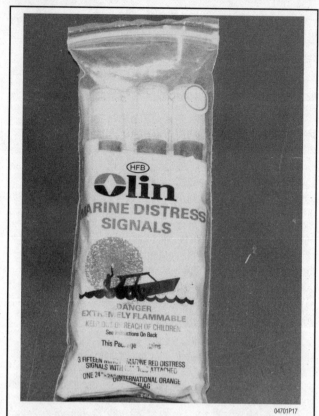

Fig. 12 Moisture protected flares should be carried onboard any vessel for use as a distress signal

A serious disadvantage of aerial flares is that they burn for only a short time. Most burn for less than 10 seconds. Most parachute flares burn for less than 45 seconds. If you use a VDS in an emergency, do so carefully. Hold hand-held flares over the side of the boat when in use. Never use a road hazard flare on a boat. It can easily start a fire. Marine type flares are carefully designed to lessen risk, but they still must be used carefully.

Aerial flares should be given the same respect as firearms since they are firearms! Never point them at another person. Don't allow children to play with them or around them. When you fire one, face away from the wind. Aim it downwind and upward at an angle of about 60" to the horizon. If there is a strong wind, aim it somewhat more vertically. Never fire it straight up. Before you discharge a flare pistol, check for overhead obstructions. These might be damaged by the flare. They might deflect the flare to where it will cause damage.

Disposal of VDS

Keep outdated flares when you get new ones. They do not meet legal requirements, but you might need them sometime, and they may work. It is illegal to fire a VDS on federal navigable waters unless an emergency exists. Many states have similar laws.

Emergency Position Indicating Radio Beacon (EPIRB)

There is no requirement for recreational boats to have EPIRBs. Some commercial and fishing vessels, though, must have them if they operate beyond the three mile limit. Vessels carrying six or fewer passengers for hire must have EPIRBs under some circumstances when operating beyond the three mile limit. If you boat in a remote area or offshore, you should have an EPIRB. An EPIRB is a

small (about 6 to 20 inches high), battery-powered, radio transmitting buoy-like device. It is a radio transmitter and requires a license or an endorsement on your radio station license by the Federal Communications Commission (FCC). EPIRBs are activated by being immersed in water or by a manual switch.

Equipment Not Required But Recommended

Although not required by law, there is other equipment that is good to have onboard.

SECOND MEANS OF PROPULSION

▶ **See Figure 13**

All boats less than 16 feet long should carry a second means of propulsion. A paddle or oar can come in handy at times. For most small boats, a spare trolling or outboard motor is an excellent idea. If you carry a spare motor, it should have its own fuel tank and starting power. If you use an electric trolling motor, it should have its own battery.

Fig. 13 A typical wooden oar should be kept onboard as an auxiliary means of propulsion. It can also function as a grab hook for someone fallen overboard

BAILING DEVICES

All boats should carry at least one effective manual bailing device in addition to any installed electric bilge pump. This can be a bucket, can, scoop, hand operated pump, etc. If your battery "goes dead" it will not operate your electric pump.

FIRST AID KIT

▶ **See Figure 14**

All boats should carry a first aid kit. It should contain adhesive bandages, gauze, adhesive tape, antiseptic, aspirin, etc. Check your

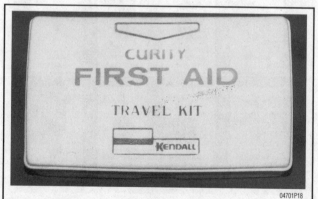

Fig. 14 Always carry an adequately stocked first aid kit on board for the safety of the crew and guests

first aid kit from time to time. Replace anything that is outdated. It is to your advantage to know how to use your first aid kit. Another good idea would be to take a Red Cross first aid course.

ANCHORS

▶ **See Figure 15**

All boats should have anchors. Choose one of suitable size for your boat. Better still, have two anchors of different sizes. Use the smaller one in calm water or when anchoring for a short time to fish or eat. Use the larger one when the water is rougher or for overnight anchoring.

Carry enough anchor line of suitable size for your boat and the waters in which you will operate. If your engine fails you, the first thing you usually should do is lower your anchor. This is good advice in shallow water where you may be driven aground by the wind or water. It is also good advice in windy weather or rough water. The anchor will usually hold your bow into the waves.

Fig. 15 All boats should have anchors. Choose and anchor of sufficient weight to secure the boat without dragging

VHF-FM RADIO

Your best means of summoning help in an emergency or in case of a breakdown is a VHF-FM radio. You can use it to get advice or assistance from the Coast Guard. In the event of a serious illness or injury aboard your boat, the Coast Guard can have emergency medical equipment meet you ashore.

TOOLS & SPARE PARTS

▶ **See Figures 16 and 17**

Carry a few tools and some spare parts, and learn how to make minor repairs. Many search and rescue cases are caused by minor

Fig. 16 A flashlight with a fresh set of batteries is handy when repairs are needed at night. It can also double as a signaling device

Fig. 17 A spare fuel jug can keep you from becoming stranded on the water. Make sure the container is approved for marine use

breakdowns that boat operators could have repaired. If your engine is an inboard or stern drive, carry spare belts and water pump impellers and the tools to change them.

Courtesy Marine Examinations

One of the roles of the Coast Guard Auxiliary is to promote recreational boating safety. This is why they conduct thousands of Courtesy Marine Examinations each year. The auxiliarists who do these examinations are well-trained and knowledgeable in the field.

These examinations are free and done only at the consent of boat owners. To pass the examination, a vessel must satisfy federal equipment requirements and certain additional requirements of the coast guard auxiliary. If your vessel does not pass the Courtesy Marine Examination, no report of the failure is made. Instead, you will be told what you need to correct the deficiencies. The examiner will return at your convenience to redo the examination.

If your vessel qualifies, you will be awarded a safety decal. The decal does not carry any special privileges. It attests to your interest in safe boating.

SAFETY IN SERVICE

It is virtually impossible to anticipate all of the hazards involved with maintenance and service, but care and common sense will prevent most accidents.

The rules of safety for mechanics range from "don't smoke around gasoline," to "use the proper tool(s) for the job." The trick to avoiding injuries is to develop safe work habits and to take every possible precaution. Whenever you are working on your boat, pay attention to what you are doing. The more you pay attention to details and what is going on around you, the less likely you will be to hurt yourself or damage your boat.

Do's

• Do keep a fire extinguisher and first aid kit handy.
• Do wear safety glasses or goggles when cutting, drilling, grinding or prying, even if you have 20–20 vision. If you wear glasses for the sake of vision, wear safety goggles over your regular glasses.
• Do shield your eyes whenever you work around the battery. Batteries contain sulfuric acid. In case of contact with the eyes or skin, flush the area with water or a mixture of water and baking soda, then seek immediate medical attention.
• Do use adequate ventilation when working with any chemicals or hazardous materials.
• Do disconnect the negative battery cable when working on the electrical system. The secondary ignition system contains EXTREMELY HIGH VOLTAGE. In some cases it can even exceed 50,000 volts.
• Do follow manufacturer's directions whenever working with potentially hazardous materials. Most chemicals and fluids are poisonous if taken internally.
• Do properly maintain your tools. Loose hammerheads, mushroomed punches and chisels, frayed or poorly grounded electrical cords, excessively worn screwdrivers, spread wrenches (open end), cracked sockets, or slipping ratchets can cause accidents.
• Likewise, keep your tools clean; a greasy wrench can slip off a bolt head, ruining the bolt and often harming your knuckles in the process.
• Do use the proper size and type of tool for the job at hand. Do select a wrench or socket that fits the nut or bolt. The wrench or socket should sit straight, not cocked.

• Do, when possible, pull on a wrench handle rather than push on it, and adjust your stance to prevent a fall.
• Do be sure that adjustable wrenches are tightly closed on the nut or bolt and pulled so that the force is on the side of the fixed jaw. Better yet, avoid the use of an adjustable if you have a fixed wrench that will fit.
• Do strike squarely with a hammer; avoid glancing blows. But, we REALLY hope you won't be using a hammer much in basic maintenance.

Don'ts

• Don't run the engine in an enclosed area or anywhere else without proper ventilation—EVER! Carbon monoxide is poisonous; it takes a long time to leave the human body and you can build up a deadly supply of it in your system by simply breathing in a little every day. You may not realize you are slowly poisoning yourself.
• Don't work around moving parts while wearing loose clothing. Short sleeves are much safer than long, loose sleeves. Hard-toed shoes with neoprene soles protect your toes and give a better grip on slippery surfaces. Jewelry, watches, large belt buckles, or body adornment of any kind is not safe working around any vehicle. Long hair should be tied back under a hat.
• Don't use pockets for toolboxes. A fall or bump can drive a screwdriver deep into your body. Even a rag hanging from your back pocket can wrap around a spinning shaft.
• Don't smoke when working around gasoline, cleaning solvent or other flammable material.
• Don't smoke when working around the battery. When the battery is being charged, it gives off explosive hydrogen gas. Actually, you shouldn't smoke anyway. Save that cigarette money and trick out your ride!
• Don't use gasoline to wash your hands; there are excellent soaps available. Gasoline contains dangerous additives which can enter the body through a cut or through your pores. Gasoline also removes all the natural oils from the skin so that bone dry hands will suck up oil and grease.
• Don't use screwdrivers for anything other than driving screws! A screwdriver used as an prying tool can snap when you least expect it, causing injuries. At the very least, you'll ruin a good screwdriver.

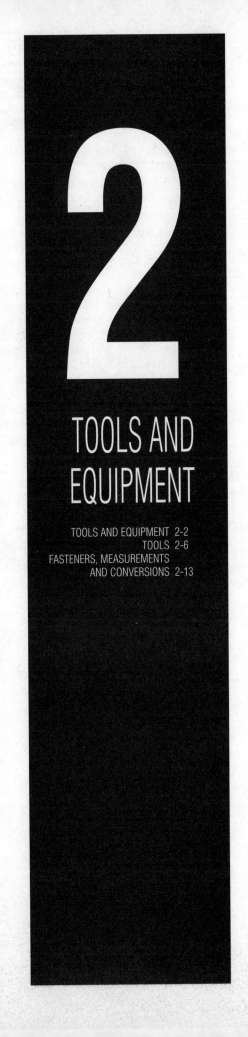

2

TOOLS AND
EQUIPMENT

TOOLS AND EQUIPMENT

Safety Tools

WORK GLOVES

▶ **See Figure 1**

Unless you think scars on your hands are cool, enjoy pain and like wearing bandages, get a good pair of work gloves. Canvas or leather are the best. And yes, we realize that there are some jobs involving small parts that can't be done while wearing work gloves. These jobs are not the ones usually associated with hand injuries.

A good pair of rubber gloves (such as those usually associated with dish washing) or vinyl gloves is also a great idea. There are some liquids such as solvents and penetrants that don't belong on your skin. Avoid burns and rashes. Wear these gloves.

And lastly, an option. If you're tired of being greasy and dirty all the time, go to the drug store and buy a box of disposable latex gloves like medical professionals wear. You can handle greasy parts, perform small tasks, wash parts, etc. all without getting dirty! These gloves take a surprising amount of abuse without tearing and aren't expensive. Note however, that it has been reported that some people are allergic to the latex or the powder used inside some gloves, so pay attention to what you buy.

EYE AND EAR PROTECTION

Don't begin any job without a good pair of work goggles or impact resistant glasses! When doing any kind of work, it's all too easy to avoid eye injury through this simple precaution. And don't just buy eye protection and leave it on the shelf. Wear it all the time! Things have a habit of breaking, chipping, splashing, spraying, splintering and flying around. And, for some reason, your eye is always in the way!

If you wear vision correcting glasses as a matter of routine, get a pair made with polycarbonate lenses. These lenses are impact resistant and are available at any optometrist.

Often overlooked is hearing protection. Power equipment is noisy! Loud noises damage your ears. It's as simple as that! The simplest and cheapest form of ear protection is a pair of noise-reducing ear plugs. Cheap insurance for your ears. And, they may even come with their own, cute little carrying case.

More substantial, more protection and more money is a good pair of noise reducing earmuffs. They protect from all but the loudest sounds. Hopefully those are sounds that you'll never encounter since they're usually associated with disasters.

WORK CLOTHES

Everyone has "work clothes." Usually these consist of old jeans and a shirt that has seen better days. That's fine. In addition, a

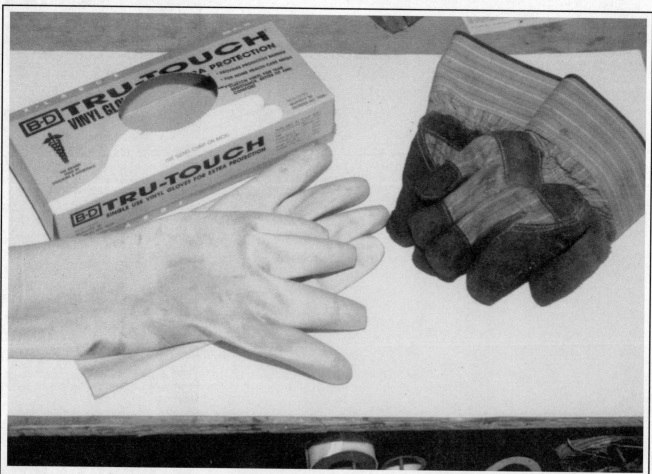

Fig. 1 Three different types of work gloves. The box contains latex gloves

87933518

denim work apron is a nice accessory. It's rugged, can hold some spare bolts, and you don't feel bad wiping your hands or tools on it. That's what it's for.

When working in cold weather, a one-piece, thermal work outfit is invaluable. Most are rated to below zero (Fahrenheit) temperatures and are ruggedly constructed. Just look at what the boat mechanics are wearing and that should give you a clue as to what type of clothing is good.

Chemicals

There is a whole range of chemicals that you'll find handy for maintenance work. The most common types are, lubricants, penetrants and sealers. Keep these handy onboard. There are also many chemicals that are used for detailing or cleaning.

When a particular chemical is not being used, keep it capped, upright and in a safe place. These substances may be flammable, may be irritants or might even be caustic and should always be stored properly, used properly and handled with care. Always read and follow all label directions and be sure to wear hand and eye protection!

LUBRICANTS & PENETRANTS

▶ **See Figure 2**

Anti-seize is used to coat certain fasteners prior to installation. This can be especially helpful when two dissimilar metals are in contact (to help prevent corrosion that might lock the fastener in place). This is a good practice on a lot of different fasteners, BUT, NOT on any fastener which might vibrate loose causing a problem. If anti-seize is used on a fastener, it should be checked periodically for proper tightness.

Lithium grease, chassis lube, silicone grease or a synthetic brake caliper grease can all be used pretty much interchangeably. All can be used for coating rust-prone fasteners and for facilitating the assembly of parts that are a tight fit. Silicone and synthetic greases are the most versatile.

➡ **Silicone dielectric grease is a non-conductor that is often used to coat the terminals of wiring connectors before fastening them. It may sound odd to coat metal portions of a terminal with something that won't conduct electricity, but here is it how it works. When the connector is fastened the metal-to-metal contact between the termi-**

Fig. 2 A variety of penetrants and lubricants is a staple of any DIYer's inventory

87933516

nals will displace the grease (allowing the circuit to be completed). The grease that is displaced will then coat the non-contacted surface and the cavity around the terminals, SEALING them from atmospheric moisture that could cause corrosion.

Silicone spray is a good lubricant for hard-to-reach places and parts that shouldn't be gooped up with grease.

Penetrating oil may turn out to be one of your best friends when taking something apart that has corroded fasteners. Not only can they make a job easier, they can really help to avoid broken and stripped fasteners. The most familiar penetrating oils are Liquid Wrench® and WD-40®. A newer penetrant, PB Blaster® also works well. These products have hundreds of uses. For your purposes, they are vital!

Before disassembling any part (especially on an exhaust system), check the fasteners. If any appear rusted, soak them thoroughly with the penetrant and let them stand while you do something else. This simple act can save you hours of tedious work trying to extract a broken bolt or stud.

SEALANTS

▶ See Figure 3

Sealants are an indispensable part for certain tasks, especially if you are trying to avoid leaks. The purpose of sealants is to establish a leak-proof bond between or around assembled parts. Most sealers are used in conjunction with gaskets, but some are used instead of conventional gasket material.

The most common sealers are the non-hardening types such as Permatex®No.2 or its equivalents. These sealers are applied to the mating surfaces of each part to be joined, then a gasket is put in place and the parts are assembled.

➡A sometimes overlooked use for sealants like RTV is on the threads of vibration prone fasteners.

One very helpful type of non-hardening sealer is the "high tack" type. This type is a very sticky material that holds the gasket in place while the parts are being assembled. This stuff is really a good idea when you don't have enough hands or fingers to keep everything where it should be.

The stand-alone sealers are the Room Temperature Vulcanizing (RTV) silicone gasket makers. On some engines, this material is used instead of a gasket. In those instances, a gasket may not be available or, because of the shape of the mating surfaces, a gasket shouldn't be used. This stuff, when used in conjunction with a conventional gasket, produces the surest bonds.

RTV does have its limitations though. When using this material, you will have a time limit. It starts to set-up within 15 minutes or so, so you have to assemble the parts without delay. In addition, when squeezing the material out of the tube, don't drop any glops into the engine. The stuff will form and set and travel around the oil gallery, possibly plugging up a passage. Also, most types are not fuel-proof. Check the tube for all cautions.

CLEANERS

▶ See Figures 4 and 5

You'll have two types of cleaners to deal with: parts cleaners and hand cleaners. The parts cleaners are for the parts; the hand cleaners are for you.

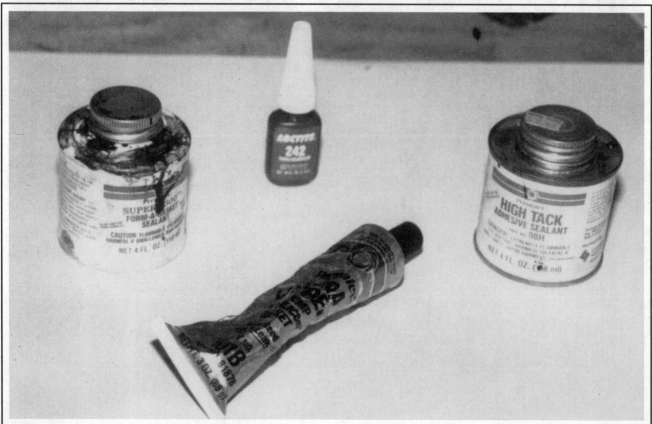

Fig. 3 Sealants are essential for preventing leaks

87933507

87933508

Fig. 4 Always read and follow label instructions. Remember, when using cleaners, they may eventually end up in the bilge

87933018

Fig. 5 This is one type of hand cleaner that not only works well but smells pretty good too

There are many good, non-flammable, biodegradable parts cleaners on the market. These cleaning agents are safe for you, the parts and the environment. Therefore, there is no reason to use flammable, caustic or toxic substances to clean your parts or tools.

As far as hand cleaners go, the waterless types are the best. They have always been efficient at cleaning, but leave a pretty smelly odor. Recently though, just about all of them have eliminated the odor and added stuff that actually smells good. Make sure that you pick one that contains lanolin or some other moisture-replenishing additive. Cleaners not only remove grease and oil but also skin oil.

One other note: most women know this already but most men don't. Use a hand lotion when you're all cleaned up. It's okay. Real men DO use hand lotion! Believe it or not, using hand lotion **before** your hands are dirty will actually make them easier to clean when you're finished with a dirty job. Lotion seals your hands, and keeps dirt and grease from sticking to your skin.

TOOLS

▶ See Figure 6

Tools; this subject could require a completely separate manual. Again, the first thing you will need to ask yourself, is just how involved do you plan to get. If you are serious about your maintenance you will want to gather a quality set of tools to make the job easier, and more enjoyable. BESIDES, TOOLS ARE FUN!!!

Almost every do-it-yourselfer loves to accumulate tools. Though most find a way to perform jobs with only a few common tools, they tend to buy more over time, as money allows. So gathering the tools necessary for maintenance does not have to be an expensive, overnight proposition.

When buying tools, the saying "You get what you pay for . . ." is absolutely true! Don't go cheap! Any hand tool that you buy should be drop forged and/or chrome vanadium. These two qualities tell you that the tool is strong enough for the job. With any tool, go with a name that you've heard of before, or, that is recommended buy your local professional retailer. Let's go over a list of tools that you'll need.

Most of the world uses the metric system. However, some American-built engines and aftermarket accessories use standard fasteners. So, accumulate your tools accordingly. Any good DIYer should have a decent set of both U.S. and metric measure tools.

Don't be confused by terminology. Most advertising refers to "SAE and metric", or "standard and metric." Both are misnomers. The Society of Automotive Engineers (SAE) did not invent the English system of measurement; the English did. The SAE likes metrics just fine. Both English (U.S.) and metric measurements are SAE approved. Also, the current "standard" measurement IS metric. So, if it's not metric, it's U.S. measurement.

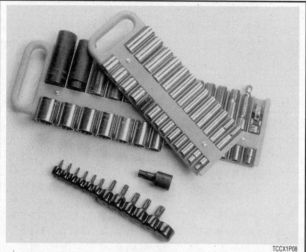

Fig. 6 Socket holders, especially the magnetic type, are handy items to keep tools in order

Hand Tools

▶ See Figure 7

SOCKET SETS

▶ See Figures 8, 9 and 10

Socket sets are the most basic, necessary hand tools repair and maintenance work. For our purposes, socket sets come in three

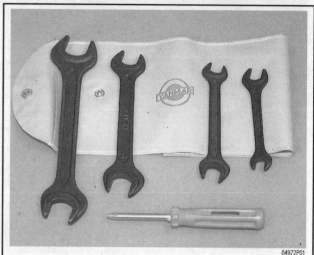

Fig. 7 Some manufacturers supply a small emergency tool kit with their engines, such as this one from Yanmar

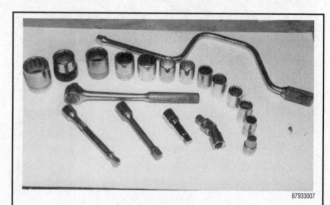

Fig. 8 A good half inch drive socket set

drive sizes: ¼ inch, ⅜ inch and ½ inch. Drive size refers to the size of the drive lug on the ratchet, breaker bar or speed handle.

A ⅜ inch set is probably the most versatile set in any mechanics tool box. It allows you to get into tight places that the larger drive ratchets can't and gives you a range of larger sockets that are still strong enough for heavy duty work. The socket set that you'll need should range in sizes from ⅜ inch through 1 inch for standard fasteners, and a 6mm through 19mm for metric fasteners.

You'll need a good ½ inch set since this size drive lug assures that you won't break a ratchet or socket on large or heavy fasteners. Also, torque wrenches with a torque scale high enough for larger fasteners are usually ½ inch drive.

¼ inch drive sets can be very handy in tight places. Though they usually duplicate functions of the ⅜ inch set, ¼ inch drive sets are easier to use for smaller bolts and nuts.

As for the sockets themselves, they come in standard and deep lengths as well as 6 or 12 point. 6 and 12 points refers to how many sides are in the socket itself. Each has advantages. The 6 point socket is stronger and less prone to slipping which would strip a bolt head or nut. 12 point sockets are more common, usually less expensive and can operate better in tight places where the ratchet handle can't swing far.

Standard length sockets are good for just about all jobs, how-

Fig. 9 A swivel (U-joint) adapter, and two types of drive adapters

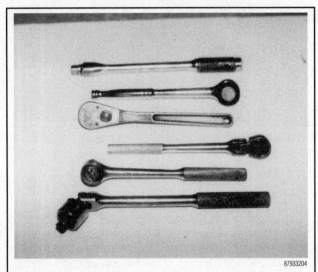

Fig. 10 Ratchets come in all sizes from rigid to swivel-headed

ever, some stud-head bolts, hard-to-reach bolts, nuts on long studs, etc., require the deep sockets.

Most manufacturers use recessed hex-head fasteners to retain many of the engine parts. These fasteners require a socket with a hex shaped driver or a large sturdy hex key. To help prevent torn knuckles, we would recommend that you stick to the sockets on any tight fastener and leave the hex keys for lighter applications. Hex driver sockets are available individually or in sets just like conventional sockets.

More and more, manufacturers are using Torx® head fasteners, which were once known as tamper resistant fasteners (because many people did not have tools with the necessary odd driver shape). They are still used where the manufacturer would prefer

only knowledgeable technicians or advanced Do-It-Yourselfers (DIYers) to work.

Torque Wrenches

▶ See Figure 11

In most applications, a torque wrench can be used to assure proper installation of a fastener. Torque wrenches come in various designs and most stores will carry a variety to suit your needs. A torque wrench should be used any time you have a specific torque value for a fastener. Keep in mind that because there is no worldwide standardization of fasteners, the charts at the end of this section are a general guideline and should be used with caution. If you

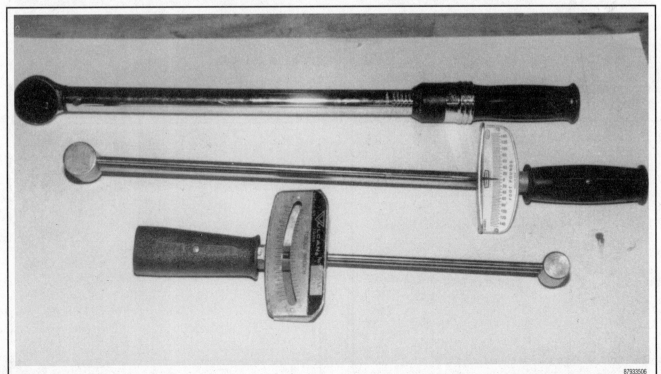

Fig. 11 Three types of torque wrenches. Top to bottom: a ½ inch drive clicker type, a ½ inch drive beam type and a ⅜ inch drive beam type that reads in inch lbs.

are using the right tool for the job, you should not have to strain to tighten a fastener.

BEAM TYPE

▶ See Figure 12

The beam type torque wrench is one of the most popular types. It consists of a pointer attached to the head that runs the length of the flexible beam (shaft) to a scale located near the handle. As the wrench is pulled, the beam bends and the pointer indicates the torque using the scale.

CLICK (BREAKAWAY) TYPE

▶ See Figure 13

Another popular torque wrench design is the click type. To use the click type wrench you pre-adjust it to a torque setting. Once the torque is reached, the wrench has a reflex signaling feature that causes a momentary breakaway of the torque wrench body, sending an impulse to the operator's hand.

Breaker Bars

▶ See Figure 14

Breaker bars are long handles with a drive lug. Their main purpose is to provide extra turning force when breaking loose tight bolts or nuts. They come in all drive sizes and lengths. Always wear gloves when using a breaker bar.

WRENCHES

▶ See Figures 15, 16 and 17

Basically, there are 3 kinds of fixed wrenches: open end, box end, and combination.

Open end wrenches have 2-jawed openings at each end of the wrench. These wrenches are able to fit onto just about any nut or bolt. They are extremely versatile but have one major drawback. They can slip on a worn or rounded bolt head or nut, causing bleeding knuckles and a useless fastener.

Box-end wrenches have a 360° circular jaw at each end of the wrench. They come in both 6 and 12 point versions just like sockets

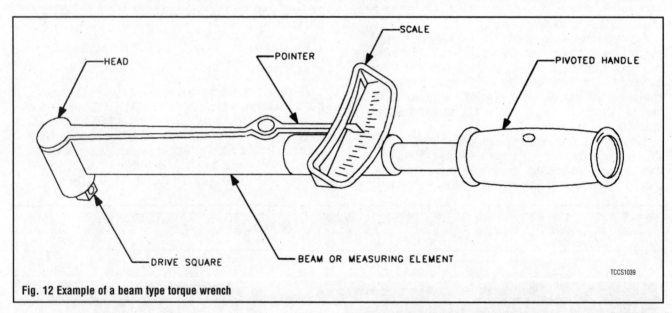

Fig. 12 Example of a beam type torque wrench

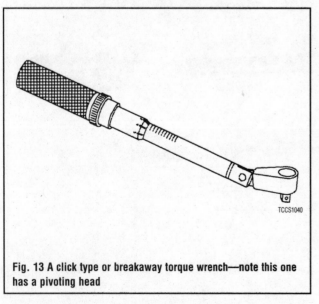

Fig. 13 A click type or breakaway torque wrench—note this one has a pivoting head

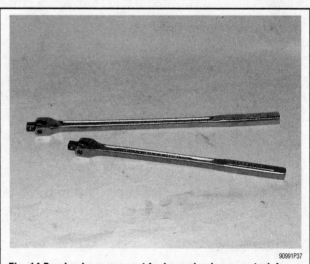

Fig. 14 Breaker bars are great for loosening large or stuck fasteners

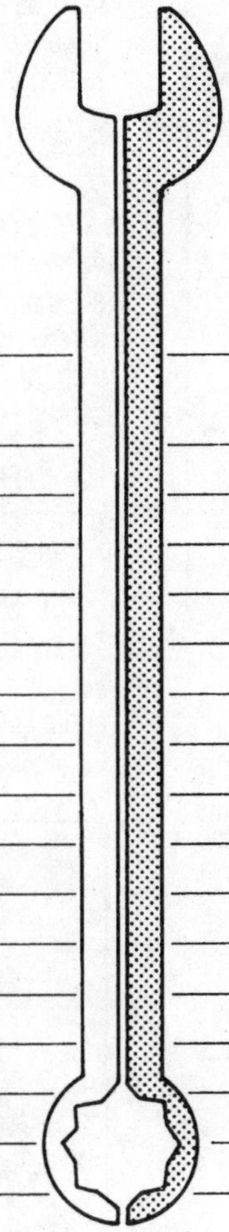

INCHES	DECIMAL		DECIMAL	MILLIMETERS
1/8''	.125		.118	3mm
3/16''	.187		.157	4mm
1/4''	.250		.236	6mm
5/16''	.312		.354	9mm
3/8''	.375		.394	10mm
7/16''	.437		.472	12mm
1/2''	.500		.512	13mm
9/16''	.562		.590	15mm
5/8''	.625		.630	16mm
11/16''	.687		.709	18mm
3/4''	.750		.748	19mm
13/16''	.812		.787	20mm
7/8''	.875		.866	22mm
15/16''	.937		.945	24mm
1''	1.00		.984	25mm

87933106

Fig. 15 Comparison of U.S. measure and metric wrench sizes

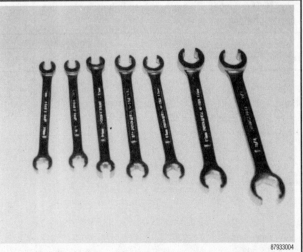

87933004

Fig. 16 Flarenut wrenches are critical to ensure tube fittings do not become rounded

and each type has the same advantages and disadvantages as sockets.

Combination wrenches have the best of both. They have a 2-jawed open end and a box end. These wrenches are probably the most versatile.

As for sizes, you'll probably need a range similar to that of the sockets, about 1/4 inch through 1 inch for standard fasteners, or 6mm through 19mm for metric fasteners. As for numbers, you'll need 2 of each size, since, in many instances, one wrench holds the nut while the other turns the bolt. On most fasteners, the nut and bolt are the same size.

➡**Although you will typically just need the sizes we specified, there are some exceptions. Occasionally you will find a nut which is larger. For these, you will need to buy ONE expensive wrench or a very large adjustable. Or you can always just convince the spouse that we are talking about safety here and buy a whole, expensive, large wrench set.**

One extremely valuable type of wrench is the adjustable wrench. An adjustable wrench has a fixed upper jaw and a moveable lower

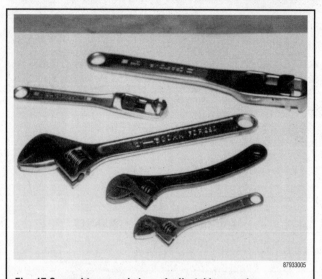

Fig. 17 Several types and sizes of adjustable wrenches

jaw. The lower jaw is moved by turning a threaded drum. The advantage of an adjustable wrench is its ability to be adjusted to just about any size fastener.

The main drawback of an adjustable wrench is the lower jaw's tendency to move slightly under heavy pressure. This can cause the wrench to slip if it is not facing the right way. Pulling on an adjustable wrench in the proper direction will cause the jaws to lock in place. Adjustable wrenches come in a large range of sizes, measured by the wrench length.

PLIERS

▶ **See Figure 18**

At least 2 pair of standard pliers is an absolute necessity. Pliers are simply mechanical fingers. They are, more than anything, an extension of your hand.

In addition to standard pliers there are the slip-joint, multi-position pliers such as ChannelLock® pliers and locking pliers, such as Vise Grips®.

Slip joint pliers are extremely valuable in grasping oddly sized parts and fasteners. Just make sure that you don't use them instead

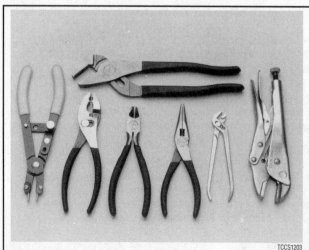

Fig. 18 Pliers and cutters come in many shapes and sizes. You should have an assortment on hand

of a wrench too often since they can easily round off a bolt head or nut.

Locking pliers are usually used for gripping bolts or studs that can't be removed conventionally. You can get locking pliers in square jawed, needle-nosed and pipe-jawed. Locking pliers can rank right up behind duct tape as the handy-man's best friend.

SCREWDRIVERS

You can't have too many screwdrivers. They come in 2 basic flavors, either standard or Phillips. Standard blades come in various sizes and thicknesses for all types of slotted fasteners. Phillips screwdrivers come in sizes with number designations from 1 on up, with the lower number designating the smaller size. Screwdrivers can be purchased separately or in sets.

HAMMERS

▶ **See Figure 19**

You always need a hammer — for just about any kind of work. For most metal work, you need a ball-peen hammer for using drivers and other like tools, a plastic hammer for hitting things safely, and a soft-faced dead-blow hammer for hitting things safely and hard. Hammers are also VERY useful with impact drivers.

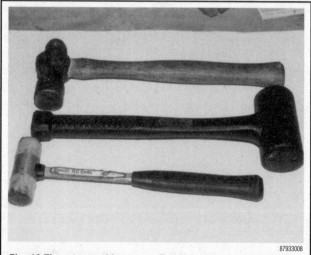

Fig. 19 Three types of hammers. Top to bottom: ball peen, rubber dead-blow, and plastic

OTHER COMMON TOOLS

There are a lot of other tools that every DIYer will eventually need (though not all for basic maintenance). They include:

- Funnels (for adding fluid)
- Chisels
- Punches
- Files
- Hacksaw
- Bench Vise
- Tap and Die Set
- Flashlight
- Magnetic Bolt Retriever
- Gasket scraper
- Putty Knife
- Screw/Bolt Extractors
- Prybar

Hacksaws have just one use—cutting things off. You may wonder why you'd need one for something as simple as maintenance, but you never know. Among other things, guide studs to ease parts installation can be made from old bolts with their heads cut off.

A tap and die set might be something you've never needed, but you will eventually. It's a good rule, when everything is apart, to clean-up all threads, on bolts, screws and threaded holes. Also, you'll likely run across a situation in which stripped threads will be encountered. The tap and die set will handle that for you.

Gasket scrapers are just what you'd think, tools made for scraping old gasket material off of parts. You don't absolutely need one. Old gasket material can be removed with a putty knife or single edge razor blade. However, putty knives may not be sharp enough for some really stubborn gaskets and razor blades have a knack of breaking just when you don't want them to, inevitably slicing the nearest body part! As the old saying goes, "always use the proper tool for the job". If you're going to use a razor to scrape a gasket, be sure to always use a blade holder.

Putty knives really do have a use in a repair shop. Just because you remove all the bolts from a component sealed with a gasket doesn't mean it's going to come off. Most of the time, the gasket and sealer will hold it tightly. Lightly driving a putty knife at various points between the two parts will break the seal without damage to the parts.

A small — 8-10 inches (20–25 centimeters) long — prybar is extremely useful for removing stuck parts.

➡**Never use a screwdriver as a prybar! Screwdrivers are not meant for prying. Screwdrivers, used for prying, can break, sending the broken shaft flying!**

Screw/bolt extractors are used for removing broken bolts or studs that have broke off flush with the surface of the part.

SPECIALTY TOOLS

Almost every marine engine around today requires at least one special tool to perform a certain task. In most cases, these tools are specially designed to overcome some unique problem or to fit on some oddly sized component.

When manufacturers go through the trouble of making a special tool, it is usually necessary to use it to assure that the job will be done right. A special tool might be designed to make a job easier, or it might be used to keep you from damaging or breaking a part.

Don't worry, MOST basic maintenance procedures can either be performed without any special tools OR, because the tools must be used for such basic things, they are commonly available for a reasonable price. It is usually just the low production, highly specialized tools (like a super thin 7-point star-shaped socket capable of 150 ft. lbs. (203 Nm) of torque that is used only on the crankshaft nut of the limited production what-dya-callit engine) that tend to be outrageously expensive and hard to find. Luckily, you will probably never need such a tool.

Special tools can be as inexpensive and simple as an adjustable strap wrench or as complicated as an ignition tester. A few common specialty tools are listed here, but check with your dealer or with other boaters for help in determining if there are any special tools for YOUR particular engine. There is an added advantage in seeking advice from others, chances are they may have already found the special tool you will need, but how to get it cheaper.

Battery Testers

The best way to test a non-sealed battery is using a hydrometer to check the specific gravity of the acid. Luckily, these are usually inexpensive and are available at most parts stores. Just be careful because the larger testers are usually designed for larger batteries and may require more acid than you will be able to draw from the battery cell. Smaller testers (usually a short, squeeze bulb type) will require less acid and should work on most batteries.

Electronic testers are available (and are often necessary to tell if a sealed battery is usable) but these are usually more than most DIYer's are willing to spend. Luckily, many parts stores have them on hand and are willing to test your battery for you.

Battery Chargers

▶ **See Figures 20 and 21**

If you are a weekend boater and take you boat out every day (or at least every week), then you will most likely want to buy a battery charger to keep your battery fresh. There are many types available,

Fig. 20 The Battery Tender® is more than just a battery charger, when left connected, it keeps your battery fully charged

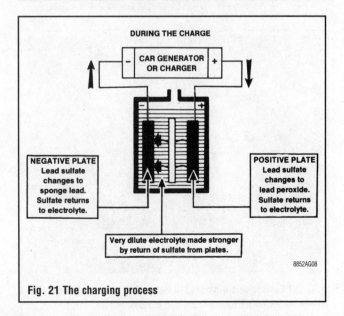

Fig. 21 The charging process

from low amperage trickle chargers to electronically controlled battery maintenance tools which monitor the battery voltage to prevent over or undercharging. This last type is especially useful if you store your boat for any length of time (such as during the severe winter months found in many Northern climates).

Even if you use your boat on a regular basis, you will eventually need a battery charger. Remember that most batteries are shipped dry and in a partial charged state. Before a new battery can be put into service it must be filled and properly charged. Failure to properly charge a battery (which was shipped dry) before it is put into service will prevent it from ever reaching a fully charged state.

Measuring Tools

Eventually, you are going to have to measure something. To do this, you will need at least a few precision tools in addition to the special tools mentioned earlier.

MICROMETERS & CALIPERS

Micrometers and calipers are devices used to make extremely precise measurements. The simple truth is that you really won't have the need for many of these items just for simple maintenance. You will probably want to have at least one precision tool such as an outside caliper to measure rotors or brake pads, but that should be sufficient to most basic maintenance procedures.

Should you decide on becoming more involved in boat engine mechanics, such as repair or rebuilding, then these tools will become very important. The success of any rebuild is dependent, to a great extent on the ability to check the size and fit of components as specified by the manufacturer. These measurements are made in thousandths and ten-thousandths of an inch.

Micrometers

▶ **See Figure 22**

A micrometer is an instrument made up of a precisely machined spindle which is rotated in a fixed nut, opening and closing the distance between the end of the spindle and a fixed anvil.

Outside micrometers can be used to check the thickness parts such as the brake rotors. They are also used during many rebuild and repair procedures to measure the diameter of components such

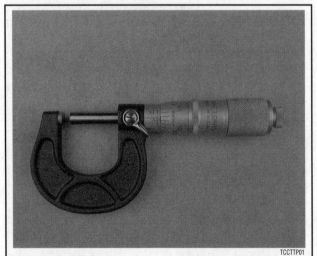

Fig. 22 Outside micrometers can be used to measure bake components including rotors, pads and pistons

as the pistons from a caliper or wheel cylinder. The most common type of micrometer reads in 1/1000 of an inch. Micrometers that use a vernier scale can estimate to 1/10 of an inch.

Inside micrometers are used to measure the distance between two parallel surfaces. For example, in engine rebuilding work, the inside mike measures cylinder bore wear and taper. Inside mikes are graduated the same way as outside mikes and are read the same way as well.

Remember that an inside mike must be absolutely perpendicular to the work being measured. When you measure with an inside mike, rock the mike gently from side to side and tip it back and forth slightly so that you span the widest part of the bore. Just to be on the safe side, take several readings. It takes a certain amount of experience to work any mike with confidence.

Metric micrometers are read in the same way as inch micrometers, except that the measurements are in millimeters. Each line on the main scale equals 1 mm. Each fifth line is stamped 5, 10, 15, and so on. Each line on the thimble scale equals 0.01 mm. It will take a little practice, but if you can read an inch mike, you can read a metric mike.

Calipers

▶ **See Figure 23**

Inside and outside calipers are useful devices to have if you need to measure something quickly and precise measurement is not necessary. Simply take the reading and then hold the calipers on an accurate steel rule.

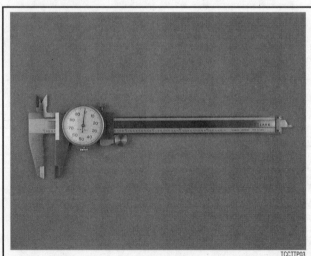

Fig. 23 Outside calipers are fast and easy ways to measure pads or rotors

DIAL INDICATORS

A dial indicator is a gauge that utilizes a dial face and a needle to register measurements. There is a movable contact arm on the dial indicator. When the arms moves, the needle rotates on the dial. Dial indicators are calibrated to show readings in thousandths of an inch and typically, are used to measure end-play and runout on various parts.

Dial indicators are quite easy to use, although they are relatively expensive. A variety of mounting devices are available so that the indicator can be used in a number of situations. Make certain that the contact arm is always parallel to the movement of the work being measured.

TELESCOPING GAUGES

A telescope gauge is used during rebuilding procedures (NOT usually basic maintenance) to measure the inside of bores. It can take the place of an inside mike for some of these jobs. Simply insert the gauge in the hole to be measured and lock the plungers after they have contacted the walls. Remove the tool and measure across the plungers with an outside micrometer.

DEPTH GAUGES

▶ **See Figure 24**

A depth gauge can be inserted into a bore or other small hole to determine exactly how deep it is. The most common use on maintenance items would be to check the depth of a rivet head (on riveted style brake pads) or to check tire depth. Some outside calipers contain a built-in depth gauge so money can be saved by just buying one tool.

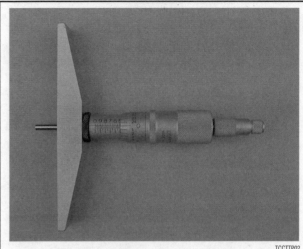

TCCTTP02

Fig. 24 Depth gauges, like this micrometer, can be used to measure the amount of pad or shoe remaining above a rivet

FASTENERS, MEASUREMENTS AND CONVERSIONS

Bolts, Nuts and Other Threaded Retainers

▶ **See Figures 25, 26, 27 and 28**

Although there are a great variety of fasteners found in the modern boat engine, the most commonly used retainer is the threaded fastener (nuts, bolts, screws, studs, etc). Most threaded retainers may be reused, provided that they are not damaged in use or during the repair. Some retainers (such as stretch bolts or torque prevailing nuts) are designed to deform when tightened or in use and should not be reinstalled.

Whenever possible, we will note any special retainers which should be replaced during a procedure. But you should always inspect the condition of a retainer when it is removed and you should replace any that show signs of damage. Check all threads for rust or corrosion which can increase the torque necessary to achieve the desired clamp load for which that fastener was originally selected. Additionally, be sure that the driver surface of the fastener has not been compromised by rounding or other damage. In some cases a driver surface may become only partially rounded, allowing the driver to catch in only one direction. In many of these occurrences, a fastener may be installed and tightened, but the driver would not be able to grip and loosen the fastener again. (This could lead to frustration down the line should that component ever need to be disassembled again).

If you must replace a fastener, whether due to design or damage, you must always be sure to use the proper replacement. In all

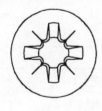
POZIDRIVE

PHILLIPS RECESS

TORX®

CLUTCH RECESS

INDENTED HEXAGON

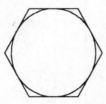
HEXAGON TRIMMED

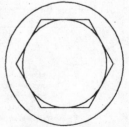
HEXAGON WASHER HEAD

TCCS1037

Fig. 25 Here are a few of the most common screw/bolt driver styles

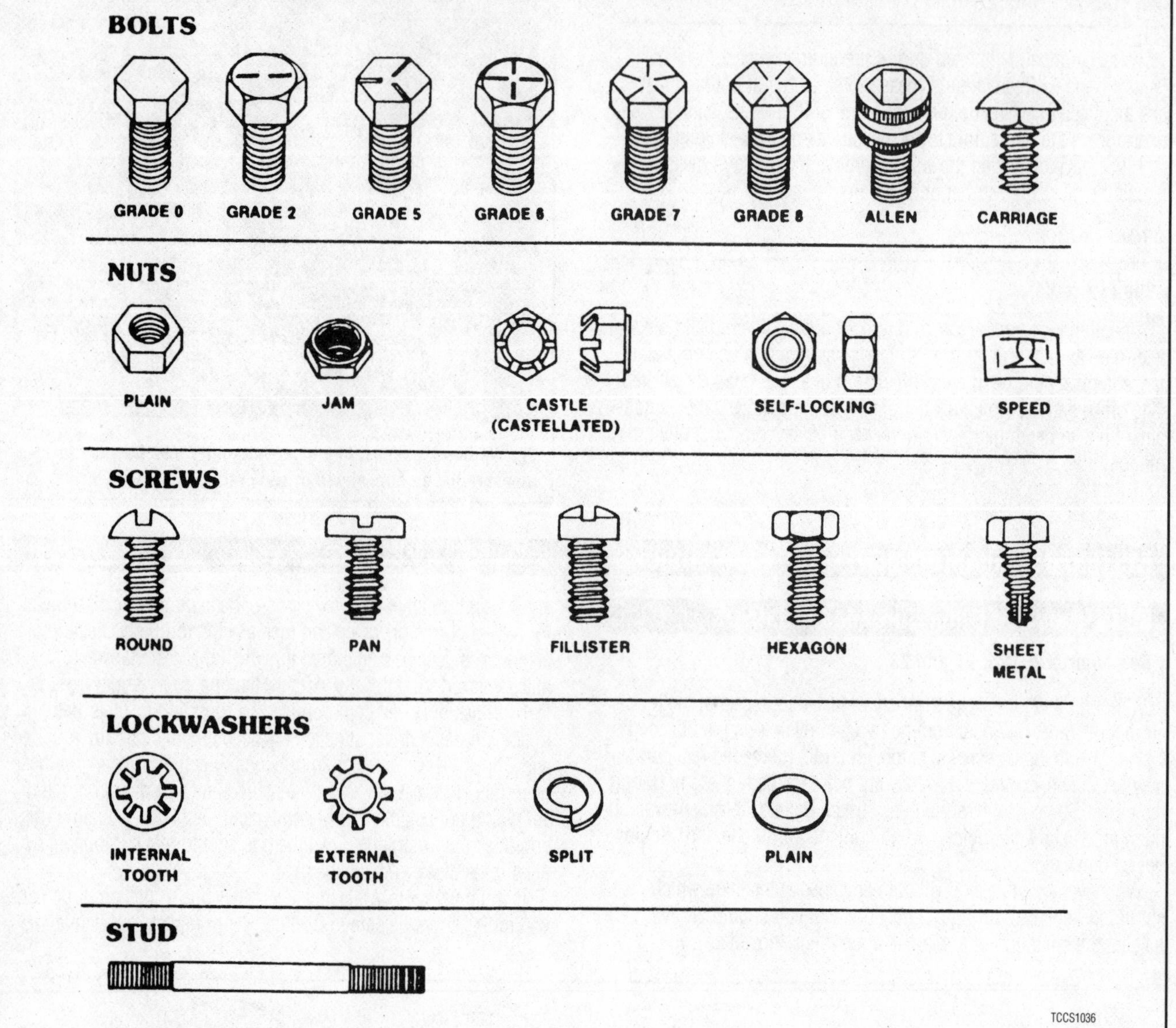

BOLTS

GRADE 0   GRADE 2   GRADE 5   GRADE 6   GRADE 7   GRADE 8   ALLEN   CARRIAGE

NUTS

PLAIN   JAM   CASTLE (CASTELLATED)   SELF-LOCKING   SPEED

SCREWS

ROUND   PAN   FILLISTER   HEXAGON   SHEET METAL

LOCKWASHERS

INTERNAL TOOTH   EXTERNAL TOOTH   SPLIT   PLAIN

STUD

TCCS1036

Fig. 26 There are many different types of threaded retainers

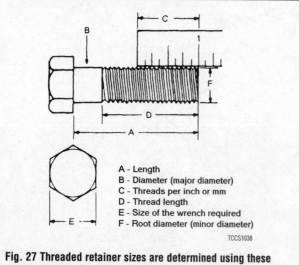

A - Length
B - Diameter (major diameter)
C - Threads per inch or mm
D - Thread length
E - Size of the wrench required
F - Root diameter (minor diameter)

TCCS1038

Fig. 27 Threaded retainer sizes are determined using these measurements

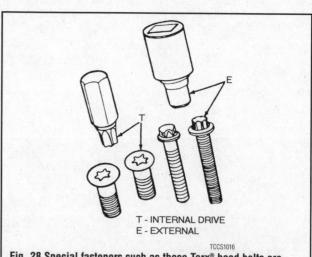

T - INTERNAL DRIVE
E - EXTERNAL

TCCS1016

Fig. 28 Special fasteners such as these Torx® head bolts are used by manufacturers to discourage people from working on vehicles without the proper tools (and knowledge)

cases, a retainer of the same design, material and strength should be used. Markings on the heads of most bolts will help determine the proper strength of the fastener. The same material, thread and pitch must be selected to assure proper installation and safe operation of the vehicle afterwards.

Thread gauges are available to help measure a bolt or stud's thread. Most part or hardware stores keep gauges available to help you select the proper size. In a pinch, you can use another nut or bolt for a thread gauge. If the bolt you are replacing is not too badly damaged, you can select a match by finding another bolt which will thread in its place. If you find a nut which threads properly onto the damaged bolt, then use that nut to help select the replacement bolt. If however, the bolt you are replacing is so badly damaged (broken or drilled out) that its threads cannot be used as a gauge, you might start by looking for another bolt (from the same assembly or a similar location) which will thread into the damaged bolt's mounting. If so, the other bolt can be used to select a nut; the nut can then be used to select the replacement bolt.

In all cases, be absolutely sure you have selected the proper replacement. Don't be shy, you can always ask the store clerk for help.

✳✳ WARNING

Be aware that when you find a bolt with damaged threads, you may also find the nut or drilled hole it was threaded into has also been damaged. If this is the case, you may have to drill and tap the hole, replace the nut or otherwise repair the threads. NEVER try to force a replacement bolt to fit into the damaged threads.

Torque

Torque is defined as the measurement of resistance to turning or rotating. It tends to twist a body about an axis of rotation. A common example of this would be tightening a threaded retainer such as a nut, bolt or screw. Measuring torque is one of the most common ways to help assure that a threaded retainer has been properly fastened.

When tightening a threaded fastener, torque is applied in three distinct areas, the head, the bearing surface and the clamp load. About 50 percent of the measured torque is used in overcoming bearing friction. This is the friction between the bearing surface of the bolt head, screw head or nut face and the base material or washer (the surface on which the fastener is rotating). Approximately 40 percent of the applied torque is used in overcoming thread friction. This leaves only about 10 percent of the applied torque to develop a useful clamp load (the force which holds a joint together). This means that friction can account for as much as 90 percent of the applied torque on a fastener.

Standard and Metric Measurements

▶ See Figure 29

Specifications are often used to help you determine the condition of various components, or to assist you in their installation. Some of the most common measurements include length (in. or cm/mm), torque (ft. lbs., inch lbs. or Nm) and pressure (psi, in. Hg, kPa or mm Hg).

In some cases, that value may not be conveniently measured with what is available in your toolbox. Luckily, many of the measuring devices which are available today will have two scales so Standard or Metric measurements may easily be taken. If any of the various measuring tools which are available to you do not contain the same scale as listed in your specifications, use the accompanying conversion factors to determine the proper value.

The conversion factor chart is used by taking the given specification and multiplying it by the necessary conversion factor. For instance, looking at the first line, if you have a measurement in inches such as "free-play should be 2 in." but your ruler reads only in millimeters, multiply 2 in. by the conversion factor of 25.4 to get the metric equivalent of 50.8mm. Likewise, if the specification was given only in a Metric measurement, for example in Newton Meters (Nm), then look at the center column first. If the measurement is 100 Nm, multiply it by the conversion factor of 0.738 to get 73.8 ft. lbs.

CONVERSION FACTORS

LENGTH–DISTANCE

Inches (in.)	x 25.4	= Millimeters (mm)	x .0394	= Inches
Feet (ft.)	x .305	= Meters (m)	x 3.281	= Feet
Miles	x 1.609	= Kilometers (km)	x .0621	= Miles

VOLUME

Cubic Inches (in3)	x 16.387	= Cubic Centimeters	x .061	= in3
IMP Pints (IMP pt.)	x .568	= Liters (L)	x 1.76	= IMP pt.
IMP Quarts (IMP qt.)	x 1.137	= Liters (L)	x .88	= IMP qt.
IMP Gallons (IMP gal.)	x 4.546	= Liters (L)	x .22	= IMP gal.
IMP Quarts (IMP qt.)	x 1.201	= US Quarts (US qt.)	x .833	= IMP qt.
IMP Gallons (IMP gal.)	x 1.201	= US Gallons (US gal.)	x .833	= IMP gal.
Fl. Ounces	x 29.573	= Milliliters	x .034	= Ounces
US Pints (US pt.)	x .473	= Liters (L)	x 2.113	= Pints
US Quarts (US qt.)	x .946	= Liters (L)	x 1.057	= Quarts
US Gallons (US gal.)	x 3.785	= Liters (L)	x .264	= Gallons

MASS–WEIGHT

Ounces (oz.)	x 28.35	= Grams (g)	x .035	= Ounces
Pounds (lb.)	x .454	= Kilograms (kg)	x 2.205	= Pounds

PRESSURE

Pounds Per Sq. In. (psi)	x 6.895	= Kilopascals (kPa)	x .145	= psi
Inches of Mercury (Hg)	x .4912	= psi	x 2.036	= Hg
Inches of Mercury (Hg)	x 3.377	= Kilopascals (kPa)	x .2961	= Hg
Inches of Water (H_2O)	x .07355	= Inches of Mercury	x 13.783	= H_2O
Inches of Water (H_2O)	x .03613	= psi	x 27.684	= H_2O
Inches of Water (H_2O)	x .248	= Kilopascals (kPa)	x 4.026	= H_2O

TORQUE

Pounds–Force Inches (in–lb)	x .113	= Newton Meters (N·m)	x 8.85	= in–lb
Pounds–Force Feet (ft–lb)	x 1.356	= Newton Meters (N·m)	x .738	= ft–lb

VELOCITY

Miles Per Hour (MPH)	x 1.609	= Kilometers Per Hour (KPH)	x .621	= MPH

POWER

Horsepower (Hp)	x .745	= Kilowatts	x 1.34	= Horsepower

FUEL CONSUMPTION*

Miles Per Gallon IMP (MPG)	x .354	= Kilometers Per Liter (Km/L)
Kilometers Per Liter (Km/L)	x 2.352	= IMP MPG
Miles Per Gallon US (MPG)	x .425	= Kilometers Per Liter (Km/L)
Kilometers Per Liter (Km/L)	x 2.352	= US MPG

*It is common to covert from miles per gallon (mpg) to liters/100 kilometers (1/100 km), where mpg (IMP) x 1/100 km = 282 and mpg (US) x 1/100 km = 235.

TEMPERATURE

Degree Fahrenheit (°F)	= (°C x 1.8) + 32
Degree Celsius (°C)	= (°F – 32) x .56

TCCS1044

Fig. 29 Standard and metric conversion factors chart

Metric Bolts

Relative Strength Marking	4.6, 4.8			8.8		
Bolt Markings						
Usage	Frequent			Infrequent		
	Maximum Torque			Maximum Torque		
Bolt Size	Ft-Lb	Kgm	Nm	Ft-Lb	Kgm	Nm
Thread Size x Pitch (mm)						
6 x 1.0	2–3	.2–.4	3–4	3–6	.4–.8	5–8
8 x 1.25	6–8	.8–1	8–12	9–14	1.2–1.9	13–19
10 x 1.25	12–17	1.5–2.3	16–23	20–29	2.7–4.0	27–39
12 x 1.25	21–32	2.9–4.4	29–43	35–53	4.8–7.3	47–72
14 x 1.5	35–52	4.8–7.1	48–70	57–85	7.8–11.7	77–110
16 x 1.5	51–77	7.0–10.6	67–100	90–120	12.4–16.5	130–160
18 x 1.5	74–110	10.2–15.1	100–150	130–170	17.9–23.4	180–230
20 x 1.5	110–140	15.1–19.3	150–190	190–240	26.2–46.9	160–320
22 x 1.5	150–190	22.0–26.2	200–260	250–320	34.5–44.1	340–430
24 x 1.5	190–240	26.2–46.9	260–320	310–410	42.7–56.5	420–550

88523G12

SAE Bolts

SAE Grade Number	1 or 2			5			6 or 7		
Bolt Markings									
Usage	Frequent			Frequent			Infrequent		
Bolt Size (inches)—(Thread)	Maximum Torque			Maximum Torque			Maximum Torque		
	Ft-Lb	kgm	Nm	Ft-Lb	kgm	Nm	Ft-Lb	kgm	Nm
1/4—20	5	0.7	6.8	8	1.1	10.8	10	1.4	13.5
—28	6	0.8	8.1	10	1.4	13.6			
5/16—18	11	1.5	14.9	17	2.3	23.0	19	2.6	25.8
—24	13	1.8	17.6	19	2.6	25.7			
3/8—16	18	2.5	24.4	31	4.3	42.0	34	4.7	46.0
—24	20	2.75	27.1	35	4.8	47.5			
7/16—14	28	3.8	37.0	49	6.8	66.4	55	7.6	74.5
—20	30	4.2	40.7	55	7.6	74.5			
1/2—13	39	5.4	52.8	75	10.4	101.7	85	11.75	115.2
—20	41	5.7	55.6	85	11.7	115.2			
9/16—12	51	7.0	69.2	110	15.2	149.1	120	16.6	162.7
—18	55	7.6	74.5	120	16.6	162.7			
5/8—11	83	11.5	112.5	150	20.7	203.3	167	23.0	226.5
—18	95	13.1	128.8	170	23.5	230.5			
3/4—10	105	14.5	142.3	270	37.3	366.0	280	38.7	379.6
—16	115	15.9	155.9	295	40.8	400.0			
7/8—9	160	22.1	216.9	395	54.6	535.5	440	60.9	596.5
—14	175	24.2	237.2	435	60.1	589.7			
1—8	236	32.5	318.6	590	81.6	799.9	660	91.3	894.8
—14	250	34.6	338.9	660	91.3	849.8			

Manufacturers' marks may vary—number of lines always two less than the grade number.

88526G10

3

MAINTENANCE
AND TUNE-UP

ENGINE MAINTENANCE

Introduction

▶ **See Figure 1**

Your decision to use a diesel engine was probably due to their reputation of being remarkably long lived and reliable. A secondary consideration could have been the small amount of regular maintenance they require. Or, the boat you purchased may have been fitted with a diesel. In any case, you are now charged with the maintenance of your investment.

It is not uncommon to see diesel engines well over 20 years of age moving a boat through the water as if the unit had recently been purchased from the current line of models. An inquiry with the proud owner will undoubtedly reveal his main credit for its performance to be regular maintenance.

It is highly uncommon to see a marine diesel wear out if properly maintained. Yanmar places a 10,000 hour life expectancy on their engines. This time period covers the engine from when it is new, to when the oil consumption is though to be too high.

Put simply, the little maintenance a diesel engine does require is absolutely essential. Not paying attention to a regular maintenance program can lead to thousands of dollars of damage, stranding or lost time out on the water.

Diesel engine maintenance centers around cleanliness. You must keep the air, fuel and oil entering the engine clean. If these items are kept clean, the result should be a happy, healthy, long lived, reliable engine.

Another thing that should be remembered when maintaining diesel engines: THERE ARE NO RULES OF THUMB. Always follow the maintenance intervals given for your specific engine.

Fig. 1 Put simply, the little maintenance a diesel engine does require is absolutely essential

04971P10

General Engine Specifications

Cylinders	Model	Years Available	Displace cu.in. (liters)	Horsepower @ RPM	Cooling System	Compression Ratio	Induction System	Combustion System
1	1GM	1980-83	19(0.293)	7.5@3600	Raw	23:1	Naturally Aspirated	Swirl Pre -Combustion
1	1GM10	1983-98	19(0.32)	9@3600	Raw	23:1	Naturally Aspirated	Swirl Pre -Combustion
2	2GM	1980-83	38(0.586)	15@3600	Raw/Fresh	23:1	Naturally Aspirated	Swirl Pre -Combustion
2	2GM20	1983-98	39(0.64)	18@3600	Raw/Fresh	23:1	Naturally Aspirated	Swirl Pre -Combustion
2	2QM20	1975-80	68(1.1)	22@2800	Raw/Fresh	20:1	Naturally Aspirated	Swirl Pre -Combustion
3	3GM	1980-83	56(0.879)	22.5@3600	Raw/Fresh	23:1	Naturally Aspirated	Swirl Pre -Combustion
3	3GM30	1983-98	58(0.95)	27@3600	Raw/Fresh	23:1	Naturally Aspirated	Swirl Pre -Combustion
3	3HM	1980-83	56(0.879)	30@3400	Raw/Fresh	22.7:1	Naturally Aspirated	Swirl Pre -Combustion
3	3HM35	1983-98	78(1.3)	34@3400	Raw/Fresh	24.8:1	Naturally Aspirated	Swirl Pre -Combustion
3	3QM30	1976-80	68(1.1)	33@2800	Raw/Fresh	20:1	Naturally Aspirated	Swirl Pre -Combustion
4	4JH-DTE	1985-98	100(1.6)	77@3600	Raw/Fresh	15.9:1	IntercooledTurbo	Direct Injection
4	4JH-E	1983-98	100(1.6)	44@3600	Raw/Fresh	17.8:1	Naturally Aspirated	Direct Injection
4	4JH-HTE	1985-98	100(1.6)	66@3600	Raw/Fresh	15.9:1	IntercooledTurbo	Direct Injection
4	4JH-TE	1983-98	100(1.6)	55@3600	Raw/Fresh	16.2:1	Turbocharged	Direct Injection
4	4JH2-DTE	1985-98	111(1.8)	88@3600	Raw/Fresh	17.2:1	IntercooledTurbo	Direct Injection
4	4JH2-E	1985-98	111(1.8)	50@3600	Raw/Fresh	17.2:1	Naturally Aspirated	Direct Injection
4	4JH2-HTE	1985-98	111(1.8)	75@3600	Raw/Fresh	17.2:1	IntercooledTurbo	Direct Injection
4	4JH2-HTE	1985-98	111(1.8)	100@3600	Raw/Fresh	17.2:1	IntercooledTurbo	Direct Injection
4	4JH2-TE	1985-98	111(1.8)	62@3600	Raw/Fresh	17.2:1	Turbocharged	Direct Injection

04974C01

Periodic Maintenance

Component	Procedure	Daily	50 Hours	100 Hours	Every 250 Hours	300 Hours	500 Hours
Fuel	Check Level	*					
	Remove Condensation	*					
	Replace Fuel Filter Element				*		
Lubricating Oil	Check Level	*					
	Change Oil in Crankcase			*			
	Change Oil Filter			*			
	Change Oil in Clutch				*		
Cooling System	Clean Thermostat				*		
	Check and Replace Zinc						*
Sea Water	Check Circulation	*					
	Check and Replace Pump Impeller						*
Fresh Water	Check and Refill	*					
	Adjust Drive Belt				*		
	Check Heat Exchanger					*	
	Check Water Pump						*
Engine Exterior	Check and Repair Leaks	*					
Engine Piping	Check and Repair Leaks	*					
Air Intake and Exhaust	Clean Air Intake Silencer						
	Clean Exhaust Water Mixing Elbow					*	
	Clean Breather Pipe					*	
	Check Exhaust Gas Condition						
	Clean Turbo Compressor			*			
Electrical System	Check Battery Electrolyte	*					
	Check Warning Lamps	*					
	Check Belt Tension				*		
Remote Control	Check Operation		*				
Fuel Injection Pump	Check Ignition Timing						*
Fuel Injection Valve	Check Spray Condition						*
	Check and Adjust Injection Pressure						*

04974C02

Note: Recommendations made here are general guidelines and are not intended to replace those recommended intervals suggested in your operators manual.

Engine Identification

♦ See Figures 2 thru 10

➡Most identification tags are difficult to see when the engine is installed in the engine compartment. You may need to use a mirror to read all the numbers off the tag.

Every engine has a model name plate and clutch name plate fitted to it. In addition, the engine serial number is stamped on the cylinder block or cylinder head.

Specifications of the engine and clutch are recorded and filed using the numbers marked on the plates. When parts are ordered, always use your engine model and number or clutch model, gear ratio or number to correctly identify the components you own.

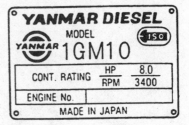

Engine model name plate

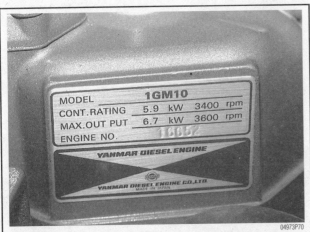

Fig. 2 The engine identification tag tells you all you need to know when ordering parts or looking up information

Clutch model name plate

Fig. 4 Engine and clutch name plates —GM and HM series engines

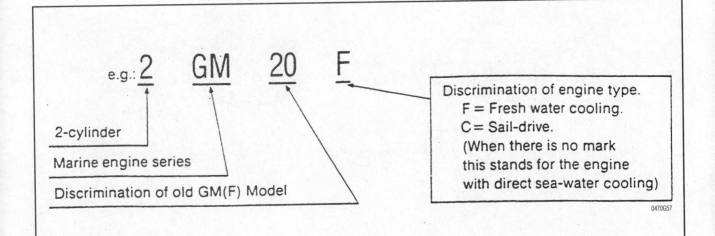

e.g.: 2 GM 20 F

2-cylinder

Marine engine series

Discrimination of old GM(F) Model

Discrimination of engine type.
F = Fresh water cooling.
C = Sail-drive.
(When there is no mark this stands for the engine with direct sea-water cooling)

Fig. 3 The engine series number can be translated as illustrated—GM and HM series engines

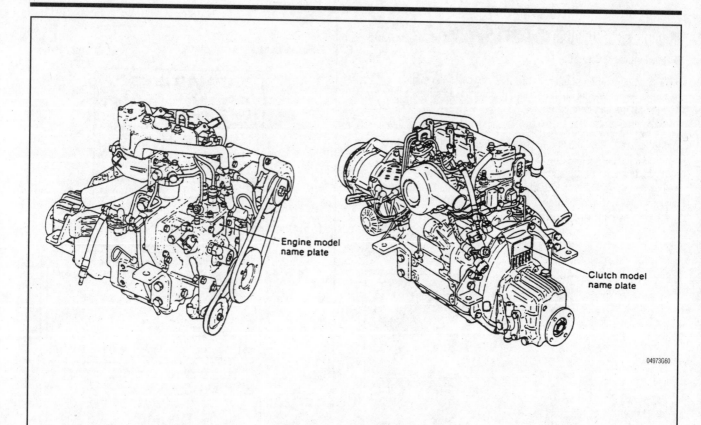

Fig. 5 Engine and clutch name plate location—2GM, 2GM20, 3GM, and 3GM30 engines

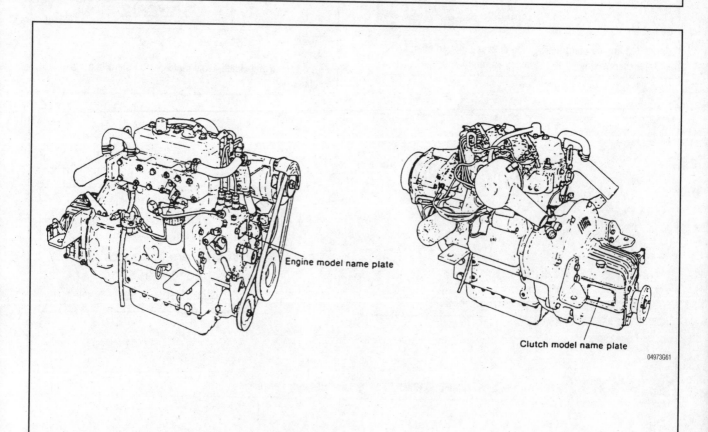

Fig. 6 Engine and clutch name plate location—3HM, 3HM35 engines

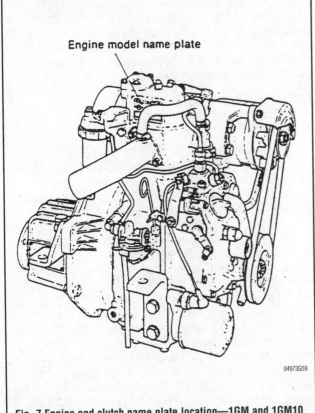

04973G59

Fig. 7 Engine and clutch name plate location—1GM and 1GM10 engines

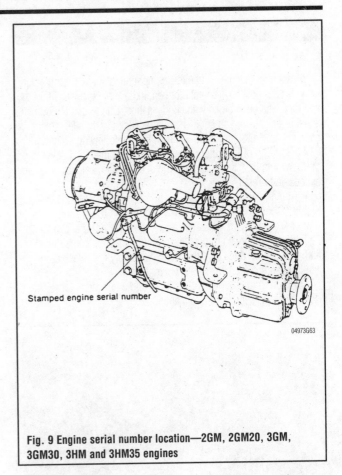

04973G63

Fig. 9 Engine serial number location—2GM, 2GM20, 3GM, 3GM30, 3HM and 3HM35 engines

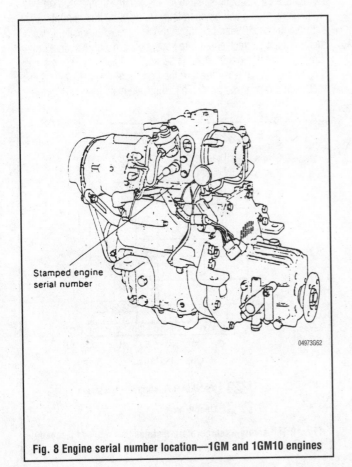

04973G62

Fig. 8 Engine serial number location—1GM and 1GM10 engines

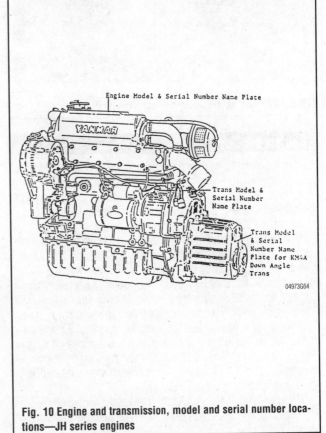

04973G64

Fig. 10 Engine and transmission, model and serial number locations—JH series engines

Fuel

▶ **See Figure 11**

Diesel fuel, due to its chemical compound, actually promotes or provides an ideal environment for certain bacterial growth within the fuel tank when combined with condensation. Due to this phenomena, a stabilizer/anti-gelling additive must be added to the fuel tank at each fill-up. Failure to do so will result in the following conditions:

• Clogging of fuel filters, lines, pumps and injectors. A symptom of this would be loss of power or no acceleration, and can be quite expensive to rectify.

• Gelling can occur at lower ambient air temperatures due to the higher viscosity of the fuel or due to chemical breakdown (destabilization) which again can cause a lack of power or no acceleration.

The bacteria does not produce a moldy off-odor to the diesel fuel. However, it does give the fuel a noticeably rich black color, which is an indicator that the problem exists.

For more information on diesel fuel and additives, refer to the "Fuel System" section of this manual.

SELSTK16

Fig. 11 Always obtain your fuel from a reputable fuel dock

Engine Oil

Nothing affects the performance and durability of a diesel engine more than the engine oil. If inferior oil is used, or if your engine oil is not changed regularly, the risk of piston seizure, piston ring sticking, accelerated wear of the cylinder walls or liners, bearings and other moving components increases significantly.

Maintaining the correct engine oil level is one of the most basic (and essential) forms of engine maintenance. Get into the habit of checking your oil on a regular basis; all engines naturally consume small amounts of oil, and if left neglected, can consume enough oil to damage the internal components of the engine. Assuming the oil level is correct because you "checked it the last time" can be a costly mistake.

If your engine has not been operated for more than 72 hours it should be pre-lubed prior to starting. This is accomplished by leaving the fuel shutoff in the **OFF** position and cranking the engine for 5 seconds. This will provide sufficient lubrication to vital engine components and prevent a dry startup.

Normally it takes 1–2 minutes to stabilize the engine oil pressure after starting. Do not be alarmed if engine oil pressure fluctuates during this period.

When shutting the engine down, always let the engine idle a few minutes to bring engine temperature down to a normal level. Since the engine is, at least in part, cooled by engine oil, it is necessary to allow engine oil temperature to stabilize prior to shutdown. Not allowing the temperature to stabilize can damage vital engine components such as the turbocharger.

ENGINE OIL RECOMMENDATIONS

▶ **See Figure 12**

Every bottle of engine oil for sale in the U.S. should have a label describing what standards it meets. Engine oil service classifications are designated by the American Petroleum Institute (API), based on the chemical composition of a given type of oil and testing of samples. The ratings include "S" (normal gasoline engine use) and "C" (commercial, fleet and diesel) applications. Over the years, the C rating has been supplemented with various letters, each one representing the latest and greatest rating available at the time of its introduction. During recent years these ratings have changed and most recently (at the time of this manual's publication), the rating is CH–4. Each successive rating usually meets all of the standards of the previous alpha designation, but also meets some new criteria, meets higher standards and/or contains newer or different additives. Since oil is so important to the life of your engine, you should obviously NEVER use an oil of questionable quality. Oils that are labeled with modern API ratings, including the "energy conserving" donut symbol, have been proven to meet the API quality standards. Always use the highest grade of oil available. The better quality of the oil, the better it will lubricate the internals of your engine.

In addition to meeting the classification of the API, your oil should be of a viscosity suitable for the outside temperature in which your engine will be operating. Oil must be thin enough to get between the close-tolerance moving parts it must lubricate. Once there, it must be thick enough to separate them with a slippery oil film. If the oil is too thin, it won't separate the parts; if it's too thick, it can't squeeze between them in the first place—either way, excess friction and wear takes place. To complicate matters, cold-morning

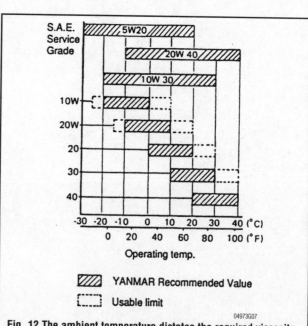

04973G07

Fig. 12 The ambient temperature dictates the required viscosity of engine oil

starts require a thin oil to reduce engine resistance, while high speed driving requires a thick oil which can lubricate vital engine parts at temperatures.

According to the Society of Automotive Engineers' (SAE) viscosity classification system, an oil with a high viscosity number (such as SAE 40 or SAE 50) will be thicker than one with a lower number (SAE 10W). The "W" in 10W indicates that the oil is desirable for use in winter operation, and does not stand for "weight". Through the use of special additives, multiple-viscosity oils are available to combine easy starting at cold temperatures with engine protection at high speeds. For example, a 10W40 oil is said to have the viscosity of a 10W oil when the engine is cold and that of a 40 oil when the engine is warm. The use of such an oil will decrease engine resistance and improve efficiency.

➡ **Yanmar does not approve synthetic oil for use in any of their engines.**

OIL LEVEL CHECK

◗ See Figures 13, 14, 15 and 16

✳✳ CAUTION

The EPA warns that prolonged contact with used engine oil may cause a number of skin disorders, including cancer! You should make every effort to minimize your exposure to used engine oil. Protective gloves should be worn when changing the oil. Wash your hands and any other exposed skin areas as soon as possible after exposure to used engine oil. Soap and water, or waterless hand cleaner should be used.

When checking the oil level, it is best that the boat be level and the oil be at operating temperature. Checking the level immediately after stopping the engine will give a false reading.

It is normal for a diesel engine to naturally consume oil during the course of operation. You should not be alarmed if the oil level in your engine drops slightly between inspections. Also the color of the oil is usually a pitch black color. Smelling the oil is a better indicator of oil condition than the color. If the oil smells burned, it should be replaced immediately.

➡ **It takes a little while for fresh oil poured into the engine to reach the crankcase. Wait for about 3 minutes and then check the oil level again.**

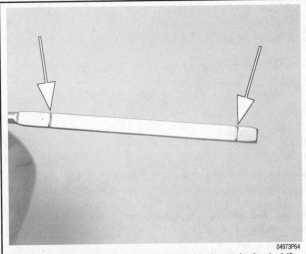

Fig. 14 The oil level on the dipstick should always be kept at the full mark on the dipstick (left)

Fig. 15 If the oil level is low, add oil through the oil cap hole in the valve cover

Fig. 13 To check the engine oil level, pull the dipstick from the tube, wipe it clean, then reinsert it to obtain a reading

Fig. 16 On JH series engines, a secondary oil fill hole is added on the front of the engine cover if access to the valve cover is limited

Oil level should be checked each day the engine is operated.

1. Locate the engine oil dipstick.

2. Clean the area around the dipstick to prevent dirt from entering the engine.

3. Remove the dipstick and note the color of the oil. Wipe the dipstick clean with a rag.

4. Insert the dipstick fully into the tube and remove it again. Hold the dipstick horizontal and read the level on the dipstick. The level should always be at the upper limit. If the oil level is below the upper limit, sufficient oil should be added to restore the proper level of oil in the crankcase.

➡**On JH series engine, slowly insert the dipstick. The seal at the top of the dipstick is quite efficient. If the dipstick is inserted too quickly, the seal will pressurize the dipstick tube and force oil out of the tube into the crankcase. This will result in an inaccurate oil level reading.**

5. See "Engine Oil Recommendations" for the proper viscosity and type of oil.

6. Oil is added through the filler port cap in the top of the valve cover. Add oil slowly and check the level frequently to prevent overfilling the engine.

❋❋ WARNING

Do not overfill the engine. If the engine is overfilled, the crankshaft will whip the engine oil into a foam causing loss of lubrication and severe engine damage.

OIL & FILTER CHANGE

▶ **See Figures 17 thru 24**

A few precautions can make the messy job of oil and filter maintenance much easier. By placing oil absorbent pads, available at industrial supply stores, into the area below the engine, you can prevent oil spillage from reaching the bilge.

It is a good idea to warm the engine oil first so it will flow better and the contaminates in the bottom of the pan are suspended in the oil. This is accomplished by starting the engine and allowing it to reach operating temperature.

Changing engine oil is sometimes complicated by the location of the drain plug. Most boats equipped with inboard engines use an evacuation pump to remove the used engine oil through the dipstick tube. If you don't have a permanently mounted oil suction pump in your engine compartment, you may want to consider installing one. This pump sucks waste oil out through either the dipstick tube or a connection on the oil drain plug. JH series engines are usually equipped with a special fitting on the side of the crankcase especially for an evacuation pump.

➡**It is recommended by Yanmar that the drain plugs on the bottom of the oil pan not be removed. Removing the plugs will break the seal installed at the factory and allow engine oil to drip into the bilge. If you have removed the plugs accidentally, purchase new plugs and install them with the proper sealant.**

The maintenance interval for oil and filter change is every 100 hours of engine operation. This interval should be strictly kept especially in the case of auxiliary sailboat engines since these engines rarely get up to operating temperature.

➡**Since you will be hanging into the engine compartment, gather all the tools and spare parts necessary for the job. Don't forget plenty of rags to clean up any spills.**

1. Connect the evacuation pump to the dipstick tube. Keep in mind that the fast flowing oil, which will spill out of the pump hose will flow with enough force that it could miss the bucket and end up in all over the deck. Position the bucket accordingly and be ready to move it if necessary.

❋❋ CAUTION

Use caution around the hot oil; when at operating temperature, it is hot enough to cause a severe burn.

2. Allow the oil to drain until nothing but a few drops come out pump. It should be noted that depending on the angle of the engine, some oil may be left in the crankcase. This is normal and should not cause the engine harm.

3. Remove evacuation pump.

4. Position a drain pan under the oil filter. Some filters are mounted horizontally and some vertically. In either case, there is usually oil left in the filter. When the filter is removed, oil will flow out of the engine. If you are not prepared, you will have a mess on

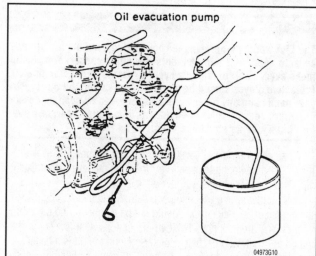

Fig. 17 If there is not enough room to place a drain pan under the engine, the oil can be removed with an oil evacuation pump

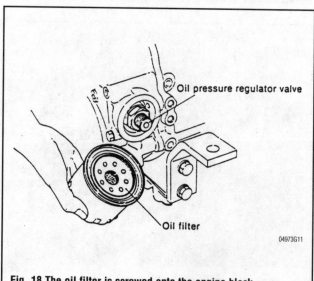

Fig. 18 The oil filter is screwed onto the engine block

Fig. 19 If there is room under the engine, place a container (like this cut out oil bottle) under the filter to catch the oil from the filter

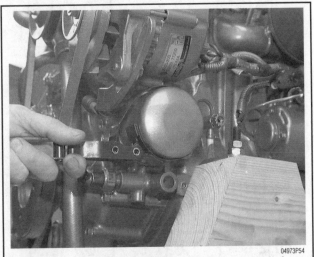

Fig. 20 Using an oil filter wrench, turn the filter counterclockwise to loosen it from the engine

Fig. 21 Once the filter is loosened, turn it a few turns and allow the oil to drain from the filter

your hands. It may also be necessary to use a funnel or fashion some type of drain shield to guide the oil into the drain pan. This can be a simple as a recycled oil bottle with the bottom cut off or an elaborate creation made of tin. In any case, it's purpose is to prevent oil from spilling all over the engine.

5. To remove the filter, you will probably need and oil filter wrench. Heat from the engine tends to tighten even a properly installed filter and makes it difficult to remove. Place the wrench on the filter as close to the engine as possible, while still leaving room to work. This will put the wrench at the strongest part of the filter (near the threaded end) and prevent crushing the filter. Loosen the filter with the wrench using a counterclockwise turning motion.

6. Once loosened, wrap a rag around the filter and unscrew it from the boss on the engine. Make sure that the drain pan and shield are positioned properly before you start unscrewing the filter. Should some of the hot oil happen to get on your hands and burn you, dump the filter into the drain pan.

7. Wipe the base of the mounting boss with a clean, dry rag. If the filter is installed vertically, you may want to fill the filter about half way with oil prior to installation. This will prevent oil starvation when you fire the engine up again. Pre-filling the oil filter is usually not possible with horizontally mounted filters. Smear a little bit of fresh oil on the filter gasket to help it seat properly on the engine. Install the filter and tighten it approximately a quarter-turn after it contacts the mounting boss (always follow the filter manufacturer's instructions). This usually equals "hand tight." Using a wrench to tighten the filter is not required.

➡Only use Yanmar or equivalent quality marine oil filters. These filters contain a bypass valve which is important in the proper operation of the lubrication system.

✱✱ WARNING

Never operate the engine without engine oil. Severe and costly engine damage will result in a matter of seconds without proper lubrication.

8. If any oil has gotten into the bilge, remove it using an oil absorbent pad. These pads are specially formulated to only absorb oil and will not soak up any water in the bilge. Perform a visual inspection and make sure all connections are tight.

9. Carefully remove the drain pan from under the oil filter and transfer the oil into a suitable container for recycling.

10. Refill the engine with the proper quantity and quality of oil immediately. You may laugh at the severity of this warning, but if you wait to refill the engine and someone unknowingly tries to start it, severe and costly engine damage will result.

11. Refill the engine crankcase slowly through the filler cap on the valve cover. Use a funnel as necessary to prevent spilling oil. Check the level often. You may notice that it usually takes less than the amount of oil listed in the Maintenance and Tune-up Specifications chart to refill the crankcase. This is only until the engine is started and the oil filter is filled with oil.

12. Since most engine wear occurs during the first few seconds after starting an engine (before the pressurized oil reaches the bearings), it is especially important to prime the oil system after an oil change. Large commercial engines have a special high pressure electric oil pump that is used to prime the engine prior to starting. However, most small engines do not have this option. Luckily, you can accomplish the same task using the starter motor. Simply pull the stop cable or disconnect the run solenoid to prevent the engine from starting and crank the engine till adequate oil pressure builds.

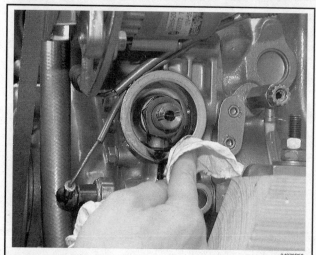

Fig. 22 Before installing the new filter, wipe the gasket mating surface clean

04973P58

Fig. 23 Dip a finger into a fresh bottle of oil, and lubricate the gasket of the new filter

04973P59

Fig. 24 Tighten the new filter ½ to ¾ of a turn after the gasket contacts the mating surface. Do not use a wrench to tighten the filter!

04973P04

Oil Capacities

Model	Capacity Gal.(liter)	
	Total	Effective
1GM	0.3 (1.3)	0.5 (2.0)
1GM10	0.3 (1.3)	0.2 (0.6)
2GM	0.5 (2.0)	0.3 (1.3)
2GM20	0.5 (2.0)	0.3 (1.3)
2QM20	1.3 (5.1)	0.9 (3.3)
3GM	0.7 (2.7)	0.5 (1.8)
3GM30	0.7 (2.7)	0.4 (1.6)
3HM	1.5 (5.5)	0.8 (3.0)
3HM35	1.4 (5.4)	0.7 (2.7)
3QM30	1.7 (6.5)	0.6 (2.2)
4JH-DTE	1.7 (6.5)	0.8 (3.0)
4JH-E	1.7 (6.5)	0.8 (3.0)
4JH-HTE	1.7 (6.5)	0.8 (3.0)
4JH-TE	1.7 (6.5)	0.8 (3.0)
4JH2-DTE	1.8 (7.0)	0.7 (2.5)
4JH2-E	1.8 (7.0)	0.7 (2.5)
4JH2-HTE	1.8 (7.0)	0.7 (2.5)
4JH2-TE	1.8 (7.0)	0.7 (2.5)

04974C03

13. To make sure the proper level is obtained, run the engine to normal operating temperature, turn the engine **OFF**, allow the oil to drain back into the oil pan and recheck the level. Top off the oil to the correct mark on the dipstick.

Transmission Oil

OIL LEVEL CHECK

▶ **See Figures 25, 26, 27, 28 and 29**

✷✷ CAUTION

The EPA warns that prolonged contact with used oil may cause a number of skin disorders, including cancer! You should make every effort to minimize your exposure to used oil. Protective gloves should be worn when changing the oil. Wash your hands and any other exposed skin areas as soon as possible after exposure to used transmission oil. Soap and water, or waterless hand cleaner should be used.

➡Each time the oil level is checked, clean the area around the transmission vent. This area tends to get clogged with oil and salt deposits. If the vent is completely clogged, the transmission will not receive proper ventilation.

When checking the transmission oil level, it is best that the boat be level and the oil be at operating temperature.

1. Locate the engine oil dipstick.

2. Clean the area around the dipstick to prevent dirt from entering the transmission.

3. Remove the dipstick and note the color of the oil. If the oil has changed color from its original tone, it is burnt and should be replaced. If the oil is a milky white color, water is present in the transmission. Wipe the dipstick clean with a rag.

➡On some dipsticks the level of oil is difficult to see. If this is the case, the dipstick can be painted with white epoxy paint. Once dry the paint will provide a background color to view the engine oil on the dipstick more clearly.

4. See "Oil Recommendations" for the proper viscosity and type of oil.

5. Insert the dipstick into the filler hole, but **DO NOT** screw it back into the transmission. The dipstick should rest on the threads.

6. Pull the dipstick out of the transmission, and read the level on the dipstick. The level should be at or near the groove on the end of the stick.

7. If the oil level is at or below the lower limit, sufficient oil should be added through the dipstick hole to restore the proper level of oil in the transmission. Add oil in small amounts to avoid overfilling the transmission.

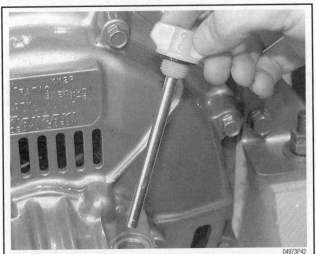

Fig. 25 To check the level of the transmission oil, remove the dipstick, wipe it clean. . .

Fig. 27 . . .without screwing it back into the hole

Fig. 26 . . .and reinsert it into the transmission. . .

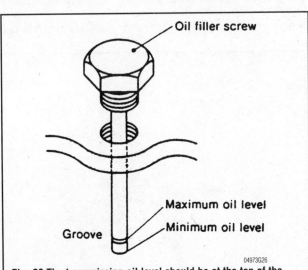

Fig. 28 The transmission oil level should be at the top of the groove on the dipstick

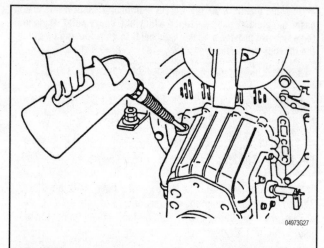

Fig. 29 A funnel or pitcher should be used when adding transmission oil to avoid spillage

Fig. 31 Tags, usually found on the outside of the transmission, identify the type and quantity of transmission oil that should be used

✳✳ WARNING

Do not overfill the transmission. If the transmission is overfilled, the gears will whip the oil into a foam causing loss of lubrication and severe damage.

OIL CHANGE

▶ **See Figures 30 and 31**

The maintenance interval changing the oil in the transmission is every 100 hours of operation.

A few precautions can make the messy job of oil and filter maintenance much easier. By placing oil absorbent pads, available at industrial supply stores, into the area below the oil drain plug, you can prevent gear oil pillage from reaching the bilge.

As with the engine oil, changing the transmission oil is sometimes complicated by the location of the drain plug. Even if you can reach the plug, on some boats there isn't enough room to fit a drain pan under the transmission. In such cases, an oil evacuation pump should be used.

1. Inspect the area under the transmission to determine if there is

adequate clearance for a drain pan. If clearance is tight, an oil evacuation pump will have to be used.

➡ **If you don't have a permanently mounted oil suction pump in your engine compartment, you may want to consider installing one. This pump sucks waste oil out through either the dipstick tube or a connection on the oil drain plug.**

2. Position the drain pan under the oil drain plug. Keep in mind that the fast flowing oil, which will spill out as you pull the plug from the transmission, will flow with enough force that it could miss the drain pan and end up in the bilge. Position the drain pan accordingly and be ready to move it more directly under the plug as the oil flow lessens to a trickle.

3. Loosen the drain plug with an appropriately sized wrench (or socket and driver), then carefully unscrew the plug with your fingers. Push in on the plug as you unscrew it, then remove the plug quickly and allow the oil to drain from the transmission.

4. Allow the oil to completely drain from the transmission.

5. Carefully thread the plug into position and tighten it to specification using a torque wrench. If a torque wrench is not available, snug the drain plug and give a slight additional turn. The transmission drain plug threads are easily stripped from overtightening (and this can be time consuming and costly to fix).

6. If any oil has gotten into the bilge, remove it using an oil absorbent pad. These pads are specially formulated to only absorb oil and will not soak up any water in the bilge. Perform a visual inspection and make sure all connections are tight.

7. Carefully remove the drain pan and transfer the oil into a suitable container for recycling.

8. Using a funnel to avoid spillage, refill the transmission with the proper quantity and quality of oil through the dipstick hole. Check the level often when adding oil to the transmission to avoid overfilling.

9. When the level of the oil in the transmission is correct, install the dipstick.

Oil Disposal

▶ **See Figure 32**

Always dispose of your used oil and other hazardous waste properly. Before draining any fluids, consult with your local authorities;

Fig. 30 As with the engine oil, an evacuation pump will need to be used if there is not sufficient clearance under the transmission

Fig. 32 Marine ecosystems are very fragile. Always dispose of your used oil and other hazardous waste properly

in many areas, waste oil is being accepted as a part of recycling programs. A number of parts stores are also accepting waste fluids for recycling.

Be sure of the recycling center's policies before draining any fluids, as many will not accept different fluids that have been mixed together.

Control Cables

♦ See Figure 33

LUBRICATION

Control cables should be lubricated at the first sign of binding. It is important to keep these cables well lubricated to prevent corrosion from forming inside the casing. If corrosion forms, it is almost impossible to remove and the cable must be replaced.

There are a couple of ways to lubricate the cable but all involve the same principle; forcing oil between the inner cable and the sheath. This can be accomplished by either gravity or pressure.

In the gravity method, the highest end of the cable is pushed through the bottom of a plastic bag and tied off with a rubber band. The bag is filled with oil and allowed to sit overnight. A container is placed at the lowest end of the cable to catch the overflowing fluid. Gravity will force the oil between the cable and the sheath.

A similar procedure uses a pressure pump and a piece of rubber hose. The highest end of the cable is attached to the pump with a piece of rubber hose and secured with hose clamps. Lubricant is then forced through the cable.

Air Filter

In a marine engine compartment, the minimal amount of dust and dirt in the air mean that a marine air filter requires less maintenance than its counterpart in the automotive world. However, the maintenance of a marine air filter is equally important.

The marine filter prevents dirt from entering the engine and scoring the cylinder walls. This lessens oil consumption and extends the engine's life. The air filter on some engines is also used as an intake silencer to quiet the intake air sound as it rushes into the cylinder head from the intake ports.

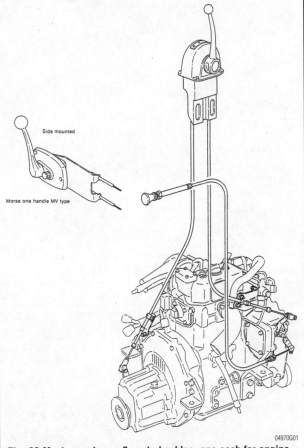

Fig. 33 Most vessels use 3 control cables, one each for engine speed, transmission range and engine stop

There are several air filter designs. Some use a coarse screen to stop larger items from being sucked into the engine. Most use either paper or foam to filter the air because it is the least restrictive. Oil soaked elements are among the most restrictive.

Over time, the air filter element will become clogged with dirt and oil, decreasing the amount of air entering the engine and lowering engine output. If an excessive amount of oil is clogging the filter, this could be an indication of worn cylinders or piston ring failure causing high pressure in the crankcase.

The maintenance interval for intake filter/silencer cleaning is every 100 hours of engine operation.

SERVICE

Naturally Aspirated

♦ See Figures 34 thru 41

Yanmar diesel engines use a foam type filter element that must be washed in a neutral detergent solution to be cleaned.

The air silencer on naturally aspirated engines is attached to the side of the cylinder head attached to the intake manifold.

1. Using a rag, wipe the intake silencer cover to prevent dirt from entering the engine when the cover is removed. Unfasten the latches securing the intake silencer cover to the base and remove the cover.

2. The silencer element is located inside the cover. Simply remove the element from the cover and soak it in a solution made from a neutral detergent and water. Agitate the element to remove all dirt and oil. Allow the element to air dry prior to installation.

Fig. 34 To remove the silencer on a GM/HM series engine, disengage the two latches on the sides of the housing

Fig. 37 Pull the screen and foam element from the housing

Fig. 35 After the latches are disengaged, unhook them from the back of the silencer base

Fig. 38 Inspect the foam for cleanliness

Fig. 36 Now the silencer can easily be removed

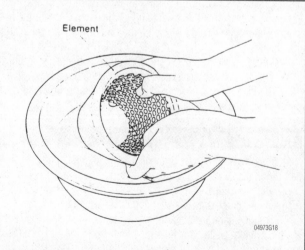

Element

Fig. 39 The silencer/filter element should be cleaned with a neutral detergent

Fig. 40 This silencer was positioned incorrectly. If water should enter the housing, the engine may be destroyed

Fig. 42 To remove the silencer on a turbocharged JH series engine, disengage the clip for the breather hose with pliers. . .

Fig. 41 The silencer should be positioned in this manner to prevent water and other debris from entering the engine

Fig. 43 . . .remove the hose, then disengage the latches. Note that this engine does not use a foam element

➡It is extremely important that the intake silencer element be completely dry prior to installation. If water enters the engine it will not compress in the combustion chamber and may cause severe internal engine damage.

3. Once the element is dry, install it in the silencer cover and place the cover back on to the base. Align the latches and fasten them securely.

Turbocharged

▶ See Figures 42, 43 and 44

The air silencer on turbocharged engines is attached to the inlet side of the turbocharger.

1. Using a rag, wipe the intake silencer cover to prevent dirt from entering the engine when the cover is removed. Note the position of the silencer for installation reference. Loosen the band clamp attaching the silencer to the turbocharger inlet and slide the silencer off the turbocharger.

2. Soak the entire silencer in a solution made from a neutral detergent and water. Agitate the element to remove all dirt and oil. Allow the element to air dry prior to installation.

Fig. 44 With the silencer removed, the impeller of the turbocharger can be viewed to check for cleanliness

➡It is extremely important that the intake silencer element be completely dry prior to installation. If water enters the engine it will not compress in the combustion chamber and may cause severe internal engine damage.

3. Once the element is dry, align and install the silencer on the turbocharger inlet with the band clamp. Tighten the clamp screw securely.

Crankcase Breather

The crankcase breather system must be kept clean to maintain good performance and durability. Periodic service is required to remove combustion products from the breather components. The components should be inspected and serviced when the air filter/silencer is serviced.

Over a period of time, deposits build up in the breather circuit. The crankcase breather system should be inspected at every oil change. Service the breather system if engine oil is being discharged into the air cleaner.

On some engines the dipstick will be blown out of the dipstick tube during operation. If this should happen, the crankcase breather is clogged and should be serviced. Do not attempt to clean the breather, rather replace it with a new one.

A similar condition will occur if the hose from the air cleaner to the valve cover is kinked during installation of the air cleaner assembly. In this case, simply reposition the air cleaner so that the hose is not kinked.

SERVICE

The crankcase breather should be serviced at least every 500–600 hours or sooner if a problem is encountered.

GM/HM Series

♦ See Figures 45, 46 and 47

1. On 2 and 3GM engines, remove the breather hose that connects the breather to the air cleaner/silencer. On the 1GM engine, this is not required, because it is equipped with a tube that vents directly into the intake manifold.

2. Remove the 4 bolts that attach the breather cover to the rocker arm cover.

Fig. 45 On GM series engines, inspect the breather hose by removing it and checking for internal obstructions

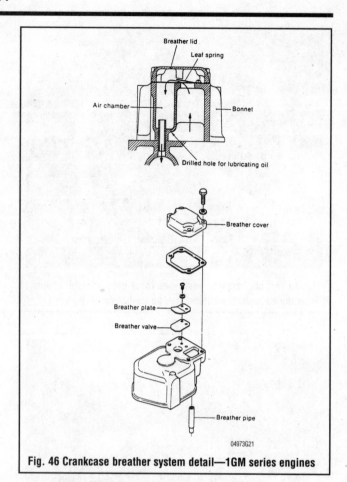

Fig. 46 Crankcase breather system detail—1GM series engines

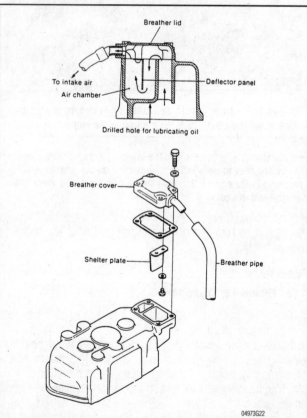

Fig. 47 Crankcase breather system detail—2GM and 3GM/HM series engines

3. Inspect the leaf spring, and the small hole at the base of the air chamber that allows excess oil to drain back into the crankcase. If the drain hole is clogged, the air chamber will fill up with oil.

4. Clean any oil deposits from inside the breather cover and air chamber.

5. Using a new gasket, install the breather cover.

6. Inspect the breather hose by blowing through it; there should not be any obstruction.

7. If the hose is in good condition, and there are no internal obstructions, it can be installed. If the hose is obstructed, cracked, or excessively soft, it should be replaced.

3HM and JH Series

▶ See Figures 48, 49, 50, 51 and 52

1. Remove the breather hose that connects the breather to the air cleaner/silencer.

2. Remove the 2 bolts that attach the breather cover to the rocker arm cover.

3. Lift the shelter plate(s) and the mesh net from the rocker cover.

4. If the mesh net is dirty, it should be washed or replaced.

5. Clean any hardened oil deposits from the shelter plate(s).

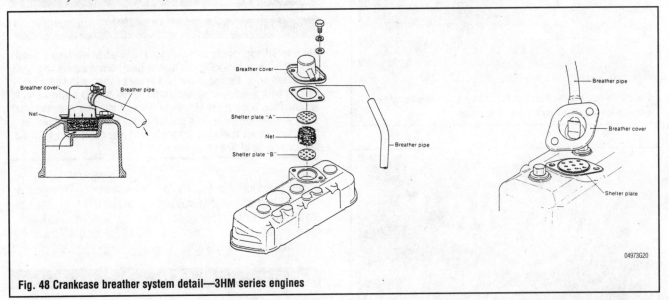

Fig. 48 Crankcase breather system detail—3HM series engines

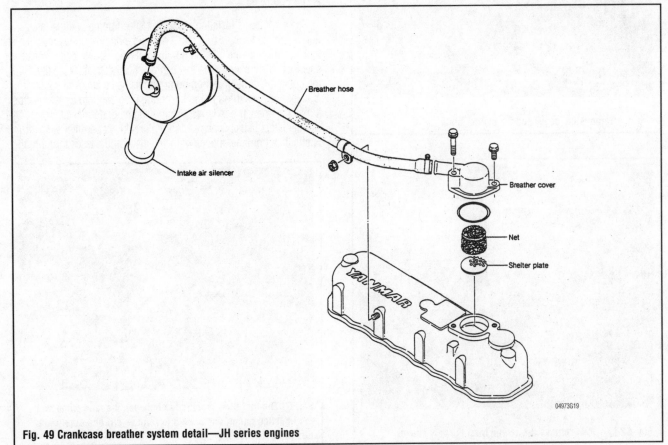

Fig. 49 Crankcase breather system detail—JH series engines

Fig. 50 To inspect the breather system on a JH series engine, remove the two bolts on the breather cover

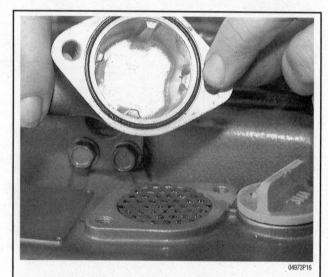

Fig. 51 Remove the cover and the shelter plate. . .

Fig. 52 . . .then remove the screen from the valve cover

6. Install the shelter plate(s) and mesh net into the rocker cover.

7. Clean any hardened oil deposits from the breather cover.

8. Using a new gasket, install the breather cover onto the rocker cover.

9. Inspect the breather hose by blowing through it; there should not be any obstruction.

10. If the hose is in good condition, and there are no internal obstructions, it can be installed. If the hose is obstructed, cracked, or excessively soft, it should be replaced.

Water Separator

✳✳ CAUTION

Observe all applicable safety precautions when working around fuel. Whenever servicing the fuel system, always work in a well ventilated area. Do not allow fuel spray or vapors to come in contact with a spark or open flame. Do not smoke while working around fuel. Keep a dry chemical fire extinguisher near the work area. Always keep fuel in a container specifically designed for fuel storage; also, always properly seal fuel containers to avoid the possibility of fire or explosion.

In addition to the fuel filter mounted on the engine, a filter is usually found inside or near the fuel tank. Because of the large variety of differences in both portable and fixed fuel tanks, it is impossible to give a single procedure to cover all applications. Check with the boat manufacturer or the marina who rigged the boat for details on replacing the fuel tank filter.

Fuel Filter

▶ See Figure 53

The fuel filter is installed between the feed pump and the injection pump, and serves to remove dirt and other impurities from the fuel tank through the feed pump. If a fuel filter is not used, a small speck of dirt or sand in the fuel can drastically affect the ability of the operation of the injection pump. If that same speck of sand manages to pass through the injection pump and into a fuel injector, it may not spray properly, decreasing engine performance.

Fuel filters usually consist of a replaceable paper filter element. Fuel enters from the outside of the element, and passes through, fil-

Fig. 53 The fuel filter is installed between the feed pump and the injection pump, and serves to remove dirt and other impurities from the fuel

tering out dirt, sand, and other impurities, allowing only clean fuel to enter the interior of the element. The fuel then exits the filter housing from an outlet at the top of the filter body, and is sent to the injection pump. Since this paper degrades over time, it is recommended that the fuel filter be replaced at the beginning of each season.

Regular replacement of the fuel filter will decrease the risk of blocking the flow of fuel to the engine, which could leave you stranded on the water. Fuel filters are usually inexpensive, and replacement is a simple task.

REMOVAL & INSTALLATION

QM and GM/HM Series Engines

▶ See Figures 54 thru 59

✳✳ CAUTION

Observe all applicable safety precautions when working around fuel. Whenever servicing the fuel system, always work in a well ventilated area. Do not allow fuel spray or vapors to come in contact with a spark or open flame. Do not smoke while working around fuel. Keep a dry chemical fire extinguisher near the work area. Always keep fuel in a container specifically designed for fuel storage; also, always properly seal fuel containers to avoid the possibility of fire or explosion.

1. Place a receptacle under fuel filter to contain any excess fuel.
2. Using a rag, thoroughly clean the filter and filter body.
3. Loosen the retaining ring that holds the fuel filter cover in place.
4. While pushing upward on the fuel filter cover, unscrew the retaining ring that holds the cover in place.
5. Once the retaining ring is free of the threads on the body, carefully lower the fuel filter cover, and the element.
6. While holding the element inside the housing in place, tip it over and allow the fuel to drain.
7. Remove the element from the cover. Be sure to note the direction in which the element is situated in the cover.

To install:

8. Clean the inside of the fuel filter cover.

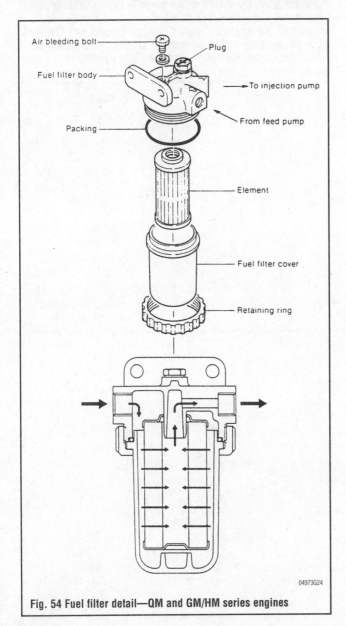

Fig. 54 Fuel filter detail—QM and GM/HM series engines

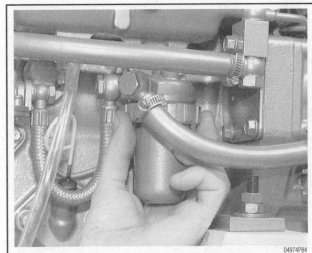

Fig. 55 To remove the fuel filter, turn the retaining ring counterclockwise

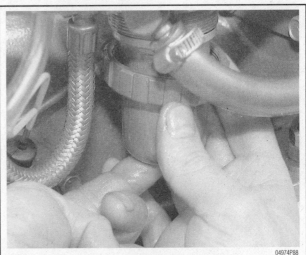

Fig. 56 While removing the retaining ring, push upwards on the filter cover until the ring is free from the housing. . .

Fig. 57 . . .then lower the housing along with the filter. Keep in mind the filter cover is filled with fuel

Fig. 58 Fill the filter cover approximately halfway with fresh fuel, and carefully lower the new element into the housing

Fig. 59 Always replace the O-ring on the filter cover when the element is replaced to avoid air from entering the system

9. Install a new filter element (in the proper direction) in the fuel filter cover.

10. Using fresh, clean fuel, fill the fuel filter cover.

11. Using a new O-ring, install the element and filter cover to the fuel filter body.

12. Tighten the retaining ring snugly by hand.

✳✳ WARNING

After installing a new fuel filter, excess air must be bled from the fuel system, or the engine may not start. Repeated attempts of starting the engine with air in the fuel system may starve the injection pump of fuel, causing damage due to lack of lubrication.

JH Series Engines

▶ See Figures 60 and 61

✳✳ CAUTION

Observe all applicable safety precautions when working around fuel. Whenever servicing the fuel system, always work in a well ventilated area. Do not allow fuel spray or vapors to come in contact with a spark or open flame. Do not smoke while working around fuel. Keep a dry chemical fire extinguisher near the work area. Always keep fuel in a container specifically designed for fuel storage; also, always properly seal fuel containers to avoid the possibility of fire or explosion.

1. Place a receptacle under fuel filter to contain any excess fuel.

2. Using a rag, thoroughly clean the filter and filter body.

3. Using an oil filter wrench, loosen the fuel filter from the filter body.

4. While pushing upward on the fuel filter, unscrew it from the filter body.

5. Once the threads are free, tip the filter over and allow the fuel to drain into the receptacle.

To install:

6. Using fresh, clean fuel, fill the fuel filter.

7. Lightly lubricate the gasket on the filter by applying a light coating of fuel.

8. Carefully raise the filter and thread it onto the filter body, tightening it according to the filter manufacturer's specifications. Most

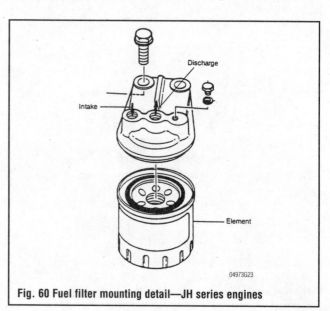

Fig. 60 Fuel filter mounting detail—JH series engines

Fig. 61 A plastic soda bottle trimmed down and placed under the filter will contain any excess fuel

filters only require ¼ to ½ turn of the filter after the gasket touches the sealing surface. This usually translates into "hand tight." A filter wrench should not be used to tighten the filter.

⁜ WARNING

After installing a new fuel filter, excess air must be bled from the fuel system, or the engine may not start. Repeated attempts of starting the engine with air in the fuel system may starve the injection pump of fuel, causing damage due to lack of lubrication.

FUEL SYSTEM BLEEDING

▶ **See Figures 62 thru 67**

Any time the fuel system is exposed to air (e.g. changing a fuel filter) the accumulated air will have to be bled from the system to allow for proper delivery of fuel to the engine.

Besides changing the fuel filter, air can be drawn into the fuel system by several routes. Typically, air is sucked into the fuel system through a poor seal upstream of the lift pump. Air can also be

Fig. 62 Although the fuel filter is slightly different between models, the air bleed screw is always the 10mm bolt with the Phillips head

Fig. 63 Loosening the air bleed screw on a 3GM engine

Fig. 64 Loosening the air bleed screw on a JH series engine

Fig. 65 After the bleed screw has been loosened, operate the priming lever on the feed pump to push the air out of the system

Fig. 66 If air has entered the injector pump, loosen the lines on the injectors ONE TURN, and crank the engine to bleed the air

Fig. 67 If your engine is equipped with spacers on the fuel injectors, use a wrench to hold the spacer stationary while loosening the line nut

drawn into the fuel system when the fuel tank is low, and the motor is being operated in rough seas. A small amount of air (about a teaspoon) in the fuel system is enough air to shut down the engine, and require fuel system bleeding.

To bleed air from the fuel system, proceed as follows:

1. Loosen the air bleed screw at the top of the secondary fuel filter (between the lift pump and injector pump).

➡When operating the priming lever on the lift pump, make sure to push down completely to ensure full operation of the diaphragm within the pump.

2. Operate the priming lever on the lift pump. If the lever is only moving slightly, rotate the crankshaft by hand to reposition the lobe on the camshaft that operates the pump. This will allow for a full stroke of the priming lever.

3. Operate the priming lever until fuel free of air flows from the air bleed screw, then tighten the screw.

➡Air can be drawn into the fuel system by a faulty bleed screw washer. After the bleeder screw has been loosened a few times, it is a good idea to replace it to ensure a leak-free seal. Also, do not overtighten the bleed screw. The screw is hollow and can be sheared very easily.

4. If air remains in the system, check for leaks in the lift pump or primary filter.

5. If the engine has stalled from air in the system, (or if the high pressure side of the system has been opened) it will be necessary to bleed the injector pump, injector lines, and injectors.

6. Loosen the bleed screw on the injector pump and operate the priming lever on the lift pump until fuel free of air flows from the bleed screw, then tighten the screw.

7. Loosen the injector line nut(s) on the injector(s) **one** turn.

✳✳ WARNING

If the injector has a hex spacer on the inlet, make sure to hold it stationary with a wrench while loosening the line nut. Do NOT attempt to loosen the line by turning the spacer. Turning the spacer will twist the fuel line, requiring replacement of the line.

8. Place the engine speed lever to the full throttle position.

9. Using the starter, crank the engine until all air is free from the lines until fuel runs from the connections.

10. Tighten the line nuts securely.

11. If air remains in the system, locate the source of the air, and perform and repair procedures necessary.

12. Repeat the procedure until the entire fuel system is free of air.

Belts

INSPECTION

▸ See Figures 68, 69, 70 and 71

Belts should be inspected on a regular basis for signs of glazing or cracking. A glazed belt will be perfectly smooth from slippage, while a good belt will have a slight texture of fabric visible. Cracks will usually start at the inner edge of the belt and run outward. All worn or damaged drive belts should be replaced immediately. It is best to replace all drive belts at one time, as a preventive maintenance measure, during this service operation.

Inspect the alternator and water pump belts every 250 hours for evidence of wear such as cracking, fraying, and incorrect tension.

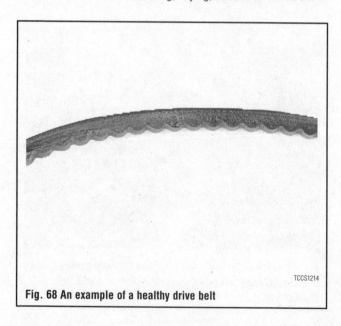

Fig. 68 An example of a healthy drive belt

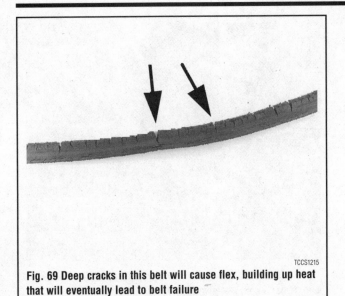

Fig. 69 Deep cracks in this belt will cause flex, building up heat that will eventually lead to belt failure

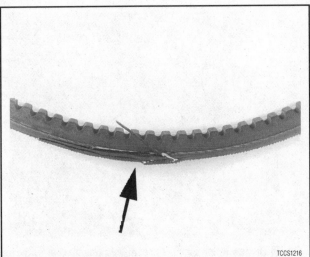

Fig. 70 The cover of this belt is worn, exposing the critical reinforcing cords to excessive wear

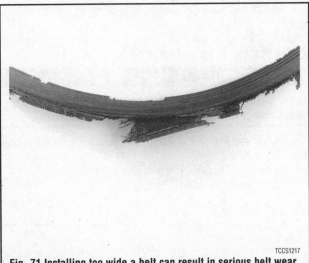

Fig. 71 Installing too wide a belt can result in serious belt wear and/or breakage

Determine belt tension at a point halfway between the pulleys by pressing on the belt with moderate thumb pressure. If the distance between the pulleys (measured at the center of the pulley) is 13–16 in. (330–400mm), the belt should deflect ½ in. (13mm) at the halfway point of its longest straight run; ¼ in. (6mm) if the distance is 7–10 in. (178–300mm). If the defection is found to be too much or too little, loosen the mounting bolts and make adjustments as necessary.

➡When replacing belts, it is recommended to clean the inside of the belt pulleys to extend the service life of the belts.

ADJUSTMENT

▶ **See Figures 72 thru 78**

Before you attempt to adjust any of your engine's belts, apply penetrating oil to the alternator bracket fasteners to make them easier to loosen.
1. Disconnect the negative battery cable.
2. Loosen the alternator or water pump pivot bolt.

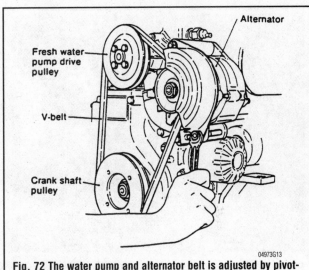

Fig. 72 The water pump and alternator belt is adjusted by pivoting the alternator

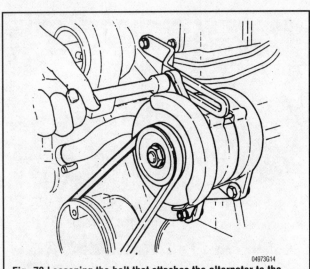

Fig. 73 Loosening the bolt that attaches the alternator to the slotted bracket will allow for belt adjustment

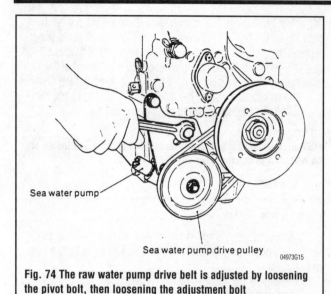

Fig. 74 The raw water pump drive belt is adjusted by loosening the pivot bolt, then loosening the adjustment bolt

Fig. 75 To adjust the alternator/fresh water pump belt, loosen the alternator pivot bolt . . .

Fig. 76 . . . followed by the adjustment bracket bolt. The alternator should be loose

Fig. 77 Once the alternator is loosened, install a belt tensioner, and adjust it until the tension on the belt is correct

Fig. 78 The raw water pump adjusts in the same manner as the alternator (note the slot on the bracket that allows the pump to pivot)

3. Loosen the alternator or water pump pulley adjustment bracket bolt.

4. Using a wooden lever, pry the alternator or water pump toward or away from the engine until the proper tension is achieved.

⁑ WARNING

Do not overtighten the drive belts; the water pump and/or alternator bearings can be damaged.

5. Tighten the alternator or water pump pulley adjustment bracket bolt.

6. Tighten the alternator or water pump pivot bolt.

7. Reconnect the negative battery cable.

REMOVAL & INSTALLATION

▶ See Figure 79

The replacement of the inner belt on multi-belted engines may require the removal of the outer belts.

Fig. 79 Once the alternator is loosened, the belt simply slips off the pulleys

Fig. 81 Visually inspect the exhaust outlet to check for proper cooling system circulation

To replace a drive belt, loosen the pivot and mounting bolts of the component which the belt is driving, then, using a wooden lever or equivalent, pry the component inward to relieve the tension on the drive belt; always be careful where you locate the prybar, or damage to components may result. Slip the belt off the component pulley, and match the new belt with the old belt for length and width. These measurements must be equal. It is normal for an old belt to be slightly longer than a new one. After a new belt is installed correctly, properly adjust the tension.

➥When removing more than one belt, be sure to mark them for identification. This will help avoid confusion when replacing the belts.

Raw Water Cooling

CHECKING CIRCULATION

▶ See Figures 80 and 81

Every time the engine is started, check for proper flow through the cooling system. The best way to check this is to visually inspect the straight through valve on dry exhaust systems or the exhaust outlet on wet exhaust engines. If the engine shows high operating temperature or overheats, the cooling system is almost certainly blocked and should be inspected immediately.

CHECKING THE RAW WATER STRAINER

▶ See Figures 82, 83 and 84

A filter or strainer is used to prevent large particles of sand, mud, and other debris from entering the cooling circuit and damaging the raw water pump. The water in which the vessel is operated dictates how frequently the filter will have to be cleaned. Most strainers have sight glasses or other means of visual inspection to determine when maintenance is necessary.

To service the strainer:
1. Close the raw water intake petcock to prevent entry of sea water once the strainer basket is open.
2. Loosen the clamp or wingnut that secures the top of the strainer.
3. Remove the strainer basket and clean it thoroughly.
4. If sediment remains in the bottom of the strainer housing, it

Fig. 80 Flappers are used to prevent seawater from backing up into the wet exhaust

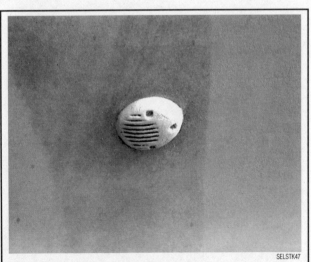

Fig. 82 As you can see, the raw water inlet can easily become clogged without proper maintenance

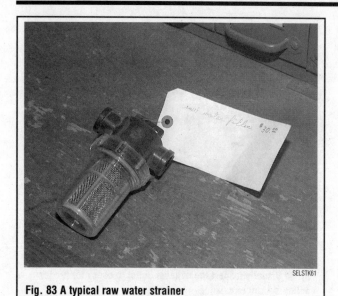

SELSTK61

Fig. 83 A typical raw water strainer

can be removed by using a turkey baster to suck out the liquid or by removing the drain plug at the bottom of the strainer and allowing the water to drain.

5. Replace the strainer basket and secure the strainer top.
6. Open the raw water intake petcock.

DRAINING & FLUSHING

▶ **See Figures 85, 86, 87 and 88**

The following procedure covers general draining and flushing procedures for raw water cooled engines. This procedure may also be used for the raw water portion of heat exchanger cooled engines. If the material presented here differs from those in your operator's manual, follow the procedure in the operator's manual the manual.

1. The boat must be out of the water to perform this procedure.
2. Disconnect the hose at the inlet petcock and secure it to a garden hose. This can usually be accomplished by using the hose clamp on the inlet hose to produce a water tight seal around the garden hose fitting.
3. Turn the garden hose **ON** and start the engine.
4. Allow the engine to run for approximately 20 minutes to flush silt, sediment and salt from the cooling system. This also allows the thermostat enough time to open fully.
5. Turn the engine **OFF** and then stop the flow of water from the garden hose. Disconnect the garden hose and allow the water to drain from the inlet fitting.
6. At this point, the water has drained from the inlet hose and probably partially drained from the water pump and cylinder head. You must now drain the water from all the low lying areas of the engine.
7. Remove all the engine drains and zincs. Have a bucket handy to catch all the water. Allow the engine to continue draining until no water is left.
8. Using a piece of stiff wire, probe all the zinc and engine drain holes. Using the wire to scrape any silt, rust, salt and scale from the engine. If this debris is left in place, it will harden and prevent the normal flow of water through the engine. Once hardened, it cannot be removed.

➡**Do not leave zincs and drain plugs out over the winter storage period. Corrosion and rust will form on the threads in the block making installation in the spring very difficult.**

Breathing plug

Element

Body

To sea water pump

04973G32

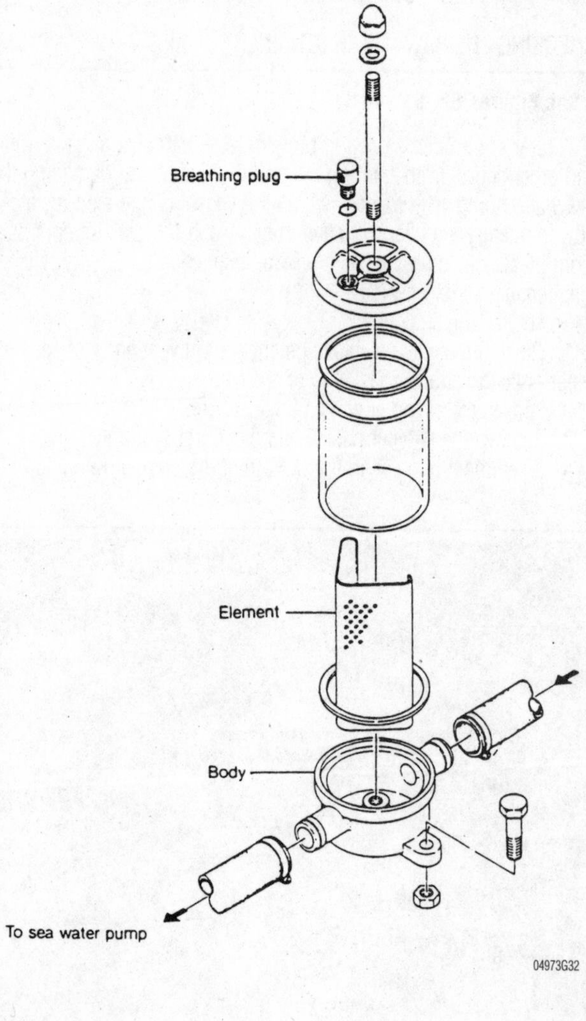

Fig. 84 Exploded view of a typical raw water strainer

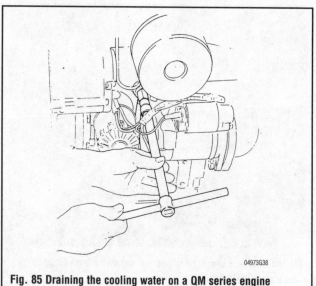

04973G38

Fig. 85 Draining the cooling water on a QM series engine

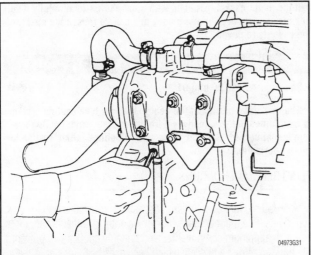

Fig. 86 The exhaust manifold also has a drain cock to make draining the cooling water easier

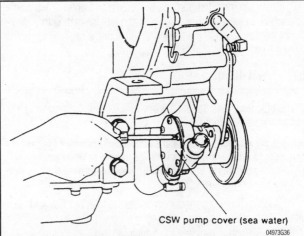

CSW pump cover (sea water)

Fig. 87 To ensure that all of the water is drained from the raw water pump, loosen the screws on the cover and allow the water to escape

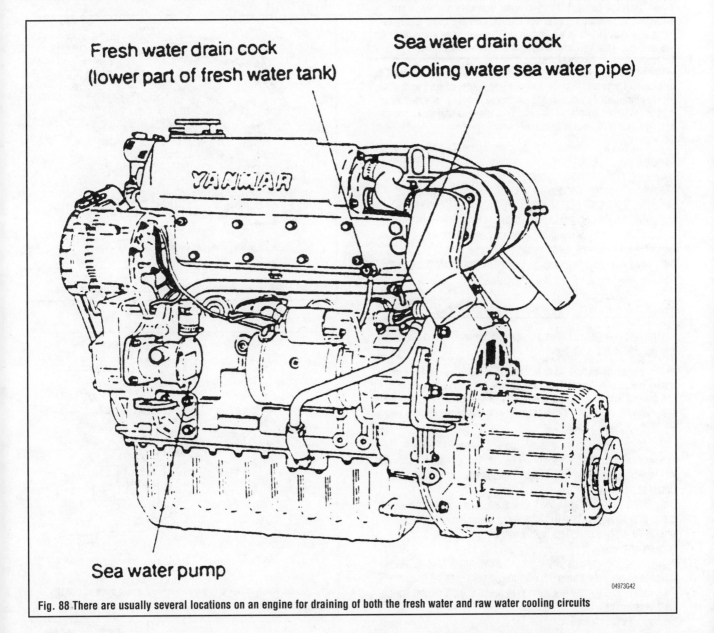

Fresh water drain cock (lower part of fresh water tank)

Sea water drain cock (Cooling water sea water pipe)

Sea water pump

Fig. 88 There are usually several locations on an engine for draining of both the fresh water and raw water cooling circuits

9. Clean the threaded portions of the zinc and drain holes with a wire brush. Inspect the zincs for wear and replace then with new ones as needed. Install the zincs and reinstall the drain plugs using Teflon® tape.

REFILLING WITH ANTIFREEZE

The next few steps outline a procedure for filling the cooling system with nontoxic antifreeze.

➡There are a few types of water pump impellers which may react adversely to antifreeze. Check with the engine or water pump impeller manufacturer for compatibility with antifreeze prior to performing this procedure.

1. Once again, disconnect the hose at the inlet petcock and secure it to a short hose submerged in a bucket of nontoxic antifreeze. You should have approximately 10–15 gallons of antifreeze available for the entire job.

✳✳ CAUTION

When draining coolant, keep in mind that cats and dogs are attracted by ethylene glycol antifreeze, and are quite likely to drink any that is left in an uncovered container or in puddles on the ground. This will prove fatal in sufficient quantity.

➡Used fluids such as antifreeze are hazardous wastes and must be disposed of properly. Before draining any fluids, consult with your local authorities; in many areas antifreeze is being accepted as a part of recycling programs. A number of marinas and parts stores are also accepting waste fluids for recycling.

2. Position a catch bucket at the point where the raw water would normally exit the boat. This will vary depending on whether you have a wet or dry exhaust system.

3. Start the engine and allow the antifreeze to run through the engine unit it comes out the exit clean. This should sufficiently coat the cooling system and leave a good bit of antifreeze in the engine. If you wish to completely fill the system, connect the inlet hose and close the inlet petcock. Remove the thermostat (which should be the high point in the system) and fill the engine manually through the thermostat housing.

LEAVING THE ENGINE DRY

The next few steps outline a procedure for leaving the cooling system dry over the winter.

➡There is some debate as to whether this is the best way to over-winter an engine. You will have to decide for yourself whether to fill the cooling system or allow it to go dry. Bear in mind that any water (not coolant) left in the block could freeze and cause serious engine damage.

1. Carefully loosen the water pump impeller cover screws. These screws are often corroded and may be difficult to remove. Use a liquid penetrating oil if you encounter difficulty. Remove the water pump impeller cover.

2. Remove the water pump impeller and inspect it for damage. If damage is noted, replace the impeller. If the impeller is intact, grease the blades impellers lightly with petroleum jelly and place it back in its bore.

3. Install the impeller cover but leave the pump cover screws loose to prevent the pump impeller from sticking to the housing. This would damage the impeller upon initial startup during spring commissioning.

➡If the paper gasket is rubbed with grease or petroleum jelly it will not stick to the cover or pump body and can be reused the next time the cover is removed.

Fresh Water Cooling

▸ See Figures 89, 90 and 91

➡The raw water portion of the heat exchanger type cooling system is maintained in the same way a traditional raw water cooling system is maintained. Refer to "Raw water Cooling" in this section for more information.

FLUID RECOMMENDATIONS

A 50/50 mixture of antifreeze and water is recommended for year round use. Use a good quality non-toxic antifreeze that is safe for use with aluminum components. Good quality coolant will prevent the buildup of rust and scale which decreases heat transfer in the heat exchanger by as much as 95 percent.

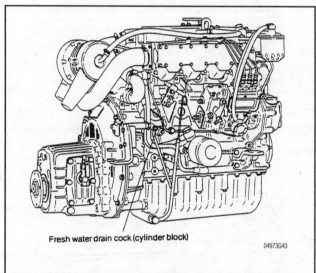

Fresh water drain cock (cylinder block)

04973G43

Fig. 89 Fresh water drain cock location on JH series engines

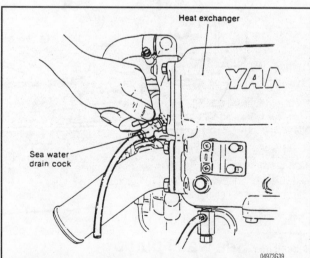

Heat exchanger

Sea water drain cock

04973G39

Fig. 90 On fresh water cooled engines, a drain cock is usually located on the header tank

Fig. 91 The top drain cock on this JH engine is for the header tank the bottom is for the aftercooler

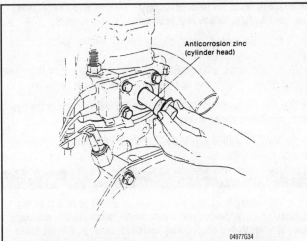

Anticorrosion zinc (cylinder head)

Fig. 92 This is what happens when the header tank is used as a step ladder out of the engine compartment. This pressure cap and housing must be replaced

LEVEL CHECK

Check the coolant level prior to starting the engine for the day. If the coolant level is low, fill the system with a proper mixture of coolant. If the coolant level in the system is allowed is drop 1.5–5 quarts, the engine will overheat.

❄❄ CAUTION

Never remove the pressure cap under any conditions while the engine is running! Failure to follow these instructions could result in damage to the cooling system or engine and/or personal injury. To avoid having scalding hot coolant or steam blow out of the expansion tank, use extreme care when removing the pressure cap. Wait until the engine has cooled, then wrap a thick cloth around the pressure cap and turn it slowly to the first stop. Step back while the pressure is released from the cooling system. When you are sure the pressure has been released, press down on the pressure cap (still have the cloth in position) turn and remove the pressure cap.

Most engines are equipped with a coolant reservoir tank (expansion tank). With the engine **COLD**, check the level of the fluid. Fluid should be visible at the bottom of the filler neck. On most engines, access to the filler neck is poor. However, you should be able to feel the level of the fluid in the expansion tank with your finger.

After check the level of coolant in the header tank, next check the level in the expansion tank and adjust as necessary.

➡**Never rely strictly on the amount of fluid in the expansion tank. This tank will some times give a false indication of the coolant level in the engine**

An adequate coolant mixture should have a strong color (other than brown). Internal corrosion and overheating are indicated by brown coolant. Inexpensive hydrometers are available to quickly test the coolant's protection level.

CHECKING THE PRESSURE CAP

◆ **See Figure 92**

While you are checking the coolant level, check the pressure cap for a worn or cracked gasket. If the cap doesn't seal properly, fluid

will be lost and the engine will overheat. Worn caps should be replaced with a new one.

CHECKING THE HOSES

Coolant hoses should be checked for deterioration, leaks and loose hose clamps. It is wise to check the hoses periodically in early spring and at the beginning of the fall or winter when you are performing other maintenance. A quick visual inspection could discover a weakened hose which might have left you stranded if it had remained unrepaired.

Whenever you are checking the hoses, make sure the engine and cooling system are cold. Visually inspect for cracking, rotting or collapsed hoses, and replace as necessary. Run your hand along the length of the hose. If a weak or swollen spot is noted when squeezing the hose wall, the hose should be replaced.

DRAINING, FLUSHING & REFILLING

At least once a year, the fresh water cooling system should be inspected, flushed, and refilled with fresh coolant. If the coolant is left in the system too long, it loses its ability to prevent rust and corrosion. If the coolant has too much water, it won't protect against freezing.

Completely draining and refilling the cooling system will remove accumulated rust, scale and other deposits. Locations of coolant drains vary with each engine. However, most engines have the following drains:

• Engine Block—usually at the rear (aft) of the engine above the crankshaft centerline
• Expansion Tank—at the bottom of the tank
• Header Tank—at the low point of the tank
• Header tank—at the low point of the exchanger

1. Drain the existing coolant. Open the drain cocks or disconnect the hoses to completely drain the system.

❄❄ CAUTION

When draining the coolant, keep in mind that cats and dogs are attracted by the ethylene glycol antifreeze, and are quite likely to drink any that is left in an uncovered container or in puddles on the ground. This will prove fatal in sufficient quantity. Always drain the coolant into a sealable container for recycling.

➡Before opening the petcocks, spray them with some penetrating lubricant to dissolve any rust or corrosion on the threads.

2. Once the system is completely drained, tighten petcocks and install all hoses to seal the system.
3. Fill the system with a mixture of water and an acid based coolant circuit cleaner. The cleaner will remove rust, scale and corrosion.

➡The acid in the cleaner will also find any weak points in the system, possibly creating a leak.

4. Start the engine and allow it to idle until it reaches proper operating temperature.

CAUTION

To avoid having scalding hot coolant or steam blow out of the expansion tank, use extreme care when removing the pressure cap. Wait until the engine has cooled, then wrap a thick cloth around the pressure cap and turn it slowly to the first stop. Step back while the pressure is released from the cooling system. When you are sure the pressure has been released, press down on the pressure cap (still have the cloth in position) turn and remove the pressure cap.

5. Stop the engine and allow it to cool. Drain the cooling system again.
6. Repeat this process until the drained water is clear and free of scale.
7. Flush the expansion tank with water and leave empty.
8. Prepare a 50/50 mixture of mix of quality antifreeze (ethylene glycol) and water in a suitable container. Fill the system the correct level and install the pressure cap.
9. Start the engine and allow it to idle to operating temperature. Stop the engine and allow it to cool. Check and adjust the coolant level, as necessary.

BOAT MAINTENANCE

Inside The Boat

➧ See Figures 93, 94 and 95

Probably the biggest surprise for boat owners is the extent to which mold and mildew developed in a boats interior. Preventing this growth is a two fold process. First, the boat's interior should be thoroughly cleaned. Second, ventilation should be provided to allow adequate air circulation.

Properly cleaning the boat includes removing as much as possible from it. The less there is on a boat, the less there is to attract mildew. Clothing, foul weather gear, shoes, books, charts, paper goods, leather, bedding, curtains, food stuffs, first aid supplies, odds and ends, etc., should be taken home. They're all fertile soil for mildew, as is dirt, grease, soap scum, etc.

The next step is to vacuum, scrub and polish every surface, especially galley and head surfaces. Use a polish that leaves a protective coating on which it's hard for mold to get a foothold. Mildew-preventive sprays are available for carpets and furniture. The cleaner and more highly polished the insides of lockers, cabi-

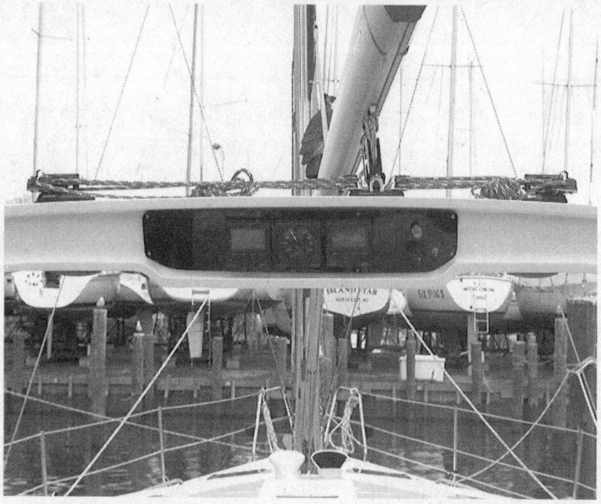

SELSTK23

Fig. 94 . . . gauges . . .

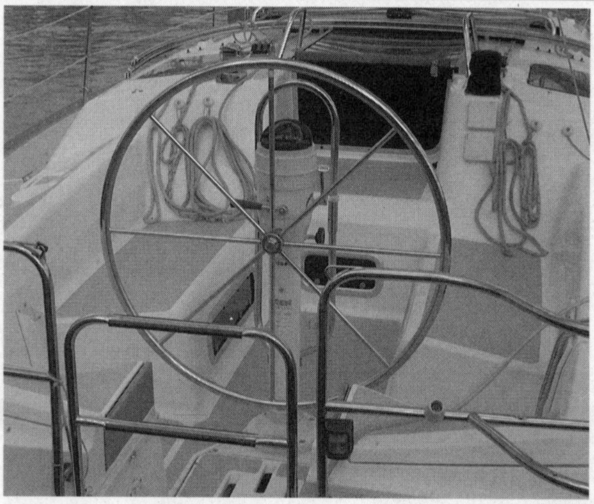

SELSTK22

Fig. 93 Items such as compasses . . .

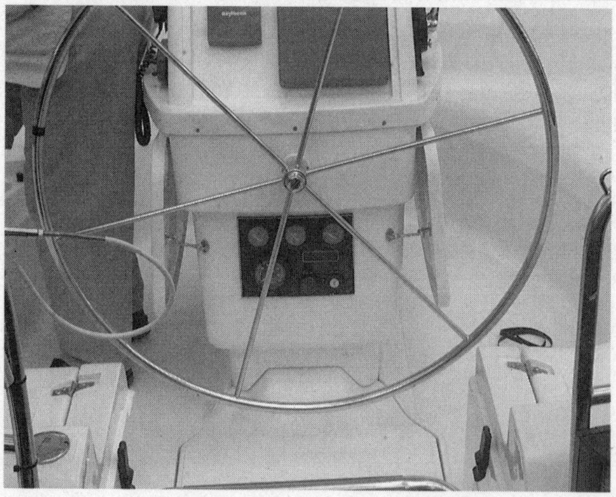

SELSTK24

Fig. 95 . . . and control panels should be protected from the elements

nets, refrigerators, icemakers, drawers and shower stalls are when you leave them, the less likely it is you'll find mold. Marine stores sell cleaners and polishes specifically for boats and for the marine environment. Many people find that regular household products are satisfactory. It's the elbow grease that counts.

If dirt and grease are the growing medium, stagnant air is the fertilizer for mold and mildew (mildew is a form of mold). The more freely air can circulate in the interior, the less conducive conditions will be for the growth of mold. Openings or vents at each end near the top can be fashioned to let air in. Flaps or stove pipe elbows can be used to keep rain out. Hatches and windows can then be left partially open. Visiting the boat on dry, sunny days and opening it up as much as possible is a great way to let fresh air in. Do this whenever possible.

The best way to protect navigation and communications equipment from deteriorating is to remove the units from the boat. Wrap them in towels, not plastic, and take them home. Television sets, VCRs and stereos should also be removed to keep them safe and to keep cold and dampness from affecting them adversely. Before storing electronics equipment, clean off the terminals and the connecting plugs and spray them with a moisture-displacing lubricant that specifically says it's intended for use on electronics. Antennas should also be stored in a safe, dry place to protect them from both the elements and accidental damage. All terminal blocks, junction blocks, fuse holders and the back of electrical panels should be cleaned and sprayed with an appropriate protectant. Remove any corrosion that has developed. Anything that could easily stolen, such as anchors, flare guns, binoculars, etc., is best taken home.

The Boat's Exterior

▶ **See Figure 96**

Fiberglass reinforced plastic hulls are tough, durable, and highly resistant to impact. However, like any other material they can be damaged. One of the advantages of this type of construction is the relative ease with which it may be repaired. Because of its break characteristics, and the simple techniques used in restoration, these hulls have gained popularity throughout the world. From the most congested urban marina, to isolated lakes in wilderness areas, to the severe cold of northern seas, and in sunny tropic remote rivers

of primitive islands or continents, fiberglass boats can be found performing their daily task with a minimum of maintenance.

A fiberglass hull has almost no internal stresses. Therefore, when the hull is broken or stove-in, it retains its true form. It will not dent to take an out-of-shape set. When the hull sustains a severe blow, the impact will be either absorbed by deflection of the laminated panel or the blow will result in a definite, localized break. In addition to hull damage, bulkheads, stringers, and other stiffening structures attached to the hull may also be affected and therefore, should be checked. Repairs are usually confined to the general area of the rupture.

The best way to care for a fiberglass hull is to first wash it thoroughly. Immediately after hauling the boat, while the bottom is still wet, is best, if possible. Remove any growth that has developed on the bottom. Use a pressure cleaner or a stiff brush to remove barnacles, grass, and slime. Pay particular attention to the waterline area. A scraper of some sort may be needed to attack tenacious barnacles. Pot scrubbers made of metal work well. Attend to any blisters; don't wait.

Remove cushions and all weather curtains and enclosures and take them home. Make sure they are clean before storing them, and don't store them tightly rolled. After washing the topsides, remove any stains that have developed with one of the fiberglass stain removers sold in marine stores. For stubborn stains, wet-sanding with 600-grit paper may be necessary. Remove oxidation and stains from metal parts. Apply a coat of wax to everything.

BELOW WATERLINE

▶ **See Figures 97, 98, 99 and 100**

A foul bottom can seriously affect boat performance. This is one reason why racers, large and small, both powerboat and sail, are constantly giving attention to the condition of the hull below the waterline.

In areas where marine growth is prevalent, a coating of vinyl, anti-fouling bottom paint should be applied. If growth has developed on the bottom, it can be removed with a solution of Muriatic acid applied with a brush or swab and then rinsed with clear water. Always use rubber gloves when working with Muriatic acid and take extra care to keep it away from your face and hands. The fumes are toxic. Therefore, work in a well-ventilated area, or if outside, keep your face on the windward side of the work.

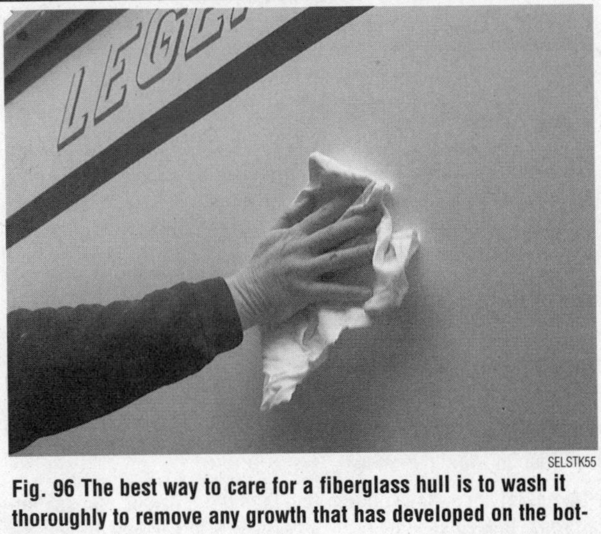

SELSTK55

Fig. 96 The best way to care for a fiberglass hull is to wash it thoroughly to remove any growth that has developed on the bottom

SELSTK00

Fig. 97 In areas where marine growth is prevalent, a coating of vinyl, anti-fouling bottom paint should be applied

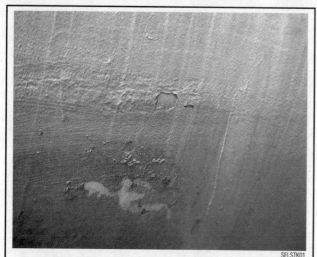

Fig. 98 This anti-foul paint has seen better days and must be replaced

Fig. 99 This hull is in even worse condition and should be sand blasted and repainted

Fig. 100 This beautiful new fiberglass hull will not stay this good looking for long if it is not protected with anti-foul paint

Barnacles have a nasty habit of making their home on the bottom of boats which have not been treated with anti-fouling paint. Actually they will not harm the fiberglass hull, but can develop into a major nuisance.

If barnacles or other crustaceans have attached themselves to the hull, extra work will be required to bring the bottom back to a satisfactory condition. First, if practical, put the boat into a body of fresh water and allow it to remain for a few days. A large percentage of the growth can be removed in this manner. If this remedy is not possible, wash the bottom thoroughly with a high-pressure fresh water source and use a scraper. Small particles of hard shell may still hold fast. These can be removed with sandpaper.

Batteries

Difficulty in starting accounts for almost half of the service required on boats each year. A survey by a major engine parts company indicated that roughly one third of all boat owners experienced a "won't start" condition in a given year. When an engine won't start, most people blame the battery when, in fact, it may be that the battery has run down in a futile attempt to start an engine with other problems.

Maintaining your battery in peak condition may be thought of as either tune-up or maintenance material. Most wise boaters will consider it to be both. A complete check up of the electrical system in your boat at the beginning of the boating season is a wise move. Continued regular maintenance of the battery will ensure trouble free starting on the water.

A complete battery service procedure is included in the here. The following are a list of basic electrical system service checks that should be performed.

- Check the battery for solid cable connections
- Check the battery and cables for signs of corrosion damage
- Check the battery case for damage or electrolyte leakage
- Check the electrolyte level in each cell
- Check to be sure the battery is fastened securely in position
- Check the battery's state of charge and charge as necessary
- Check battery voltage while cranking the starter. Voltage should remain above 9.5 volts
- Clean the battery, terminals and cables
- Coat the battery terminals with dielectric grease or terminal protector
- Check the tension on the alternator belt.

Batteries which are not maintained on a regular basis can fall victim to parasitic loads (small current drains which are constantly drawing current from the battery, like clocks, small lights, etc.). Normal parasitic loads may drain a battery on boat that is in storage and not used frequently. Boats that have additional accessories with increased parasitic load may discharge a battery sooner. Storing a boat with the negative battery cable disconnected or battery switch turned **OFF** will minimize discharge due to parasitic loads.

CLEANING

Keep the battery clean, as a film of dirt can help discharge a battery that is not used for long periods. A solution of baking soda and water mixed into a paste may be used for cleaning, but be careful to flush this off with clear water.

➡ **Do not let any of the solution into the filler holes on non-sealed batteries. Baking soda neutralizes battery acid and will de-activate a battery cell.**

CHECKING SPECIFIC GRAVITY

The electrolyte fluid (sulfuric acid solution) contained in the battery cells will tell you many things about the condition of the battery. Because the cell plates must be kept submerged below the fluid level in order to operate, maintaining the fluid level is extremely important. In addition, because the specific gravity of the acid is an indication of electrical charge, testing the fluid can be an aid in determining if the battery must be replaced. A battery in a boat with a properly operating charging system should require little maintenance, but careful, periodic inspection should reveal problems before they leave you stranded.

✳✳ CAUTION

Battery electrolyte contains sulfuric acid. If you should splash any on your skin or in your eyes, flush the affected area with plenty of clear water. If it lands in your eyes, get medical help immediately.

As stated earlier, the specific gravity of a battery's electrolyte level can be used as an indication of battery charge. At least once a year, check the specific gravity of the battery. It should be between 1.20 and 1.26 on the gravity scale. Most parts stores carry a variety of inexpensive battery testing hydrometers. These can be used on any non-sealed battery to test the specific gravity in each cell.

Conventional Battery

◆ See Figure 101

A hydrometer is required to check the specific gravity on all batteries that are not maintenance-free. The hydrometer has a squeeze bulb at one end and a nozzle at the other. Battery electrolyte is sucked into the hydrometer until the float or pointer is lifted from its seat. The specific gravity is then read by noting the position of the float/pointer. If gravity is low in one or more cells, the battery should be slowly charged and checked again to see if the gravity has come up. Generally, if after charging, the specific gravity of any two cells varies more than 50 points (0.50), the battery should be replaced, as it can no longer produce sufficient voltage to guarantee proper operation.

Check the battery electrolyte level at least once a month, or more often in hot weather or during periods of extended operation. Electrolyte level can be checked either through the case on translucent batteries or by removing the cell caps on opaque-case types. The electrolyte level in each cell should be kept filled to the split ring inside each cell, or the line marked on the outside of the case.

➡ **Never use mineral water or water obtained from a well. The iron content in these types of water is too high and will shorten the life of or damage the battery.**

If the level is low, add only distilled water through the opening until the level is correct. Each cell is separate from the others, so each must be checked and filled individually. Distilled water should be used, because the chemicals and minerals found in most drinking water are harmful to the battery and could significantly shorten its life.

If water is added in freezing weather, the battery should be warmed to allow the water to mix with the electrolyte. Otherwise, the battery could freeze.

Maintenance-Free Batteries

Although some maintenance-free batteries have removable cell caps for access to the electrolyte, the electrolyte condition and level is usually checked using the built-in hydrometer "eye." The exact

Fig. 101 The best way to determine the condition of a battery is to test the electrolyte with a battery hydrometer

type of eye varies between battery manufacturers, but most apply a sticker to the battery itself explaining the possible readings. When in doubt, refer to the battery manufacturer's instructions to interpret battery condition using the built-in hydrometer.

The readings from built-in hydrometers may vary, however a green eye usually indicates a properly charged battery with sufficient fluid level. A dark eye is normally an indicator of a battery with sufficient fluid, but one that may be low in charge. In addition, a light or yellow eye is usually an indication that electrolyte supply has dropped below the necessary level for battery (and hydrometer) operation. In this last case, sealed batteries with an insufficient electrolyte level must usually be discarded.

BATTERY TERMINALS

At least once a season, the battery terminals and cable clamps should be cleaned. Loosen the clamps and remove the cables, negative cable first. On batteries with top mounted posts, the use of a puller specially made for this purpose is recommended. These are inexpensive and available in most parts stores.

Clean the cable clamps and the battery terminal with a wire brush, until all corrosion, grease, etc., is removed and the metal is shiny. It is especially important to clean the inside of the clamp thoroughly (a wire brush is useful here), since a small deposit of foreign material or oxidation there will prevent a sound electrical connection and inhibit either starting or charging. It is also a good idea to apply some dielectric grease to the terminal, as this will aid in the prevention of corrosion.

After the clamps and terminals are clean, reinstall the cables, negative cable last; do not hammer the clamps onto battery posts. Tighten the clamps securely, but do not distort them. Give the clamps and terminals a thin external coating of clear polyurethane paint after installation, to retard corrosion.

Check the cables at the same time that the terminals are cleaned. If the insulation is cracked or broken, or if its end is frayed, that cable should be replaced with a new one of the same length and gauge.

BATTERY & CHARGING SAFETY PRECAUTIONS

Always follow these safety precautions when charging or handling a battery.

1. Wear eye protection when working around batteries. Batteries contain corrosive acid and produce explosive gas a byproduct of their operation. Acid on the skin should be neutralized with a solution of baking soda and water made into a paste. In case acid contacts the eyes, flush with clear water and seek medical attention immediately..

2. Avoid flame or sparks that could ignite the hydrogen gas produced by the battery and cause an explosion. Connection and disconnection of cables to battery terminals is one of the most common causes of sparks.

3. Always turn a battery charger **OFF**, before connecting or disconnecting the leads. When connecting the leads, connect the positive lead first, then the negative lead, to avoid sparks.

4. When lifting a battery, use a battery carrier or lift at opposite corners of the base.

5. Ensure there is good ventilation in a room where the battery is being charged.

6. Do not attempt to charge or load-test a maintenance-free battery when the charge indicator dot is indicating insufficient electrolyte.

7. Disconnect the negative battery cable if the battery is to remain in the boat during the charging process.

8. Be sure the ignition switch is **OFF** before connecting or turning the charger **ON**. Sudden power surges can destroy electronic components.

9. Use proper adapters to connect charger leads to batteries with non-conventional terminals.

BATTERY CHARGERS

Before using any battery charger, consult the manufacturer's instructions for its use. Battery chargers are electrical devices that change Alternating Current (AC) to a lower voltage of Direct Current (DC) that can be used to charge a marine battery. There are two types of battery chargers—manual and automatic.

A manual battery charger must be physically disconnected when the battery has come to a full charge. If not, the battery can be overcharged, and possibly fail. Excess charging current at the end of the charging cycle will heat the electrolyte, resulting in loss of water and active material, substantially reducing battery life.

➡As a rule, on manual chargers, when the ammeter on the charger registers half the rated amperage of the charger, the battery is fully charged. This can vary, and it is recommended to use a hydrometer to accurately measure state of charge.

Automatic battery chargers have an important advantage—they can be left connected (for instance, overnight) without the possibility of overcharging the battery. Automatic chargers are equipped with a sensing device to allow the battery charge to taper off to near zero as the battery becomes fully charged. When charging a low or completely discharged battery, the meter will read close to full rated output. If only partially discharged, the initial reading may be less than full rated output, as the charger responds to the condition of the battery. As the battery continues to charge, the sensing device monitors the state of charge and reduces the charging rate. As the rate of charge tapers to zero amps, the charger will continue to supply a few milliamps of current—just enough to maintain a charged condition.

REPLACING BATTERY CABLES

Battery cables don't go bad very often, but like anything else, they can wear out. If the cables on your boat are cracked, frayed or broken, they should be replaced.

When working on any electrical component, it is always a good idea to disconnect the negative (-) battery cable. This will prevent potential damage to many sensitive electrical components

Always replace the battery cables with one of the same length, or you will increase resistance and possibly cause hard starting. Smear the battery posts with a light film of dielectric grease, or a battery terminal protectant spray once you've installed the new cables. If you replace the cables one at a time, you won't mix them up.

➡Any time you disconnect the battery cables, it is recommended that you disconnect the negative (-) battery cable first. This will prevent you from accidentally grounding the positive (+) terminal when disconnecting it, thereby preventing damage to the electrical system.

Before you disconnect the cable(s), first turn the ignition to the **OFF** position. This will prevent a draw on the battery which could cause arcing. When the battery cable(s) are reconnected (negative cable last), be sure to check all electrical accessories are all working correctly.

TUNE-UP

Introduction

A proper tune-up is the key to long and trouble-free engine life, and the work can yield its own rewards. Studies have shown that a properly tuned and maintained engine can achieve better fuel economy than an out-of-tune engine. As a conscientious boater, set aside a Saturday morning, say once a month, to check or replace items which could cause major problems later. Keep your own personal log to jot down which services you performed, how much the parts cost you, the date, and the number of hours on the engine at the time. Keep all receipts for such items as engine oil and filters, so

that they may be referred to in case of related problems or to determine operating expenses. As a do-it-yourselfer, these receipts are the only proof you have that the required maintenance was performed. In the event of a warranty problem, these receipts will be invaluable.

The efficiency, reliability, fuel economy and enjoyment available from engine performance are all directly dependent on having your engine tuned properly. The importance of performing service work in the proper sequence cannot be over emphasized. Before making any adjustments, check the specifications. Never rely on memory when making critical adjustments.

Before beginning to tune any engine, ensure the engine has satisfactory compression. An engine with worn or broken piston rings, burned pistons, or scored cylinder walls, will not perform properly no matter how much time and expense is spent on the tune-up. Poor compression must be corrected or the tune-up will not give the desired results.

A regular maintenance program that is followed throughout the year, is one of the best methods of ensuring the engine will give satisfactory performance. As they say, you can spend a little time now or a lot of time and money later.

The extent of the engine tune-up is usually dependent on the time lapse since the last service. A complete tune-up of the entire engine would entail almost all of the work outlined in this manual. However, this is usually not necessary in most cases.

In this section, a logical sequence of tune-up steps will be presented in general terms. If additional information or detailed service work is required, refer to the section containing the appropriate instructions.

Tune-Up Sequence

During a tune-up, a definite sequence of procedures should be followed to return the engine to its maximum performance level. This type of work should not be confused with troubleshooting (attempting to locate a problem when the engine is not performing satisfactorily). In many cases, these two areas will overlap, because many times a minor or major tune-up will correct a malfunction and return the system to normal operation.

The following list is a suggested sequence of tasks to perform during a tune-up.
* Perform a compression check of each cylinder.
* Perform a valve adjustment
* Start the engine in a body of water and check the water flow through the engine.
* Check the injection pump for adequate performance and delivery.
* Test the starting and charging systems.
* Check the internal wiring.

Cylinder Compression

Cylinder compression test results are extremely valuable indicators of internal engine condition. The best marine mechanics automatically check an engine's compression as the first step in a comprehensive tune-up. A compression test will uncover many mechanical problems that can cause rough running or poor performance.

A compression gauge for diesel engines consists of a dummy injector connected to a gauge capable of reading 600 psi.

CHECKING COMPRESSION

1. Make sure that the proper amount and viscosity of engine oil is in the crankcase, then ensure the battery is fully charged.
2. Warm-up the engine to normal operating temperature, then shut the engine **OFF**.
3. Remove the injector lines and remove the injectors from each cylinder.
4. Install a diesel compression gauge into the No. 1 cylinder injector hole until the fitting is snug. When fitting the compression gauge adapter to the cylinder head, make sure the bleeder of the gauge (if equipped) is closed.

5. According to the tool manufacturer's instructions, connect a remote starting switch to the starting circuit.
6. With the ignition switch in the **OFF** position, use the remote starting switch to crank the engine through at least five compression strokes (approximately 5 seconds of cranking) and record the highest reading on the gauge.
7. Repeat the test on each cylinder, cranking the engine approximately the same number of compression strokes and/or time as the first.
8. Compare the highest readings from each cylinder to that of the others. The indicated compression pressures are considered within specifications if the lowest reading cylinder is within 75 percent of the pressure recorded for the highest reading cylinder. For example, if your highest reading cylinder pressure was 150 psi (1034 kPa), then 75 percent of that would be 113 psi (779 kPa). So the lowest reading cylinder should be no less than 113 psi (779 kPa).
9. Compression readings that are generally low indicate worn, broken, or sticking piston rings, scored pistons or worn cylinders.
10. If a cylinder exhibits an unusually low compression reading, squirt a tablespoon of clean engine oil into the cylinder through the injector hole and repeat the compression test. If the compression rises after adding oil, it means that the cylinder's piston rings and/or cylinder bore are damaged or worn. If the pressure remains low, the valves may not be seating properly (a valve job is needed), or the head gasket may be blown near that cylinder.
11. If compression in any two adjacent cylinders is low (with normal compression in the other cylinders), and if the addition of oil doesn't help raise compression, there is leakage past the head gasket. Oil and coolant in the combustion chamber, combined with blue or constant white smoke from the tailpipe, are symptoms of this problem. However, don't be alarmed by the normal white smoke emitted from the tailpipe during engine warm-up during cold weather. There may be evidence of water droplets on the engine oil dipstick and/or oil droplets in the cooling system if a head gasket is blown.
12. When reinstalling the injector assemblies, install new washers and/or gaskets as appropriate.

Valve Adjustment

◆ See Figures 102 thru 108

Four-stroke diesel engines use valves to admit the fuel/air mixture into the combustion chamber, to seal the combustion chamber for compression, and to allow the spent exhaust gases to escape. All of these functions occur using the valve train (camshaft, lifters/shims and rocker arms and pushrods.)

In order for the valves to operate properly, they must be adjusted to assure that the full benefit of the camshaft lobe lift is realized, but they also must be able to close fully once the lobe of the camshaft has gone by. Valves are adjusted by increasing or decreasing their lash, which is the amount of free-play in the valve train when the valve is closed (meaning the camshaft lobe is not actuating the pushrod or rocker arm). Valve lash therefore, is basically a gap that exists between components when the valve is fully closed.

Since valves open and close with every turn of the crankshaft, their movement becomes a blur at engine speeds, creating a pounding on the entire valve train. As the engine is operated, internal components slowly wear, affecting distances between the components in the valve train. Valve lash will tend to change (increase or decrease) depending on the model. On some engines, the valve seats and heads will wear slowly, causing the valve to come further

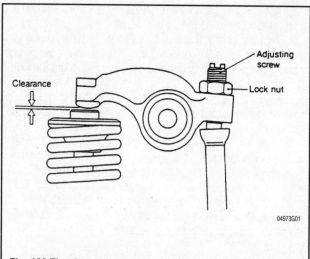

Fig. 102 The clearance between the rocker arm and the valve tip is known as valve lash

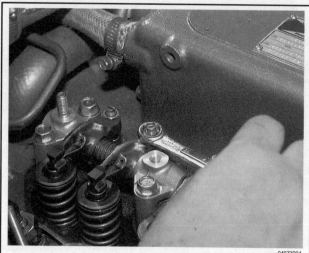

Fig. 105 If the lash is not correct, loosen the locknut on the adjuster. . .

Fig. 103 To access the rocker arms for adjustment, the valve cover must be removed

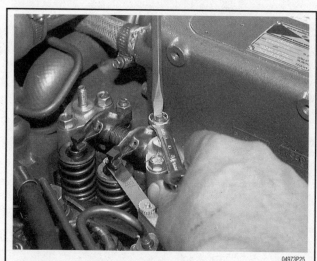

Fig. 106 . . .and hold it stationary while turning the adjuster with a screwdriver

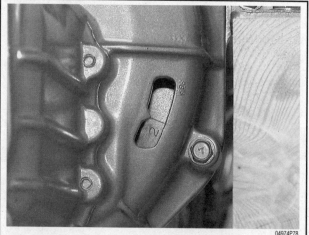

Fig. 104 To properly adjust the valves for each cylinder, the marks on the flywheel must be aligned to the view port on the transmission

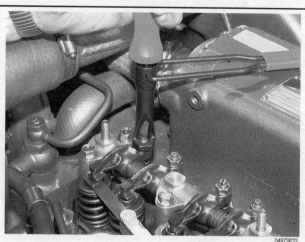

Fig. 107 Here is an example of a special valve lash adjuster tool. A screwdriver and wrench are combined, making the job easier

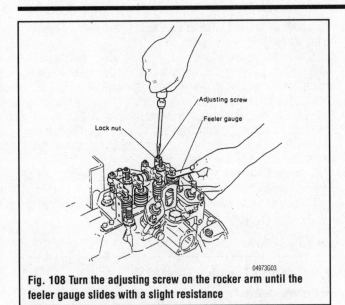

Fig. 108 Turn the adjusting screw on the rocker arm until the feeler gauge slides with a slight resistance

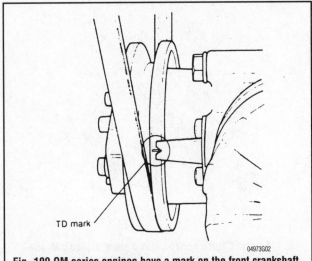

Fig. 109 QM series engines have a mark on the front crankshaft pulley for locating Top Dead Center (TDC) for valve adjustment

into the cylinder head (moving the stem closer to the shim or rocker arm and decreasing valve lash). On other models, the stem, shim or other valve train components will wear, causing the gap to increase.

Increased valve lash will not allow a valve to fully open since some of the camshaft lobe lift will be wasted on taking up the excess lash. If an intake valve does not open sufficiently, the full air charge will not make it into the cylinder and power will be lost during combustion. If an exhaust valve does not fully open, some exhaust gases will be left in the cylinder. Again, the result will be a reduction in engine power.

Decreased valve lash will have a less noticeable effect on engine power, but could have a more devastating effect on your engine. As valve lash is decreased beyond specification, the valve train components may be "too tight" and not allow the valve to fully come into contact with the seat. This will prevent the valve from cooling through heat transfer with the valve seat. The term "burnt valve" which you have likely heard someone mention before means that a valve was ruined by heat. A burnt valve will not properly seal the combustion chamber, and can also break, destroying your piston and cylinder wall. As valve lash decreases, the engine could lose power if valves are held partially open by the valve train (not allowing for proper compression).

Intake and exhaust valves usually have different specifications, because of the differences in their sizes and jobs. The exhaust valve is usually set a little looser than the intake, because of the harsh environment it lives in (superheated gases pass over it into the exhaust system when it is open). Intake valves have it easy (for a valve) as they are in contact with the cylinder head when they are exposed to combustion gases and temperatures (when they are open, relatively cool air/fuel mixture passes over their surface).

To identify an exhaust or intake valve, look at its position in relation to the rest of the cylinder head. In most cases, the intake valve will be closest to and in alignment with the intake manifold, while the exhaust valve is closest to and adjacent to the exhaust pipe.

QM SERIES ENGINES

◆ See Figures 102 thru 109

➡The valve lash should be adjusted when the engine is cold.

1. Remove the rocker cover from the engine.
2. Turn the engine over by hand, and set the number one cylin-

der to Top Dead Center (TDC). By aligning the mark on the crankshaft pulley with the mark on the engine. Both the intake and exhaust valves for the number one cylinder should be in a resting position (not depressed). The rocker arms should not move when the engine is slightly turned back and forth from the TDC mark on the flywheel. If the rocker arms move back and forth, and the flywheel is on the correct mark, rotate the engine one full revolution of the timing mark. This should be the compression stroke.

✳✳ CAUTION

It is advisable to loosen the injector fuel lines before attempting to turn the engine over by hand. It IS possible for the engine to start when turning it by hand! When turning the engine, do so very slowly to prevent accidental starting of the engine.

3. Insert a flat feeler gauge of the proper dimension between the tappet and the rocker lever for the number 1 cylinder. The lash should be 0.0059 in. (0.15mm) The gauge should slide between these two parts with a slight pull.
4. If the gauge will not slide readily, or there is no resistance when pulling it through, loosen the locknut on the rocker arm, and use a screwdriver to rotate the adjusting screw. When a slight pull is obtained, hold the adjusting screw in place with a screwdriver and use a box or open end wrench to tighten the locknut. If the adjustment has tightened up, change the setting of the adjusting screw. If a great deal of effort is required to pull the gauge (adjustment too tight), burned valves may result.
5. On two cylinder engines, adjust the valves on the number 2 cylinder by rotating the crankshaft by 180 degrees. On three cylinder engines, rotate the crankshaft pulley by 120 degrees. The number three cylinder can be adjusted by turning the number three cylinder an additional 120 degrees.
6. Once all of the valves are adjusted to the proper specification, install the rocker arm cover.

GM/HM SERIES ENGINES

◆ See Figures 102 thru 108, 110 and 111

➡The valve lash should be adjusted when the engine is cold.

1. Remove the rocker cover from the engine.

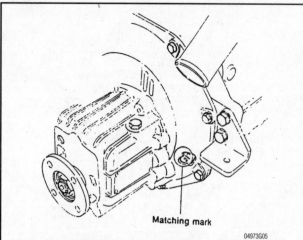

Fig. 110 Some GM/HM series engines are equipped with an opening in the flywheel housing on the for viewing of the timing marks

Fig. 111 On GM series engines, the starter can be removed to view the timing marks if the marks on the transmission cannot be seen

2. If necessary, remove the starter motor from the engine. This will allow for viewing of the timing mark(s) on the flywheel. (Some transmissions do not have a viewing port for the timing marks on the flywheel)

✳✳ CAUTION

It is advisable to loosen the injector fuel lines before attempting to turn the engine over by hand. It IS possible for the engine to start when turning it by hand! When turning the engine, do so very slowly to prevent accidental starting of the engine.

3. Turn the engine over by hand, and set the number one cylinder to Top Dead Center (TDC). Both the intake and exhaust valves should be in a resting position (not depressed). The rocker arms should not move when the engine is slightly turned back and forth from the TDC mark on the flywheel. If the rocker arms move back and forth, and the flywheel is on the correct mark, rotate the engine one full revolution of the timing mark. This will be the compression stroke.
4. Insert a flat feeler gauge of the proper dimension between the tappet and the rocker lever for the No. 1 cylinder. The lash should

be 0.0079 in. (0.2 mm). The gauge should slide between these two parts with a slight pull.
5. If the gauge will not slide readily, or there is no resistance when pulling it through, loosen the locknut on the rocker arm, and use a screwdriver to rotate the adjusting screw. When a slight pull is obtained, hold the adjusting screw in place with a screwdriver and use a box or open end wrench to tighten the locknut. If the adjustment has tightened up, change the setting of the adjusting screw. If a great deal of effort is required to pull the gauge (adjustment too tight), burned valves may result.
6. Adjust the remaining valves on the engine by aligning the timing mark for each of the remaining cylinders.
7. Once all of the valves are adjusted to the proper specification, install the starter, and the rocker arm cover.

JH SERIES ENGINES

▶ See Figures 102 thru 108, 112 and 113

➡The valve lash should be adjusted with the engine cold.

1. Remove the rocker (valve) cover from the engine.

✳✳ CAUTION

It is advisable to loosen the injector fuel lines before attempting to turn the engine over by hand. It IS possible for the engine to start when turning it by hand! When turning the engine, do so very slowly to prevent accidental starting of the engine.

2. Turn the engine over by hand, and set the number 1 cylinder to Top Dead Center (TDC) by aligning the mark on the flywheel with the mark on the flywheel housing. Both the intake and exhaust valves for the number one cylinder should be in a resting position (not depressed). The rocker arms should not move when the engine is slightly turned back and forth from the TDC mark on the flywheel. If the rocker arms move back and forth, and the flywheel is on the correct mark, rotate the engine one full revolution of the timing mark. This will be the compression stroke.
3. Insert a flat feeler gauge of the proper dimension between the tappet and the rocker lever for the number 1 cylinder. The lash should be 0.0079 in. (0.2mm). The gauge should slide between these two parts with a slight pull.

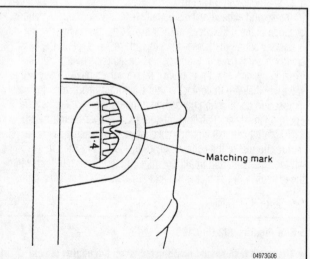

Fig. 112 The timing marks on JH series engines are viewed through an opening in the flywheel housing

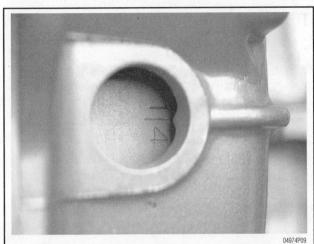

Fig. 113 On this JH engine, the TDC marks are aligned for cylinders 1 and 4 (both cylinders are NOT on the compression stroke, though)

4. If the gauge will not slide readily, or there is no resistance when pulling it through, loosen the locknut on the rocker arm, and use a screwdriver to rotate the adjusting screw. When a slight pull is obtained, hold the adjusting screw in place with a screwdriver and use a box or open end wrench to tighten the locknut. If the adjustment has tightened up, change the setting of the adjusting screw. If a great deal of effort is required to pull the gauge (adjustment too tight), burned valves may result.

5. Adjust the valves on the number 2 cylinder by rotating the crankshaft clockwise until the TDC mark on the flywheel lines up with the mark on the flywheel housing. After the number 2 cylinder valves have been adjusted, proceed with the number 3 and 4 cylinders, respectively. When aligning the timing mark, make sure the TDC mark is aligned and not the injection timing mark, which may not allow for proper adjustment.

6. Once all of the valves are adjusted to the proper specification, install the rocker arm cover.

WINTER STORAGE

▶ **See Figures 114, 115, 116 and 117**

Taking extra time to store the boat properly at the end of each season will increase the chances of satisfactory service at the next season. Remember, storage is the greatest enemy of a marine engine. In a perfect world the unit should be run on a monthly basis. The steering and shifting mechanism should also be worked through complete cycles several times each month. But who lives in a perfect world!

For most of us, if a small amount of time is spent in winterizing our beloved boats, the reward will be satisfactory performance, increased longevity and greatly reduced maintenance expenses.

Winter Storage Checklist

Proper winterizing involves adequate protection of the unit from physical damage, rust, corrosion, and dirt. The following steps provide guide to winterizing your marine diesel at the end of a season.

1. Always keep a note or a checklist of just what maintenance needs to be done prior to starting the engine in the spring.

2. Replace the oil and oil filter. Normal combustion produces corrosive acids that are absorbed by the oil. Leaving dirty oil in the engine for an extended time allows these acids to attack and damage bearing surfaces.

3. Change the engine oil and filter. Used oil contains harmful acids and contaminants that should not be allowed to go to work on the engine all winter long. The transmission oil should also be changed.

4. Replace the fuel filter elements—draining any water from the filter bowls.

5. Keeping the fuel tank full over the winter will help cut down on condensation. Using fuel additives which control bacterial growth or prevent gelling in cold climates are also popular with many boaters.

6. Bleed the fuel system of air.

7. Run the engine up to operating temperature and then turn it **OFF**.

8. With the fuel shut-off pulled, squirt some oil into the inlet manifold and turn the engine over a few times without starting. This

Fig. 114 Most sailboats are stored outside . . .

Fig. 115 . . . however, some owners prefer indoor storage

Fig. 116 Taking extra time to store the boat properly at the end of each season will increase the chances of satisfactory service at the next season

Fig. 117 Storage cradles should be adjusted to fit the hull properly and not allow the keel to rest on the ground

will spread the oil around the cylinders and prevent rust and corrosion from forming.

9. On raw water cooled engines, perform the following:

a. Flush the raw water circuit to remove corrosive salts by connecting a freshwater supply to the raw water inlet hose.

b. Drain the raw water system taking special care to empty all the low spots.

➡ At this point the system can be filled with a non-toxic antifreeze or the as the check list suggests, the water pump can be disassembled.

c. Remove the raw water pump impeller. Grease the impellers lightly with petroleum jelly and replace.

d. Leave the pump cover screws loose to prevent the pump impeller from sticking to the housing. This would damage the impeller upon initial startup in the spring.

e. If the paper gasket is rubbed with grease or petroleum jelly it will not stick to the cover or pump body next time the cover is removed.

10. On fresh water cooled engines, perform the following:

a. Drain the engine coolant, opening all engine, heat-exchanger, and oil-cooler drains. The antifreeze itself does not wear out, but it has various additives to fight corrosion that do.

b. Backflush the fresh water circuit to remove sediment.

c. Replace the coolant with a clean, fresh, 50/50 mixture of water and antifreeze. Always mix the antifreeze and water mixture prior to pouring it into the engine.

11. Loosen the exhaust manifolds and check for carbon build-up.

12. Inspect all hoses for signs of softening, cracking or bulging, especially those routinely exposed to high heat. Check hose clamps for tightness and corrosion.

13. On turbocharged engines, remove the inlet and exhaust ducting from the turbo and inspect the compressor wheel and turbine for excessive deposits or damage. Clean as necessary.

14. Fully charge the batteries. Disconnect all leads. Unattended, a battery naturally discharges over a period of several weeks. The electrolyte on a discharged battery can freeze at 20°F (−7°C), so keep the batteries fully charged or, better still, remove them to a warmer storage area. Small automatic trickle chargers work well.

15. Treat battery and cable terminals with petroleum jelly, silicone grease, or a heavy-duty corrosion inhibitor.

16. Protect external surfaces with a heavy-duty corrosion inhibitor.

17. Grease all greaseable points on the drivetrain.

18. Lightly coat the alternator and starter with a light lubricant to disperse the water. Loosen the alternator belt tension.

19. Cover the engine with a waterproof sheet in case there are any leaks from above. Some boaters will take the time to seal all openings to the engine (air inlet, breathers, exhaust), however this traps moisture and may do more harm than good.

20. Place a checklist in a handy spot to remind you of just what maintenance needs to be done prior to starting the engine in the spring.

21. If you visit the boat during the winter, additional protection can be achieved using the starter to turn the engine over and circulate oil to the bearings and cylinder walls. Remember to pull out the stop cable and keep turning until the low oil pressure light extinguishes or pressure registers on the gauge.

➡ Special inhibiting oils are available that provide greater protection. Use these to replace the standard engine oil; run the engine briefly to coat all surfaces, and then drain the oil. Protection remains good, provided the engine is not turned.

22. If the engine cannot be fully winterized, replace oil, coolant, and all filters and run the engine up to operating temperatures monthly.

SPRING COMMISSIONING

Satisfactory performance and maximum enjoyment can be realized if a little time is spent in commissioning your boat in the spring. Assuming you have followed the steps we recommended to winterize your vessel (in addition to any the manufacturer specifies) and the unit has been properly stored, a minimum amount of work should be required to prepare it for use.

After performing the spring commissioning and testing the boat on the water, it is a good idea to perform a full tune-up. Remember, you are relying on your engine to get you where you want to go when the wind is not blowing. Treat it good now and it will treat you good later.

Spring Commissioning Checklist

▶ **See Figures 118 and 119**

The following steps outline a logical sequence of tasks to be performed before starting your engine for the first time in a new season.

1. Pick up the checklist you made to remind yourself of just what maintenance needs to be done prior to starting the engine in the spring. You did remember to write yourself a checklist . . . right?

2. Remove the cover placed over the engine last winter. Unseal any engine openings (air inlet, breathers, exhaust) previously sealed.

3. Replace all zincs.

4. If you took our advice, you removed the battery for the winter. While it was in storage, you should have kept it fully charged. It should be ready to go. So install the battery and connect the battery cables. Treat battery and cable terminals with petroleum jelly, silicone grease, or a heavy-duty corrosion inhibitor. Capacity test the batteries.

5. Tighten alternator and other belts.

6. On turbocharged engines, Connect the inlet and exhaust ducting from the turbo if removed. It may also be a good idea to pre-oil the turbocharger prior to starting the engine.

7. As you did in the winter, inspect all hoses for signs of soften-ing, cracking or bulging, especially those routinely exposed to high heat. Check hose clamps for tightness and corrosion.

8. Ensure the exhaust manifolds are tight.

9. On fresh water cooled engines, check the condition of the coolant mixture with a coolant tester and adjust the mixture by adding antifreeze or water.

10. On raw water cooled engines, install a new pump cover gasket and tighten the pump cover screws.

11. Prime the cooling system.

12. Bleed the fuel system of air.

13. With the fuel shut-off pulled, crank the engine without starting it until oil pressure is established.

14. Start the engine and allow it to reach operating temperature.

15. Once running, check the oil pressure, the raw water discharge and the engine for oil and water leaks.

16. Draining water from the filter bowls. If an excessive amount of water is noted in the fuel system, you may want to consider a fuel conditioner or replacing the old fuel with fresh fuel.

17. After testing the boat on the water, it is a good idea to change the engine oil and filter. The transmission oil should also be changed. This eliminates the adverse effects of moisture in the oil.

18. After the boat has been in the water for a few days, check the engine alignment.

SELSTK37

Fig. 118 Most marinas have the capacity to store boats of all sizes

SELSTK49

Fig. 119 Moving boats is easy at the marina. This boat is ready for a maiden voyage after spring commissioning

COMPONENT LOCATIONS

Component Locations—2QM20

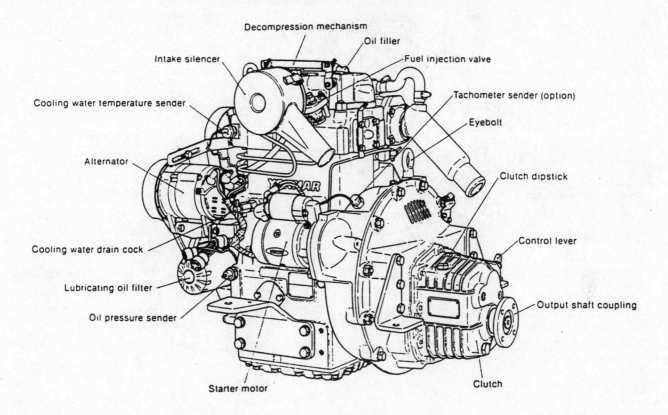

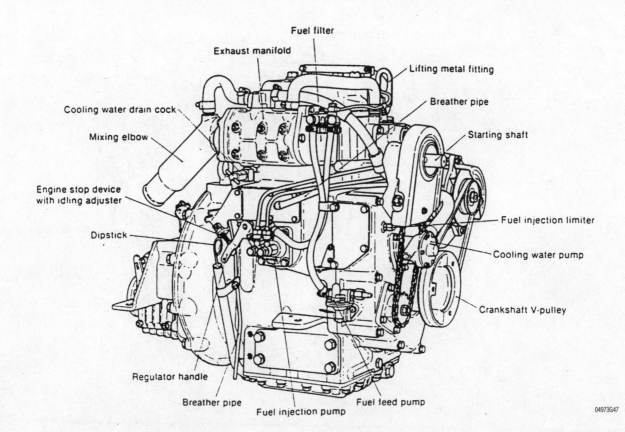

04973G47

Component Locations—3QM30H

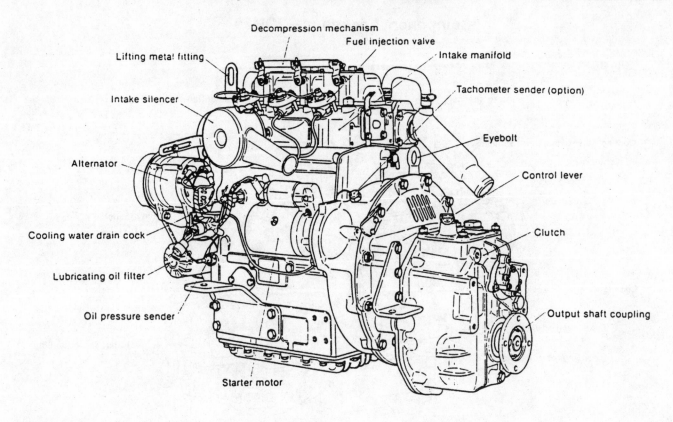

Decompression mechanism
Fuel injection valve
Lifting metal fitting
Intake manifold
Intake silencer
Tachometer sender (option)
Eyebolt
Alternator
Control lever
Cooling water drain cock
Clutch
Lubricating oil filter
Oil pressure sender
Output shaft coupling
Starter motor

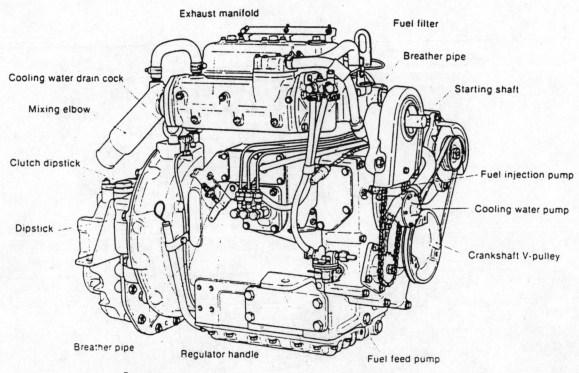

Exhaust manifold
Fuel filter
Breather pipe
Cooling water drain cock
Starting shaft
Mixing elbow
Clutch dipstick
Fuel injection pump
Cooling water pump
Dipstick
Crankshaft V-pulley
Breather pipe
Regulator handle
Fuel feed pump
Engine stop device
with idling adjuster
Fuel injection limiter

04973G48

Component Locations—3QM30Y

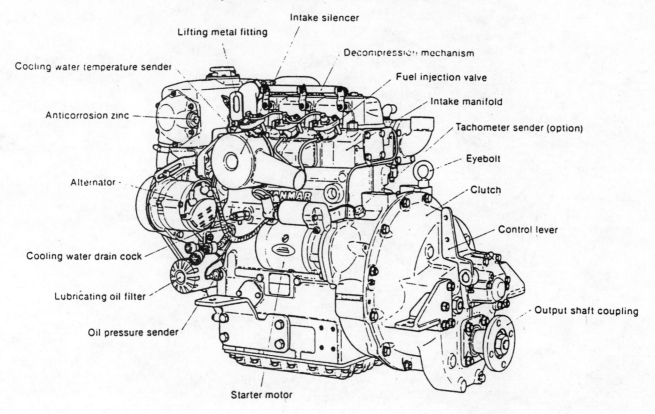

Lifting metal fitting
Intake silencer
Decompression mechanism
Cooling water temperature sender
Fuel injection valve
Anticorrosion zinc
Intake manifold
Tachometer sender (option)
Eyebolt
Alternator
Clutch
Control lever
Cooling water drain cock
Lubricating oil filter
Oil pressure sender
Output shaft coupling
Starter motor

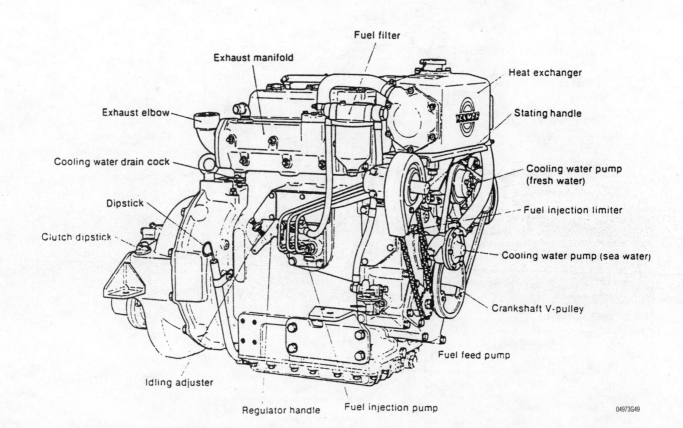

Fuel filter
Exhaust manifold
Heat exchanger
Exhaust elbow
Stating handle
Cooling water drain cock
Cooling water pump (fresh water)
Dipstick
Fuel injection limiter
Clutch dipstick
Cooling water pump (sea water)
Crankshaft V-pulley
Fuel feed pump
Idling adjuster
Regulator handle
Fuel injection pump

04973G49

Component Locations—1GM

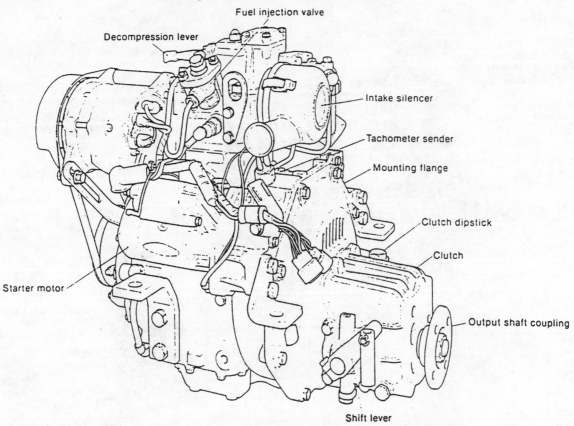

Fuel injection valve
Decompression lever
Intake silencer
Tachometer sender
Mounting flange
Clutch dipstick
Clutch
Output shaft coupling
Starter motor
Shift lever

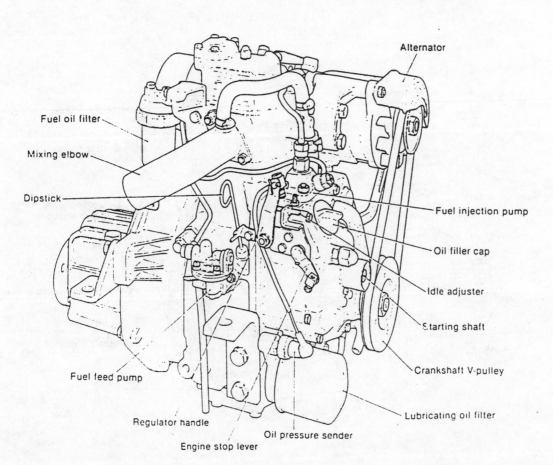

Alternator
Fuel oil filter
Mixing elbow
Dipstick
Fuel injection pump
Oil filler cap
Idle adjuster
Starting shaft
Crankshaft V-pulley
Fuel feed pump
Lubricating oil filter
Regulator handle
Oil pressure sender
Engine stop lever

04973G50

Component Locations—2GM

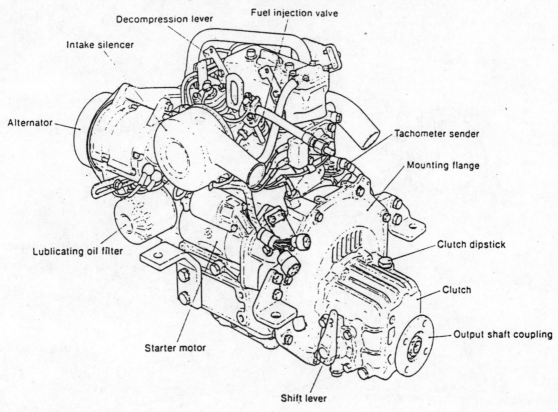

- Decompression lever
- Fuel injection valve
- Intake silencer
- Alternator
- Tachometer sender
- Mounting flange
- Lublicating oil filter
- Clutch dipstick
- Clutch
- Output shaft coupling
- Starter motor
- Shift lever

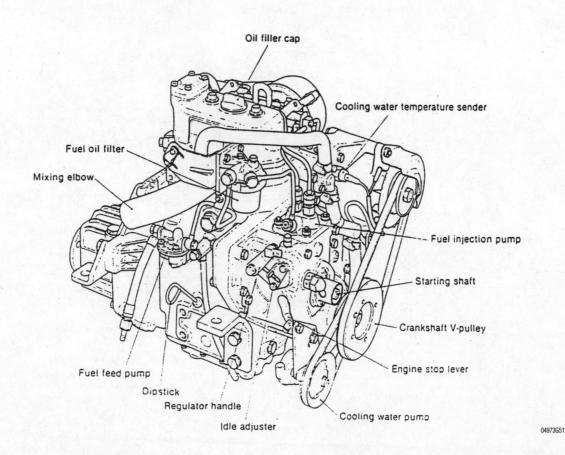

- Oil filler cap
- Cooling water temperature sender
- Fuel oil filter
- Mixing elbow
- Fuel injection pump
- Starting shaft
- Crankshaft V-pulley
- Fuel feed pump
- Engine stop lever
- Dipstick
- Regulator handle
- Cooling water pump
- Idle adjuster

04973G51

Component Locations—3HM

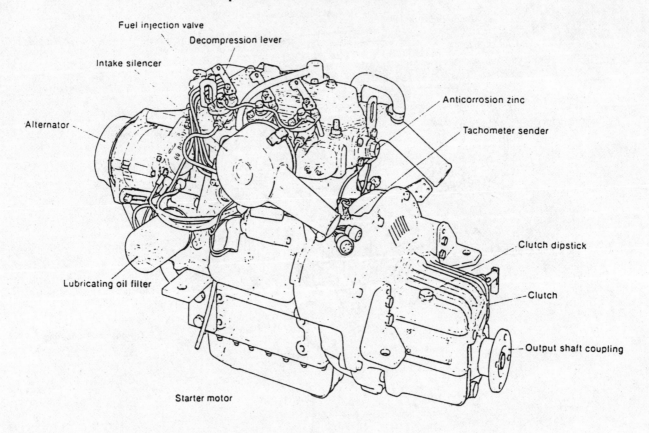

Fuel injection valve
Decompression lever
Intake silencer
Anticorrosion zinc
Alternator
Tachometer sender
Clutch dipstick
Lubricating oil filter
Clutch
Output shaft coupling
Starter motor

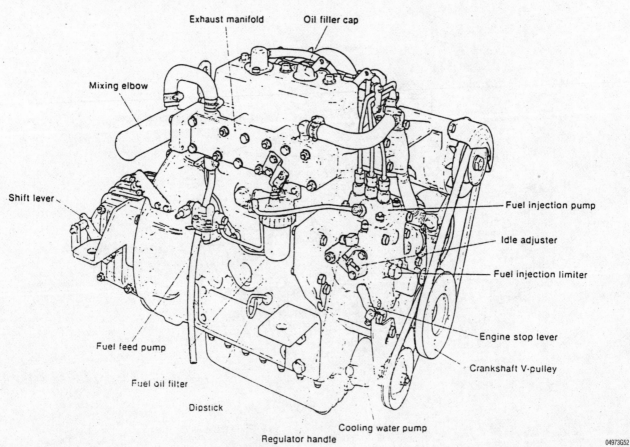

Exhaust manifold
Oil filler cap
Mixing elbow
Shift lever
Fuel injection pump
Idle adjuster
Fuel injection limiter
Fuel feed pump
Engine stop lever
Crankshaft V-pulley
Fuel oil filter
Dipstick
Cooling water pump
Regulator handle

04973G52

Component Locations—3GM

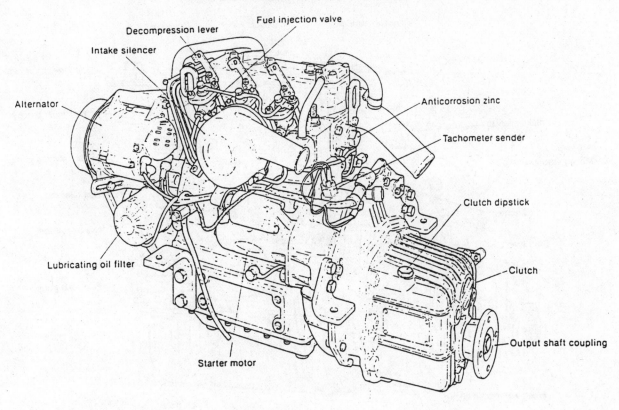

Decompression lever
Fuel injection valve
Intake silencer
Alternator
Anticorrosion zinc
Tachometer sender
Clutch dipstick
Clutch
Lubricating oil filter
Output shaft coupling
Starter motor

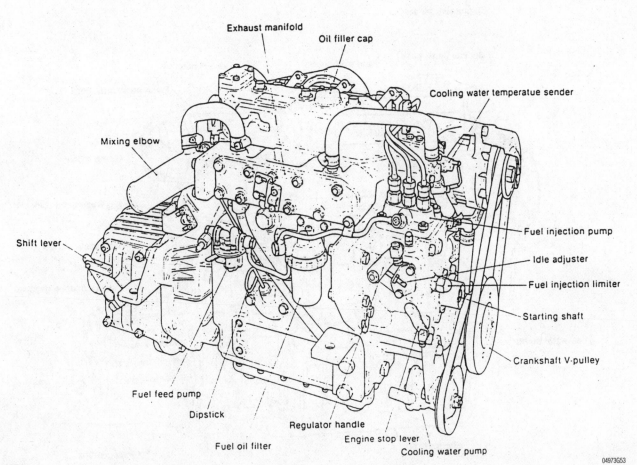

Exhaust manifold
Oil filler cap
Cooling water temperatue sender
Mixing elbow
Fuel injection pump
Shift lever
Idle adjuster
Fuel injection limiter
Starting shaft
Crankshaft V-pulley
Fuel feed pump
Dipstick
Regulator handle
Engine stop lever
Cooling water pump
Fuel oil filter

04973G53

Component Locations—4JHE and 4JH2E

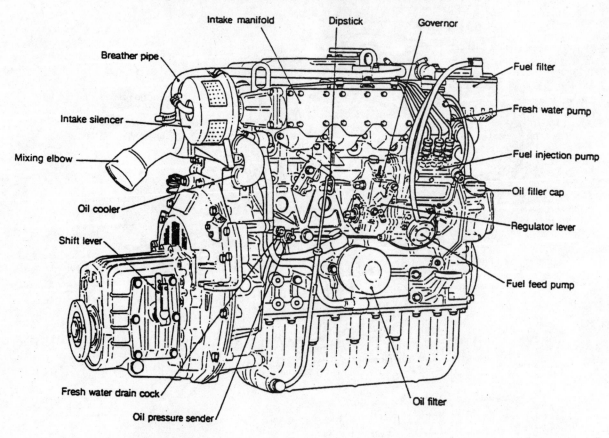

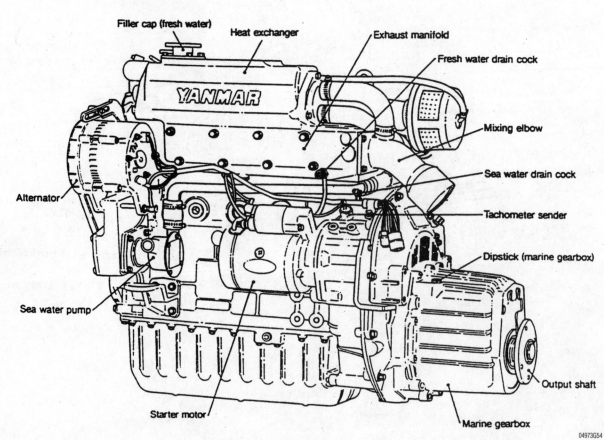

04973G54

Component Locations—4JHTE and 4JH2TE

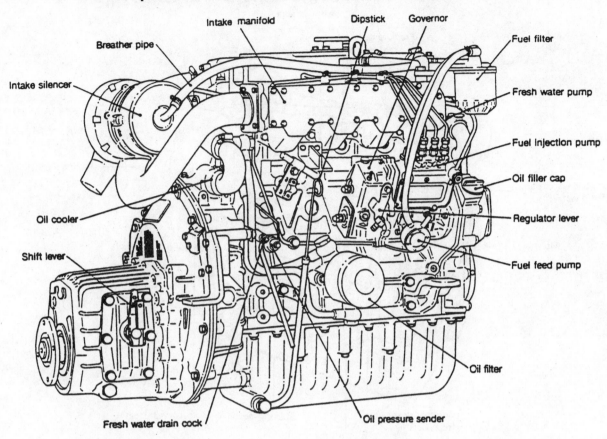

Intake manifold — Dipstick — Governor — Fuel filter — Breather pipe — Fresh water pump — Intake silencer — Fuel Injection pump — Oil filler cap — Oil cooler — Regulator lever — Shift lever — Fuel feed pump — Oil filter — Fresh water drain cock — Oil pressure sender

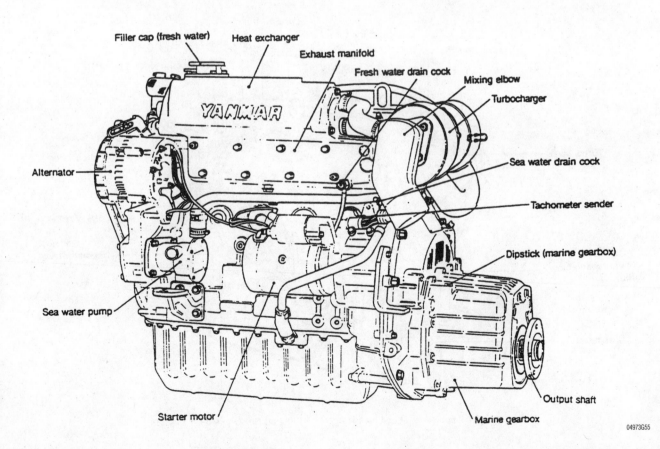

Filler cap (fresh water) — Heat exchanger — Exhaust manifold — Fresh water drain cock — Mixing elbow — Turbocharger — Sea water drain cock — Alternator — Tachometer sender — Dipstick (marine gearbox) — Sea water pump — Output shaft — Starter motor — Marine gearbox

YANMAR

04973G55

Component Locations—4JHHTE, 4JHDTE, 4JH2HTE, 4JH2DTE, and 4JH2UTE

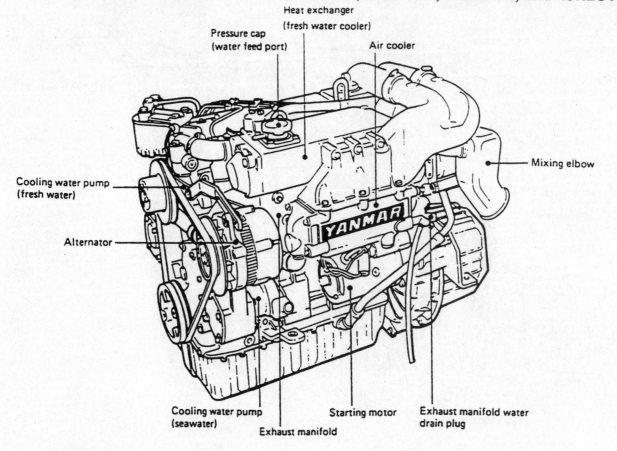

Heat exchanger
(fresh water cooler)

Pressure cap
(water feed port)

Air cooler

Mixing elbow

Cooling water pump
(fresh water)

Alternator

Cooling water pump
(seawater)

Exhaust manifold

Starting motor

Exhaust manifold water
drain plug

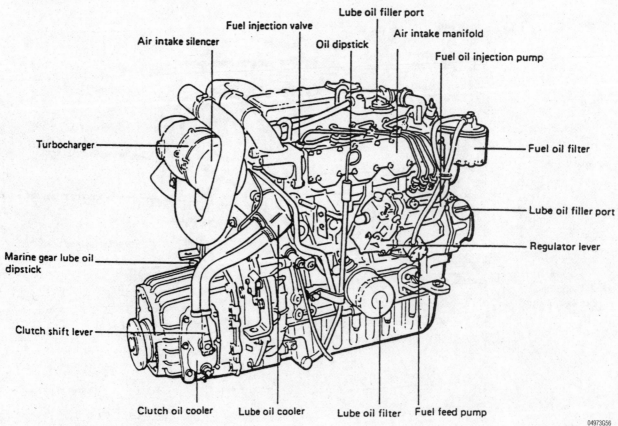

Lube oil filler port

Fuel injection valve

Oil dipstick

Air intake manifold

Air intake silencer

Fuel oil injection pump

Turbocharger

Fuel oil filter

Lube oil filler port

Regulator lever

Marine gear lube oil
dipstick

Clutch shift lever

Clutch oil cooler

Lube oil cooler

Lube oil filter

Fuel feed pump

04973G56

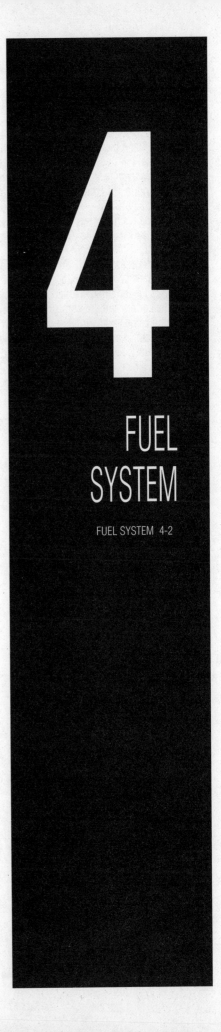

4

FUEL
SYSTEM

FUEL SYSTEM 4-2

FUEL SYSTEM

▶ **See Figures 1 and 2**

The output of a diesel engine is controlled by regulating the amount of fuel injected into the cylinders. In marine use, you normally want the engine to run at a specific speed, regardless of the load placed on it. This cannot be done by simply placing the throttle at a certain point because every time the load increases or decreases, the engine slows down or speeds up. Constant running is achieved by connecting the fuel control lever on the injector pump to a governor.

The injector pump is the heart of the fuel system. This complex component receives low-pressure fuel from the lift pump, senses the engine's fuel requirement, and then, at exactly the right moment, pumps the correct amount of fuel to the injectors at very high pressure. Both in-line and distributor-type pumps are common on small diesels. injector pumps will provide years of reliable service if supplied with clean fuel. They are factory set and, for the most part, require no maintenance. A few models have their own oil sump and require regular oil changes. Failures are rare, and most problems are due to the presence of air.

Fuel timing determines when each injector receives its shot of fuel from the injector pump. It is set by the mesh of the timing gears and the position in which the injector pump is mounted on the engine.

Timing varies with engine design and number of cylinders. Typically, a four-cylinder diesel will have a setting of 10–15 inches before top dead center (TDC). Timing is factory set and should not need adjustment unless the timing gears are disassembled or the injector pump disturbed. Rarely, the injector pump or timing-gear adjustment may work loose and affect the timing.

Governors are rpm-sensitive mechanisms that balance centrifuged flyweights against spring pressure to control the injector pump setting and maintain rpm under changing loads.

Governors, whether part of the engine or injector pump, require no maintenance. They may have external adjustments for maximum-rpm, acceleration, and idle-rpm settings, but only the idle-rpm setting can be altered. Maximum rpm and acceleration screws are factory set, and usually sealed; do not attempt to adjust them.

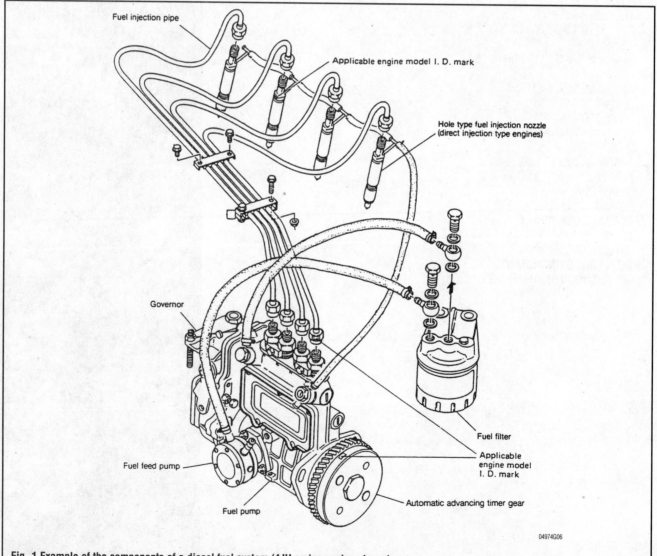

04974G06

Fig. 1 Example of the components of a diesel fuel system (4JH series engine shown)

Fig. 2 Some engines, like this JH series, use a cold start enrichment

Fuel

GENERAL INFORMATION

♦ **See Figure 3**

Diesel fuel is obtained from crude oil, which is a mixture of hydrocarbons, which are compounds consisting of Hydrogen and Carbon. These compounds are Benzine, Pentane, Hexane, Heptane, Toluene, Propane and Butane.

The compounds which form crude oil vaporize at different temperatures. That is, they have different boiling points. To separate the hydrocarbons, the crude oil is heated and the different hydrocarbons are given off as vapor. The hydrocarbon With the lowest boiling point is given off first.

When this separation is complete, the temperature is again used to obtain the hydrocarbon with the next highest boiling point and so on until all the commercial gasoline, kerosene, diesel fuel, domestic heating oil, industrial fuel oil, lubricating oil, paraffin, etc. are obtained, leaving coke and asphalt. This process of distilling diesel

Fig. 3 Always obtain your fuel from a reputable fuel dock

fuel from crude oil is a highly complicated process involving precision control of both temperatures and pressures.

As crude oil is refined, approximately 44 percent is gasoline, 36 percent is fuel oil and the balance is kerosene, lubricants, etc. The characteristic properties of diesel fuel include, heat value, specific gravity, flash point, pour point, viscosity, volatility ignition quality, carbon residue, sulfur content, oxidation and water.

HEAT VALUE

The heat value of a fuel is of primary importance as it is an indication of how much power the fuel will provide when burned. The heat value of fuel oil can be determined by burning the oil in a special device known as a calorimeter. With such equipment a measured quantity of fuel burned and the amount of heat is carefully measured in British Thermal Units (BTU) per pound of fuel.

➡**A BTU is the amount of heat required to raise the temperature of one pound of water one degree Fahrenheit.**

The energy content of diesel fuel is about 10 percent higher than that of gasoline on a volume basis, allowing it to deliver more work per gallon than an equivalent volume of gasoline. A gallon of diesel Contains some 141,000 BTU'S, a gallon of premium gasoline contains l25,000.

The diesel engine vastly multiplies this initial advantage by the way it operates and uses far less fuel to achieve performance equal to or better than that of the gasoline engine. It does this because it uses a much higher compression ratio.

SPECIFIC GRAVITY

Specific gravity of a liquid, such as diesel fuel, is the ratio of the density of the fuel to the density of water. Specific gravity may be measured by using a hydrometer. Specific gravity of a fuel affects the spray penetration as the fuel is injected into the combustion chamber. It is also, to a degree, a measure of the heat content of the fuel. Fuel with a low American Petroleum Institute (API) specific gravity usually has greater heat value per gallon than fuel with a higher rating.

FLASH POINT

The flash point of a fuel is the temperature at which it must be heated until sufficient flammable vapor is driven off to ignite when brought into contact with a flame or heat. Fire point is the higher temperature at which the oil vapors will continue to burn after being ignited. The fire point is usually 50-70°F higher than the flash point.

Flash point is also an indication of fire hazard. The lower the flash point, the greater the fire hazard. The flash point of diesel fuel is 100°F for type I-D fuel, 120°F for type 2-D fuel and 130°F for type 4-D fuel. In some states there are laws specifying the flash point of diesel fuels which may be used.

Flash point is only an index of combustion temperature and tells nothing about diesel fuel ignition quality within the engine.

CLOUD POINT & POUR POINT

The cloud point of diesel fuel is the temperature at which the components of the fuel become insoluble, cannot be dissolved and wax crystals begin to form.

Pour point is the temperature at which enough of the fuel becomes insoluble and prevents flow under specified conditions. High pour point implies that in cold weather the fuel oil will flow easily through the filters and fuel system of the engine.

VISCOSITY

Viscosity is the property of a fluid that resists the force which causes the fluid to flow. Viscosity is measured by observing the time required for a certain volume of the fluid to flow, under stated conditions, through a short tube of small bore. The flow is measured by using a device called a viscometer. The viscosity of the fuel oil is measured at 77°F and 122°F.

Viscosity of diesel fuel affects the pattern of spray in the combustion chambers. Low viscosity produces a fine mist, while high viscosity tends to result in coarse atomization.

VOLATILITY

Volatility of a liquid is its ability to change into a vapor. The volatility of a liquid fuel is indicated by the air-vapor ratio that can be formed at a specific temperature. In the case of diesel fuels, volatility is indicated by a 90 percent distillation temperature(a specific temperature at which 90 percent of the fuel is distilled off). As volatility is decreased, carbon deposits and in Some engines, wear is increased. Some engines will produce smoke as volatility is decreased.

CETANE RATING

The ignition quality of diesel fuel, which is the ease with which the fuel will ignite and the manner in which it burns, is expressed in Cetane numbers. A Cetane number rating is obtained by comparing the fuel with Cetane. Cetane is a colorless liquid hydrocarbon which has excellent ignition qualities and is rated at 100. The higher the Cetane number, the shorter the lag between the instant the fuel enters the combustion chamber and the instant it begins to burn. By comparing the performance of a diesel fuel of unknown quality with Cetane, a Cetane rating may be obtained.

➡**The Cetane rating has a similar relationship to diesel fuel as the Octane rating has to gasoline. The higher the Cetane rating, the faster the fuel burns. 55 is the highest rating normally given to diesel fuel.**

CARBON RESIDUE

The carbon residue of a diesel fuel, which is soot left over after combustion, is an indication of the amount of combustion chamber deposits that will be formed when the fuel is burned in the engine. This can be measured in the laboratory by heating a sample of the fuel in a closed container in the absence of air. The carbon residue will remain in the container. The amount of carbon residue which is considered permissible in diesel fuel oil depends somewhat on characteristics of the engine. This is more critical in small high speed engines than in large, low speed industrial engines. Standard requirements allow a maximum of 0.01% ash content.

➡**The Environmental Protection Agency (EPA) limits the amount of soot/ash particulates a diesel engine may produce for reasons of air quality.**

SULFUR CONTENT

The presence of sulfur in diesel fuel in excessive quantities is objectionable, as it increases piston ring and cylinder wear. In addition, it causes the formation of varnish on the piston skirts, and oil sludge in the engine crankcase. When fuel containing sulfur is burned in an engine, the sulfur combines with water, which results from the fuel combustion, to form corrosive acids. These acids tend to etch finished surfaces, add to the deterioration of engine oil and the production of sludge. Standard requirements for No. 1 and No. 2 diesel fuel allow a maximum of 0.5% sulfur content by weight, although better grades typically contain less then 0.16%.

➡**Fuels which have a high sulfur content often contain considerable quantities of various nitrogen compounds. There is evidence that the high wear rate of engine parts is partially caused by the nitrogen compounds.**

DIESEL FUEL GRADES

By the American Society of Testing Materials standards there are three classifications of diesel fuels available commercially. They are technically designated Grade I-D, Grade 2-D and Grade 4-D, but are more commonly referred to as simply No. 1, No. 2 or No; 4 diesel fuel. There was a Grade 3-D, but it is no longer produced.

Grade I-D is the most refined and volatile diesel fuel available, the premium quality and is used in high rpm engines requiring frequent changes in load and speed. Grade I-D diesel fuel has a lower cloud and gel point than 2-D that has been "winterized" with anti-gel additives, making it suitable for use year-round at temperatures below 20°F. Although rated as the best all-around diesel fuel, Grade I-D will deliver slightly lower fuel economy than 2-D and is not always available where ambient temperatures exceed 20°F.

Grade 2-D or No. 2 diesel fuel is the most widely used. It is a low volatility fuel suitable for use without additives or winterizing at temperatures above 20°F and delivers the best fuel economy of all the diesel grades. To expand Grade 2-D's effective use range to lower temperatures (typically to 0°F), fuel makers "winterize" 2-D with various fuel additives and blends to lower both its cloud and pour points.

Grade 4-D is a fuel oil for low and medium speed engines. It is the least refined and contains the highest levels of ash and sulfur, making it unsuitable for marine use.

As engine speed increases, cleanliness and viscosity become more critical to proper diesel engine operation. In addition to the grades of diesel fuel mentioned, there are also special fuels for railroad locomotives, buses and military uses. Chemical, physical and engine tests may be used to determine the quality of diesel fuel, which must pass tough fuel standard tests to receive a I-D or 2-D rating.

ADDITIVES

As diesel fuel flows from the pump, it is not always the same. As we have learned, diesel fuel is a mixture of over 200 compounds which make up unrefined crude oil.

Since October of 1993, diesel fuel refining included a process called hydro-treating. This change was implemented to reduce the sulfur and aromatic content of diesel fuel. On the positive side, the new low sulfur diesel fuels are less polluting and slightly more stable than previous diesel fuel. Conversely, the new low sulfur diesel has less lubricating qualities, contains more water, and has a higher paraffin wax content. These changes necessitate the use of fuel additives to insure top diesel engine performance. The reduced lubricity can be offset by using lubricity improvers.

Water in diesel fuel is a bigger problem than ever before. The water can be in several forms, free water or heavy water found usually in the bottom of tanks, emulsified water which can give the fuel a cloudy appearance, and simple dissolved water. In addition to high water content, at the pump, additional water is formed year round by condensation,

Each time a diesel engine operates, regardless of the time of year, condensation is formed. When this water reaches the diesel fuel injection system, the effect can be ruinous. The injectors depend on the diesel fuel for lubrication. Water further reduces lubricity while washing away the thin oil film coating the injector's micro-machined surfaces, causing enlargement of the spray holes, eroding of the injector tips, and destruction of the spray pattern. In a precision part like a fuel injector pump, this damage can be enormously expensive. To offset the presence of water in diesel fuel, a water dispersant chemical should be used.

In cold weather, regular diesel fuel thickens or gels to a point that it will no longer flow. This is called the pour point temperature. Prior to reaching this temperature, the diesel fuel starts to cloud up. This temperature is called the cloud point. Between the cloud point temperature, and the pour point temperature is the Cold Filter Plug Point (CFPP), or the temperature at which s fuel will not flow through the standard diesel fuel filter.

In the past, diesel engine owners have used kerosene or No. I Diesel to thin No. 2 fuel when it had gelled. With the availability of anti-gel additives on the market, this practice has all but disappeared.

Fuel filter plugging in cold weather is a major operating problem and the major cause for shut-down. Most diesel engine manufacturers have a 20 micron (as a point of reference, a human hair is about 70 micron in diameter) filter on the suction side of the lift pump. Most engines utilize a primary and secondary filter arrangement. The primary is usually a 20 micron filter, the secondary a 5 or 10 micron.

In order to prevent filters from plugging with ice crystals, and to prevent the filter media from swelling with water, or a combination of swelling then plugging with wax particles, an anti-gel product should be used.

These products also have water dispersants to guard against water related winter problems.

The best time to treat your fuel is prior to filling up your fuel tank. This insures excellent mixing of the anti-gel additive. If the fuel is already too cold, the anti-gel compound will not be as effective.

If your fuel system is already gelled up due to either wax, water or both, you can treat the system with a de-gelling compound.

Fuel Tank

Fuel tanks are one of the most overlooked components of a diesel fuel system. Since a fuel tank has virtually no moving parts, it would appear to be a maintenance-free item. But the accumulation of sediment and debris from constant refueling can build up quickly, causing fuel delivery problems. Regular draining and inspection will keep any fuel delivery problems from forming, and ensure the delivery of fresh, clean fuel to the engine.

INSPECTION

The fuel tank should be drained periodically to remove any water and debris present in the tank. This is a general guideline and actual inspecting and draining the fuel tank depends on how well fuel is filtered before it enters the tank, the age of the fuel tank, and the material from which the tank is constructed. Although your vessel most likely has two or more fuel filters to prevent water and debris from entering the engine, frequent draining and inspection of the fuel tank will help keep the fuel as clean as possible, preventing premature clogging of the filters.

Also, if your vessel has a deck mounted filler cap, it should also be inspected regularly. Faulty filler caps can allow water to enter the tank.

DRAINING

1. Place an appropriate sized container under the drain cock, and drain the fuel until the tank is empty. Look for signs of sediment or water in the drained fuel. These can be easily viewed if the fuel is set aside to allow the water and sediment to separate.

➡ **Try to plan ahead when draining the fuel tank; wait until the tank is near empty to make fuel tank draining quick and easy.**

2. Leaving the drain cock open, pour a small amount (a gallon or two) of fuel into the tank and allow it to drain; this will help flush out any remaining contaminates.

3. If the fuel tank has an inspection cover, remove it and check for signs of rust, sediment, and water.

➡ **Between inspections, keep the fuel level as high as possible to prevent condensation (water) from forming inside the fuel tank. Water in the fuel tank will cause problems with engine operation, and in some cases, fungus and algae growth.**

Fuel Filtration

♦ **See Figure 4**

Fuel filters are installed between the fuel tank and feed pump and between the feed pump and injection pump. These filters serve to remove dirt and other impurities from the fuel tank.

Regular replacement of the fuel filter will decrease the risk of blocking the flow of fuel to the engine, which could leave you stranded on the water. Fuel filters are usually inexpensive, and replacement is a simple task.

For information regarding fuel filter servicing, refer to section 3.

SELSTK63

Fig. 4 The filter is installed between the feed pump and the injection pump, and serves to remove dirt and other impurities from the fuel

Lift Pump

The fuel lift pump (also known as a transfer pump, or lift pump) is used to supply the injector pump with a steady supply of fuel at a low pressure (around 2.5 psi). A reciprocating diaphragm within the pump sucks fuel from the fuel tank with flow controlled by two small internal one way valves.

The lift pump, which is mounted to the engine block, is mechanically driven from a cam lobe on the engine camshaft. On the 4JH engines, the lift pump is mounted directly to the injector pump.

CHECKING

▶ **See Figures 5 and 6**

1. Disconnect the fuel outlet pipe from the lift pump.
2. Using the starter, crank the engine over; fuel should flow from the pump in strong bursts.
3. If the pump flow seems weak, check for blockage of the fuel supply.

➡**On some engines, it may be necessary to mark the fuel lines before they are removed from the pump. If the lines are reversed, the lift pump will not supply fuel to the injector pump.**

4. If the fuel supply to the pump is normal, remove both the inlet and outlet lines, and operate the pump priming lever while holding a finger over one of the connections. Suction and pressure should be felt from both connections, as well as the pump making a loud sucking noise.

Fig. 5 Removing the banjo fittings from the pump will allow you to check the pump for proper operation

5. The lift pump can also be tested by removing the entire pump, submersing it in kerosene, and operating the cam lever (or pump piston) with one hand while placing a finger over the outlet port (usually marked with an arrow). If bubbles emit from the pump with the outlet port plugged, there is a leak, and the pump is faulty and should be replaced.
6. If the pump is functioning properly, install the pump.
7. Loosen the air bleed screw on the secondary (between the lift pump and injector pump) fuel filter.

➡**Any time the fuel lines are disconnected, air must be bled from the fuel system.**

8. Operate the priming lever on the lift pump until fuel runs out of the air bleed screw opening, then tighten the screw.

REMOVAL & INSTALLATION

QM And GM/HM Series Engines

▶ **See Figures 7, 8, 9, 10 and 11**

1. Turn the fuel tank petcock to the **OFF** position.
2. Place a small receptacle under the lift pump to contain any excess fuel.
3. Remove the fittings that attach to the sides of the lift pump.

➡**On some engines, it may be necessary to mark the fuel lines before they are removed from the pump. If the lines are reversed, the lift pump will not supply fuel to the injector pump.**

4. Remove the two bolts that attach the lift pump to the engine block.
5. Draw the lift pump from the engine block.

To install:

6. Carefully scrape any gasket residue from the mating surfaces of the lift pump and engine block.
7. Lightly grease the arm on the lift pump that rides on the eccentric cam.
8. Using a new gasket, install the lift pump to the engine block.
9. Using new copper washers, install the fuel line fittings to the lift pump.
10. Once the lines are attached, turn the fuel petcock to the **ON** position.

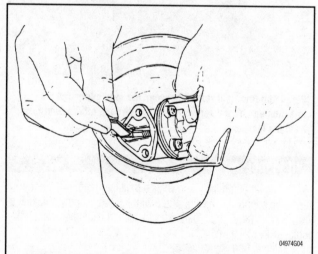

Fig. 6 Submersing the pump and operating it by hand is an accurate method of checking the pump for leaks

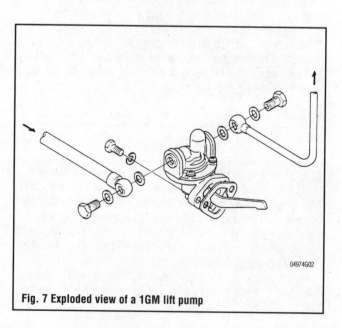

Fig. 7 Exploded view of a 1GM lift pump

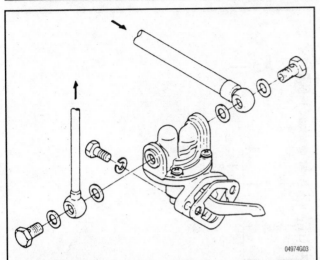

Fig. 8 Exploded view of a 2GM and 3GM/HM lift pump (note the reverse fuel flow)

Fig. 9 On GM series engines, the lift pump is attached to the engine block by two bolts

Fig. 10 Once the lines and the two bolts are removed, the pump is easily removed from the engine

Fig. 11 Before installing the pump, clean the gasket mating surface on the engine block to ensure the new gasket seals properly

11. Loosen the air bleed screw on the secondary (between the lift pump and injector pump) fuel filter.

12. Operate the priming lever on the lift pump until fuel runs out of the air bleed screw opening, then tighten the screw. If necessary, refer to the Fuel System Bleeding section for more information.

JH Series Engines
▶ See Figure 12

1. Turn the fuel tank petcock to the **OFF** position.

2. Place a small receptacle under the lift pump to contain any excess fuel.

3. Remove the fuel line fittings attached to the lift pump. Be careful not to lose the copper washers.

➡Be sure to mark the fuel lines before they are removed from the pump. If the lines are reversed, the lift pump will not supply fuel to the injector pump.

4. Remove the two bolts that attach the lift pump to the engine block.

5. Draw the lift pump from the engine block.

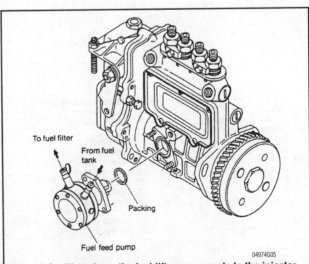

Fig. 12 On JH engines, the fuel lift pump mounts to the injector pump

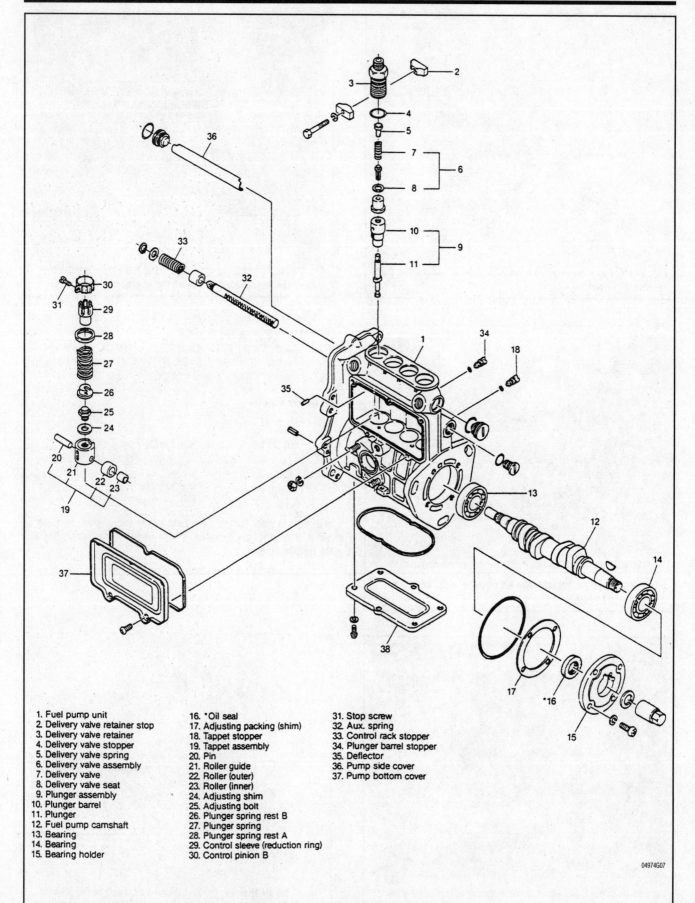

1. Fuel pump unit
2. Delivery valve retainer stop
3. Delivery valve retainer
4. Delivery valve stopper
5. Delivery valve spring
6. Delivery valve assembly
7. Delivery valve
8. Delivery valve seat
9. Plunger assembly
10. Plunger barrel
11. Plunger
12. Fuel pump camshaft
13. Bearing
14. Bearing
15. Bearing holder

16. *Oil seal
17. Adjusting packing (shim)
18. Tappet stopper
19. Tappet assembly
20. Pin
21. Roller guide
22. Roller (outer)
23. Roller (inner)
24. Adjusting shim
25. Adjusting bolt
26. Plunger spring rest B
27. Plunger spring
28. Plunger spring rest A
29. Control sleeve (reduction ring)
30. Control pinion B

31. Stop screw
32. Aux. spring
33. Control rack stopper
34. Plunger barrel stopper
35. Deflector
36. Pump side cover
37. Pump bottom cover

04974G07

Fig. 13 Diesel injector pumps are made of highly precision parts, and should only be serviced by qualified technicians

To install:

6. Carefully scrape any gasket residue from the mating surfaces of the lift pump and injector pump.

7. Lightly grease the end of the lift pump piston where it rides on the cam.

8. Using a new gasket, install the lift pump to the injector pump.

9. Install the fuel lines to the lift pump in the proper direction.

10. Once the lines are attached, turn the fuel petcock to the **ON** position.

11. Loosen the air bleed screw on the secondary (between the lift pump and injector pump) fuel filter.

12. Operate the priming lever on the lift pump until fuel runs out of the air bleed screw opening, then tighten the screw.

Injector Pump

▶ See Figure 13

The fuel injector pump is a highly precise, complex mechanical assembly that supplies the fuel injectors with the proper amount of pressurized fuel to be sprayed in the cylinders at the proper time.

On all engines, Yanmar stresses the importance of refraining from making adjustments to the injector pump. Many of the adjustment screws are sealed in place with lead-bound wire. These seals and other means of securing adjustments are placed on the injector pump at the factory, where the injector pump was properly adjusted on a testing machine. The adjustments on the injector pump and governor mechanism are set by the factory for optimum performance, and should not be changed.

❊❊ WARNING

Do NOT attempt to disassemble the injector pump. The internal components of the pump are EXTREMELY close-fitting, precision parts, and can fail quickly if contaminated with dirt and dust. Injector pumps should be serviced only by a professional diesel service facility equipped with the proper tools and equipment.

Due to the complex mechanical nature of the injector pump, no procedures for complete disassembly are given. Although a procedure for adjusting the injector pump timing is given, it is recommended that only a qualified diesel engine mechanic with proper testing equipment perform any kind of adjustments or repairs to the injector pump. Foolish tampering with the injector pump can cause serious problems with engine operation.

TESTING

▶ See Figures 14 and 15

Testing the injector pump for proper operation requires special equipment, and should be performed by an experienced diesel mechanic. If it has been determined that the injector pump is not functioning properly, the pump can be removed for testing at a qualified facility.

However, it is advisable that a qualified diesel mechanic verify that the injector pump is faulty **before** removal, since removal can be a lengthy task on some engines. Keep in mind that even if the injector pump is removed, taken to a shop and repaired, installation of the injector pump may require partial disassembly of the pump and other special tools to make timing adjustments. Unless you fully understand the operation of a diesel fuel injection system, and have the special tools and equipment, removal is not recommended.

Fig. 14 Diesel injector pumps must be tested and calibrated on special equipment for proper operation

Fig. 15 The injector pump is fitted to a bracket on the calibration machine, and operated to simulate engine conditions

CHECKING & ADJUSTING TIMING

QM Series Engines

▶ See Figures 16, 17, 18, 19 and 20

❊❊ CAUTION

Observe all applicable safety precautions when working around fuel. Whenever servicing the fuel system, always work in a well ventilated area. Do not allow fuel spray or vapors to come in contact with a spark or open flame. Keep a dry chemical fire extinguisher near the work area. Always keep fuel in a container specifically designed for fuel storage; also, always properly seal fuel containers to avoid the possibility of fire or explosion.

1. Install a measuring pipe on each of the pump outputs for viewing of the fuel.

2. Bleed the air from the injector pump.

3. Set the control rack to the middle position. Pull the lever when setting the accelerator lever.

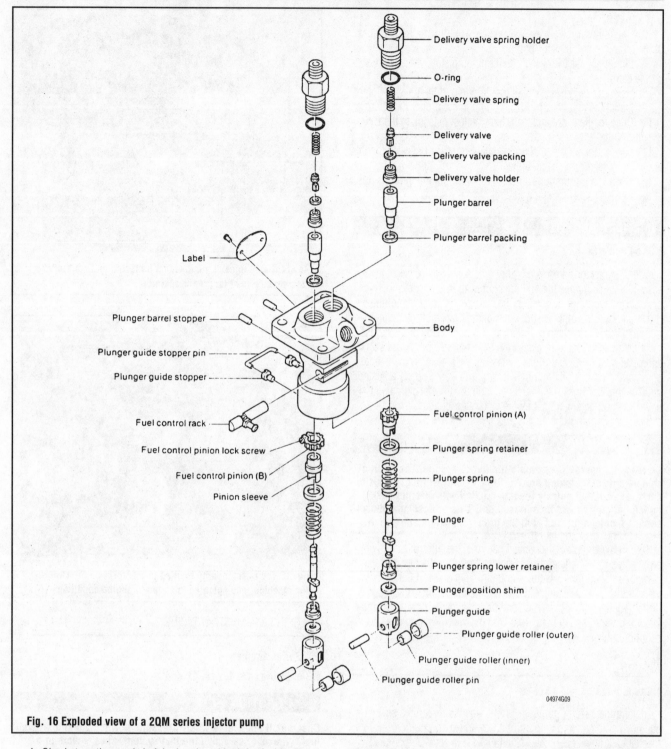

- Delivery valve spring holder
- O-ring
- Delivery valve spring
- Delivery valve
- Delivery valve packing
- Delivery valve holder
- Plunger barrel
- Plunger barrel packing

Label

Plunger barrel stopper

Plunger guide stopper pin

Plunger guide stopper

Fuel control rack

Fuel control pinion lock screw

Fuel control pinion (B)

Pinion sleeve

- Body
- Fuel control pinion (A)
- Plunger spring retainer
- Plunger spring
- Plunger
- Plunger spring lower retainer
- Plunger position shim
- Plunger guide
- Plunger guide roller (outer)
- Plunger guide roller (inner)
- Plunger guide roller pin

04974G09

Fig. 16 Exploded view of a 2QM series injector pump

4. Slowly turn the crankshaft by hand, and look for fuel in the measuring pipe of the number 1 cylinder. Stop turning the crankshaft the instant that fuel appears in the pipe.

5. Note the fuel injector timing mark on the crankshaft.

6. If the timing is incorrect, add (or subtract) shims from between the injector pump body and the engine block.

7. Once the timing is correct for the number 1 cylinder, repeat the process of turning the crankshaft by hand (in the proper direction) and noting the timing of the remaining cylinders.

8. If the injector timing of the remaining cylinders is not correct, note the deviation from specification (estimate in degrees) of each

of the cylinders on the crankshaft. Plunger shims will have to be added (or subtracted) to the individual pumps to make the timing equal between all cylinders, which requires partial disassembly of the injector pump.

※※ WARNING

When disassembling the injector pump, EXTREME cleanliness is essential. Before disassembling the pump for shim adjustment, thoroughly clean the pump, the work surface, all tools, and your hands. The injector pump components are very precise, and sensitive to contamination.

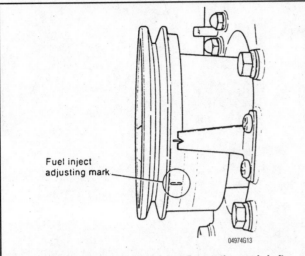

Fuel inject
adjusting mark

04974G13

Fig. 17 Injector pump timing mark location on the crankshaft pulley—QM series engines

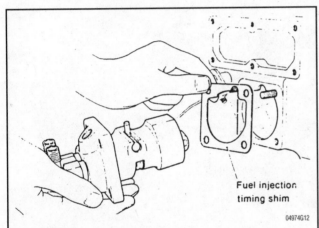

Fuel injection
timing shim

04974G12

Fig. 18 In addition to the adjustment shims in the plunger guides, shims are also used between the injector pump and the engine block

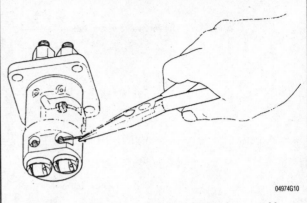

04974G10

Fig. 19 Using needlenose pliers to remove the plunger guide stopper pin

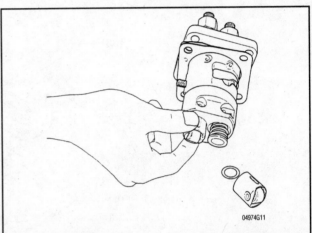

04974G11

Fig. 20 Once the stopper pin is removed, the plunger guides can be removed to access the shims

9. Mark the individual cylinders on the injector pump that are out of adjustment.

10. Remove the injector pump from the engine.

11. Starting with the number 2 cylinder, remove the plunger guide stopper pin by pushing in on the plunger guide, and removing the stopper with needlenose pliers.

12. Shims can be added (or subtracted) as necessary to correct the pump timing. Generally, each shim will change the timing by approximately 1 degree on the crankshaft. Adding shims will advance the injector timing, and removing them will retard timing.

13. Once the shims are installed and the pump is reassembled, install the injector pump to the engine and repeat the first 5 steps of this procedure to recheck the timing. If too many (or too few) shims were added to the plungers of the pump, the entire process will have to be repeated.

GM/HM Series Engines

▶ See Figures 21 thru 28

✳✳ CAUTION

Observe all applicable safety precautions when working around fuel. Whenever servicing the fuel system, always work in a well ventilated area. Do not allow fuel spray or vapors to come in contact with a spark or open flame. Keep a dry chemical fire extinguisher near the work area. Always keep fuel in a container specifically designed for fuel storage; also, always properly seal fuel containers to avoid the possibility of fire or explosion.

1. Install a measuring pipe on each of the pump outputs for viewing of the fuel.

2. Bleed the air from the injector pump.

3. Set the control rack to the middle position. Pull the lever when setting the accelerator lever.

4. If necessary, remove the starter to view the timing marks on the flywheel.

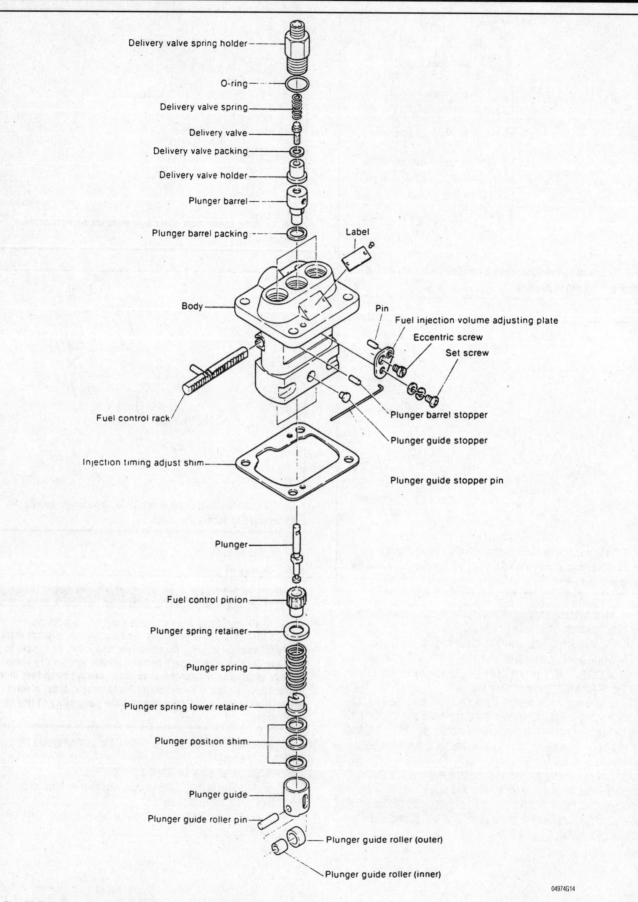

Delivery valve spring holder

O-ring

Delivery valve spring

Delivery valve

Delivery valve packing

Delivery valve holder

Plunger barrel

Plunger barrel packing

Label

Body

Pin

Fuel injection volume adjusting plate

Eccentric screw

Set screw

Fuel control rack

Plunger barrel stopper

Plunger guide stopper

Injection timing adjust shim

Plunger guide stopper pin

Plunger

Fuel control pinion

Plunger spring retainer

Plunger spring

Plunger spring lower retainer

Plunger position shim

Plunger guide

Plunger guide roller pin

Plunger guide roller (outer)

Plunger guide roller (inner)

04974G14

Fig. 21 Exploded view of a 3GM/HM series injector pump

5. Slowly turn the crankshaft by hand, and look for fuel in the measuring pipe of the number 1 cylinder. Stop turning the crankshaft the instant that fuel appears in the pipe.

6. Note the fuel injector timing mark on the flywheel.

7. If the timing is incorrect, add (or subtract) shims from between the injector pump body and the engine block.

8. Once the timing is correct for the number 1 cylinder, repeat the process of turning the crankshaft by hand (in the proper direction) and noting the timing of the remaining cylinders.

➡ On single cylinder engines, further adjustment of the timing is not necessary.

9. If the injector timing of the remaining cylinders is not correct, note the deviation from specification (estimate in degrees) of each of the cylinders on the flywheel. Plunger shims will have to be added (or subtracted) to the individual pumps to make the timing equal between all cylinders, which requires partial disassembly of the injector pump.

Fig. 24 The injection timing marks can also be seen through the view port on the transmission on some models

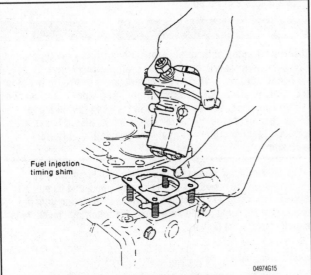
Fig. 22 As with the QM series engines, shims between the engine and injector pump are used to set the initial timing

Fig. 25 This pin retains small stoppers that hold the plunger guides in position

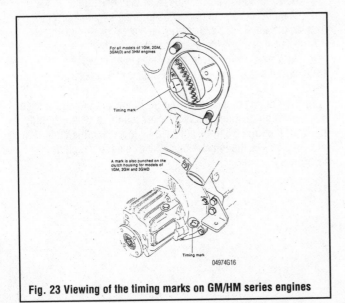
Fig. 23 Viewing of the timing marks on GM/HM series engines

Fig. 26 On 1GM engines, a circular spring clip is used instead of a pin to retain the plunger guide stopper

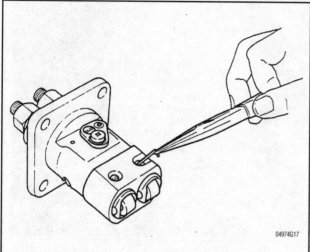

Fig. 27 To add or subtract shims from the injector pump, remove the plunger guide stopper pin . . .

Fig. 28 . . . and carefully remove the plunger guide to access the shims for adjusting the timing

10. Mark the individual cylinders on the injector pump that are out of adjustment.

11. Remove the injector pump from the engine.

12. Starting with the number 2 cylinder, remove the plunger guide stopper pin by pushing in on the plunger guide, and removing the stopper with needlenose pliers.

13. On 1GM engines, removing the plunger guide involves removing the circular clip that secures a retaining screw, and sliding the plunger guide from the body.

14. Shims can be added (or subtracted) as necessary to correct the pump timing. Generally, each shim will change the timing by approximately 1 degree on the crankshaft. Adding shims will advance the injector timing, and removing them will retard timing.

15. Once the shims are installed and the pump is reassembled, install the injector pump to the engine and repeat the first 5 steps of this procedure to recheck the timing. If too many (or too few) shims were added to the plungers of the pump, the entire process will have to be repeated.

16. Once the timing is correct, install the low pressure inlet and return lines, and the high pressure injector lines.

17. Install the starter, if removed.

18. Bleed the air from the fuel system.

JH series engines

▶ **See Figures 29 thru 36**

✳✳ **CAUTION**

Observe all applicable safety precautions when working around fuel. Whenever servicing the fuel system, always work in a well ventilated area. Do not allow fuel spray or vapors to come in contact with a spark or open flame. Keep a dry chemical fire extinguisher near the work area. Always keep fuel in a container specifically designed for fuel storage; also, always properly seal fuel containers to avoid the possibility of fire or explosion.

➡The injector pump timing on JH engines is automatically set properly when the marks are aligned on the pump and pump bracket. The marks are set by the factory, and deviation from the factory timing should not be necessary. The following procedure is to verify the correct timing.

1. Install a measuring pipe on the number one cylinder output on the pump for viewing of the fuel.

2. Bleed the air from the injector pump.

3. Set the control rack to the middle position. Pull the lever when setting the accelerator lever.

4. Slowly turn the crankshaft by hand, and look for fuel in the measuring pipe of the number 1 cylinder. Stop turning the crankshaft the instant that fuel appears in the pipe.

5. Note the fuel injector timing marks on the flywheel.

6. The timing should be correct. If the timing is out of adjustment, the automatic advance mechanism may be out of adjustment.

7. If the timing is incorrect, and the automatic timing advance mechanism on the pump drive gear is thought to be functioning correctly, the timing can be adjusted by loosening the three bolts that secure the injector pump, and rotating the pump accordingly.

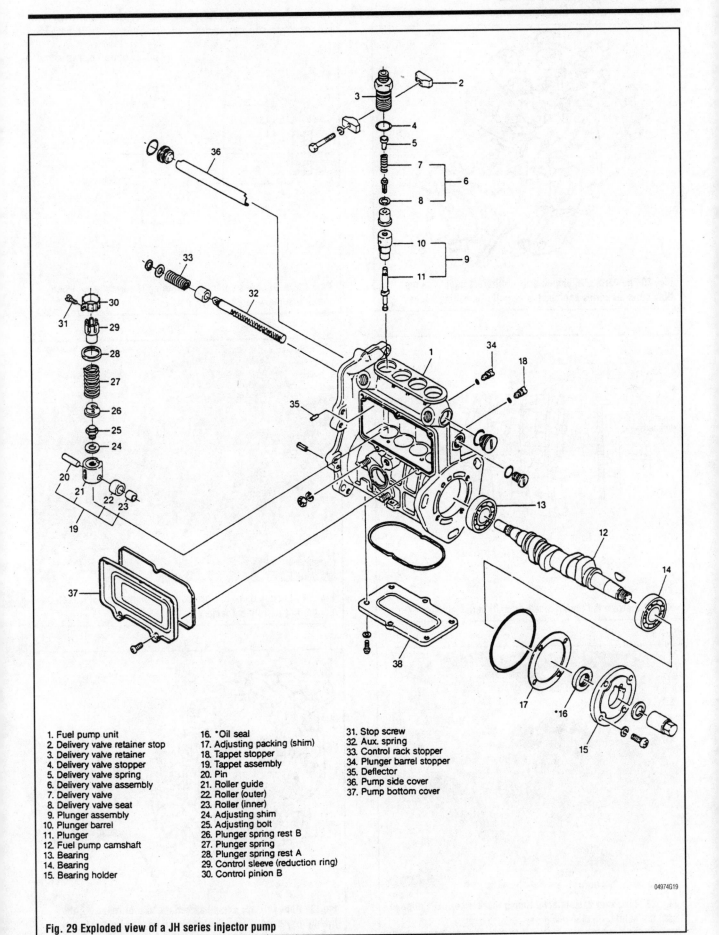

1. Fuel pump unit
2. Delivery valve retainer stop
3. Delivery valve retainer
4. Delivery valve stopper
5. Delivery valve spring
6. Delivery valve assembly
7. Delivery valve
8. Delivery valve seat
9. Plunger assembly
10. Plunger barrel
11. Plunger
12. Fuel pump camshaft
13. Bearing
14. Bearing
15. Bearing holder

16. *Oil seal
17. Adjusting packing (shim)
18. Tappet stopper
19. Tappet assembly
20. Pin
21. Roller guide
22. Roller (outer)
23. Roller (inner)
24. Adjusting shim
25. Adjusting bolt
26. Plunger spring rest B
27. Plunger spring
28. Plunger spring rest A
29. Control sleeve (reduction ring)
30. Control pinion B

31. Stop screw
32. Aux. spring
33. Control rack stopper
34. Plunger barrel stopper
35. Deflector
36. Pump side cover
37. Pump bottom cover

04974G19

Fig. 29 Exploded view of a JH series injector pump

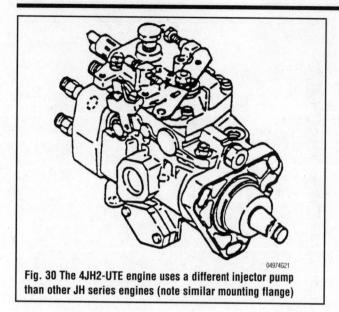

Fig. 30 The 4JH2-UTE engine uses a different injector pump than other JH series engines (note similar mounting flange)

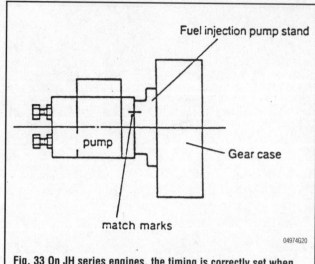

Fig. 33 On JH series engines, the timing is correctly set when the timing marks are aligned

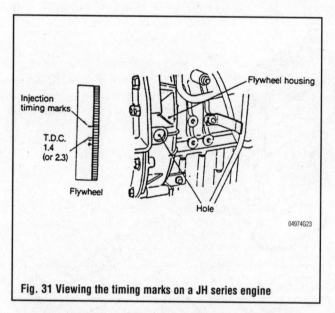

Fig. 31 Viewing the timing marks on a JH series engine

Fig. 34 Although there is a scale on this timing pump, it should not be moved from the factory setting

Fig. 32 Make sure each injector timing mark is perfectly aligned with the pointer on the view port

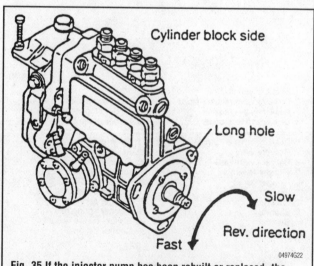

Fig. 35 If the injector pump has been rebuilt or replaced, the timing may need to be adjusted by rotating the pump slightly

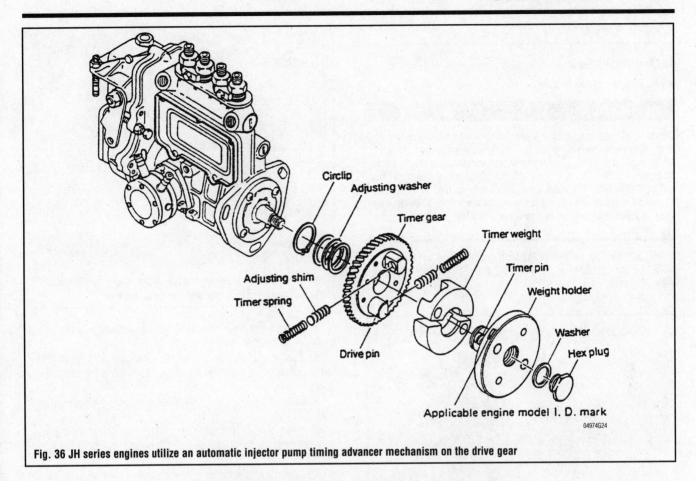

Fig. 36 JH series engines utilize an automatic injector pump timing advancer mechanism on the drive gear

Fuel Injection Specifications

Model	Fuel Injection Timing Degrees (BTDC)	Pressure (PSI)
1GM	14-16	170
1GM10	14-16	2347-2489
2GM	14-16	170
2GM20	14-16	2347-2489
2QM20	25	150-170
3GM	17-19	170
3GM30	17-19	2347-2489
3HM	17-19	160
3HM35	20-22	2204-2347
3QM30	28	150-170
4JH-DTE	11-13	2773-2915
4JH-E	11-13	2773-2915
4JH-HTE	11-13	2773-2915
4JH-TE	11-13	2773-2915
4JH2-DTE	16-18	2773-2915
4JH2-E	11-13	2773-2915
4JH2-HTE	16-18	2773-2915
4JH2-TE	16-18	2773-2915

04974C01

REMOVAL & INSTALLATION

QM And GM/HM Engines

♦ See Figures 37, 38 and 39

✳✳ CAUTION

Observe all applicable safety precautions when working around fuel. Whenever servicing the fuel system, always work in a well ventilated area. Do not allow fuel spray or vapors to come in contact with a spark or open flame. Keep a dry chemical fire extinguisher near the work area. Always keep fuel in a container specifically designed for fuel storage; also, always properly seal fuel containers to avoid the possibility of fire or explosion.

1. Turn the fuel petcock to the **OFF** position.
2. Disconnect the low pressure inlet and return hoses from the injector pump.

Fig. 37 When removing the injector pump, the control rack arm will need to be moved slightly to be removed from the engine

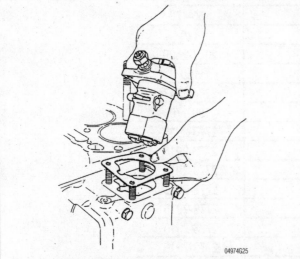

Fig. 38 Removing an injector pump on a 2GM series engine—other models similar

Fig. 39 When installing the injector pump, make sure the control rack engages properly with the governor arm

3. Remove any hoses, brackets, or other components necessary for removal op the injector pump.
4. Remove the high pressure line(s) that connect the pump to the injector(s). Be careful not to bend or distort the metal line(s).
5. Remove the gear case side cover and remove the governor lever.
6. Using an appropriately sized wrench, loosen and remove the nuts from the studs that secure the injector pump to the engine.
7. Carefully lift the pump from the engine, while aligning the cut-out in the pump base with the control rack on the injector pump.
8. Remove the timing adjustment shims, noting their thickness.
   To install:
9. Place the timing adjustment shims on the pump base.
10. Carefully lower the injector pump into the base, while aligning the control rack and the governor lever.
11. Install the nuts on the studs that secure the injector pump and tighten them securely.
12. If the injector pump was removed for inspection, and no parts were replaced or disturbed, the timing does not need adjustment.
13. If the injector pump was serviced or rebuilt, the timing should be adjusted. Refer to the timing adjustment procedure in this section.
14. After the timing has been adjusted properly, install the high pressure fuel line(s) that connect the pump to the injector(s).
15. Install the low pressure inlet and outlet hoses to the injector pump.
16. Prime and bleed the fuel system as required.

JH Engines

♦ See Figures 40, 41, 42, 43 and 44

✳✳ CAUTION

Observe all applicable safety precautions when working around fuel. Whenever servicing the fuel system, always work in a well ventilated area. Do not allow fuel spray or vapors to come in contact with a spark or open flame. Keep a dry chemical fire extinguisher near the work area. Always keep fuel in a container specifically designed for fuel storage; also, always properly seal fuel containers to avoid the possibility of fire or explosion.

1. Turn the fuel petcock to the **OFF** position.

2. Disconnect the low pressure inlet and return hoses from the injector pump.

3. Remove the high pressure lines that connect the pump to the injectors. Be careful not to bend or distort the metal lines.

4. Remove the fuel injector pump gear cover from the timing gear cover on the engine.

5. Verify the matchmarks on the injector pump and pump housing. If there are no markings, use a metal scribe or paint to **accurately** mark the relationship between the pump and housing. This is critical for proper engine operation.

6. Rotate the timing marks on the pump gear and drive gear until they are in alignment.

7. Remove the injector pump drive shaft hex nut and washer.

8. Remove the box nut and washer from the injector pump drive shaft.

Fig. 42 To access the injector drive gear, remove the 4 bolts that retain the cover on the front of the engine

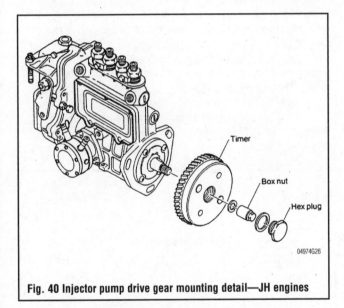

Fig. 40 Injector pump drive gear mounting detail—JH engines

Fig. 43 Once the bolts are removed, remove the cover

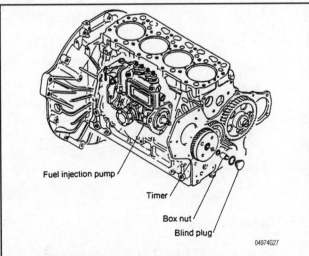

Fig. 41 Exploded view of the mounting detail of a JH series injector pump. Cylinder head and front cover removed for clarity

Fig. 44 Although the gear cannot be removed, the pump can be removed from the gear by removing the nut

9. Remove the drive gear from the injector pump shaft, and lift the pump from the engine.

➡**Use caution when removing the fasteners from the pump within the timing gear housing. If a part is dropped into the housing, the engine may have to be disassembled to retrieve lost parts.**

10. Remove the injector pump support brackets.
11. Remove the three bolts that attach the injector pump to the timing gear housing.

To install:

12. Carefully align the three bolt holes on the injector pump with the housing, and install the bolts.
13. Align the timing marks on the pump and housing, then tighten the bolts.
14. Install the injector pump support brackets.
15. Rotate the timing marks on the pump gear and drive gear until they are in alignment, and install the drive gear from the injector pump shaft.
16. Apply grease to the injector pump drive shaft box nut and washer. Install the box nut and tighten it to 43–51 ft. lbs. (58–69 Nm)
17. Install the hex plug and washer onto the drive gear.
18. Install the fuel injector pump gear cover onto the timing gear cover.
19. Install the high pressure lines that connect the pump to the injectors. Be careful not to bend or distort them upon installation.
20. Connect the low pressure inlet and return hoses from the injector pump.
21. Turn the fuel petcock to the **ON** position.
22. Prime and bleed the fuel system as required.

Governor

▶ **See Figures 45, 46, 47 and 48**

The output of a diesel engine is controlled by the amount of fuel injected into the cylinders by the injector pump. In marine use, it is most desirable for the engine to run at a specific speed, regardless of the load.

The governor is a mechanism that is used to keep engine speed constant by automatically adjusting the amount of fuel supplied to the engine regardless of load. This helps to protect the engine from abrupt changes in load, such as disengagement of the clutch, or the propeller leaving the water in rough seas.

If an engine was not equipped with a governor, simply positioning the throttle lever at full speed will cause the engine to slow down and speeds up as the vessel moves through the water due to the constant change in load. The governor helps mediate the power output of the engine by automatically operating the throttle when a load is applied.

The QM and GM/HM series engines use an all-speed governor that is activated by the centrifugal action of pivoting weights on the camshaft. The position of the governor weights changes as speed increases and decreases, which activates a governor lever. The governor lever serves two functions; movement of the fuel control rack on the injector pump, and activation of the regulator lever. The governor lever operates the regulator lever with a special spring, so movement of the fuel control rack on the injector pump is not hampered.

The JH series engines (including 4JH2-UTE) employ injector pumps with built-in governors. The means of operation are the same as the governing mechanism found on QM and GM/HM series engines, but the governing mechanism is fully contained within the injector pump.

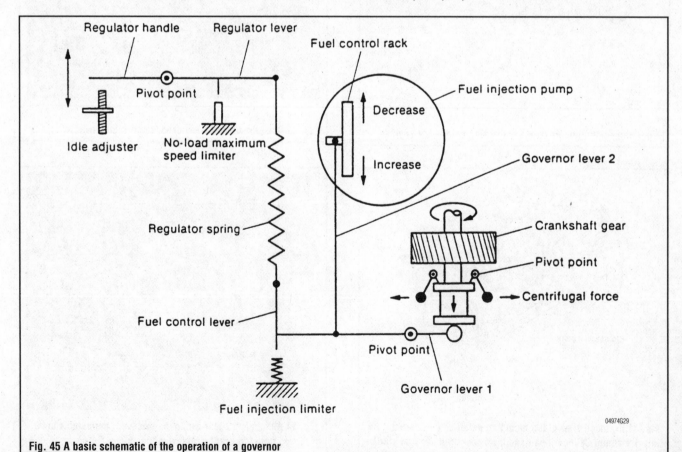

Fig. 45 A basic schematic of the operation of a governor

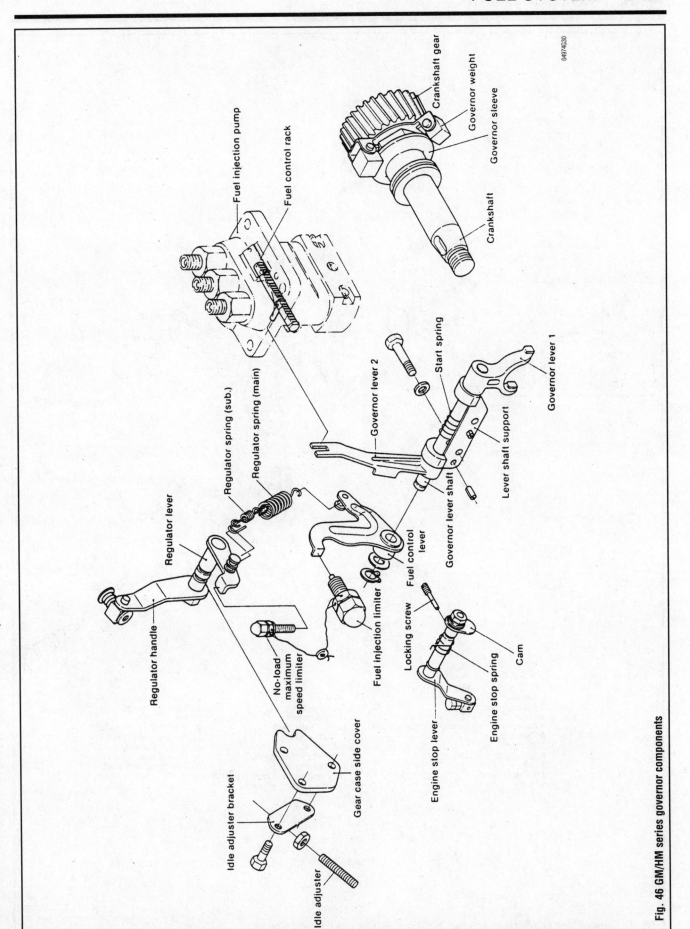

Fig. 46 GM/HM series governor components

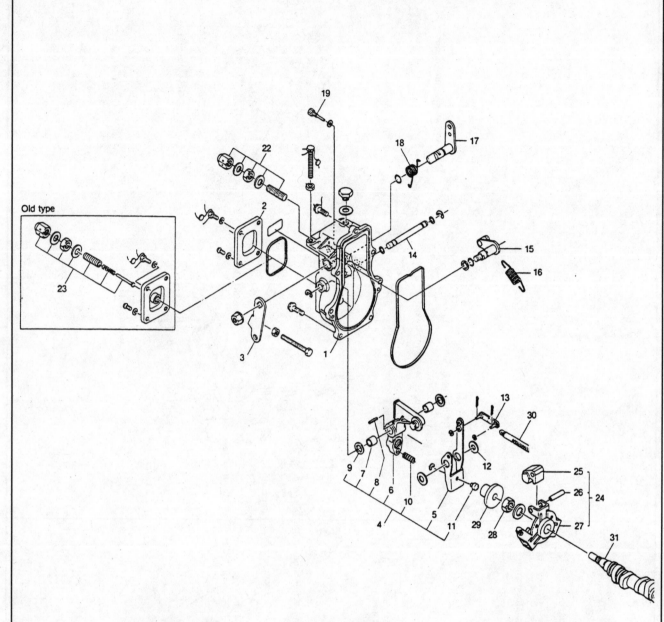

Old type

1. Governor case
2. Governor case cover
3. Control lever
4. Governor lever assembly 5. Governor lever
6. Tension lever
7. Bushing
8. Spring pin
9. Shim
10. Throttle spring
11. Shifter

12. Washer
13. Governor link
14. Governor shaft
15. Control lever shaft
16. Governor spring
17. Stop lever
18. Stop lever return spring
19. Stop lever stop pin
22. Fuel stopper (limit bolt) assembly
23. Adjusting spring assembly

24. Governor weight
25. Governor weight
26. Pin
27. Governor weight support
28. Governor weight nut
29. Governor sleeve
30. Control rack
31. Fuel pump cam shaft

04974G31

Fig. 47 Exploded view of a JH series governor, which is integrated into the injector pump

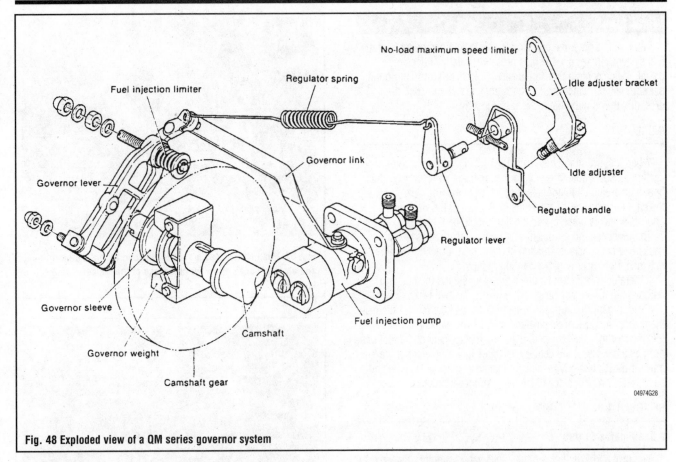

Fig. 48 Exploded view of a QM series governor system

ADJUSTMENTS

▶ **See Figures 49 and 50**

On all engines, Yanmar stresses the importance of refraining from adjusting the governor mechanisms. Many of the adjustment screws are sealed in place with lead-bound wire. These seals are placed on the adjustment mechanisms at the factory, where the injector pump was properly adjusted on a testing machine. The adjustments on the injector pump and governor mechanism are set for optimum performance, and should not be changed.

If you have concluded that the injector pump or governor needs adjustment, it is highly recommended that a qualified diesel engine mechanic inspect your engine to confirm your diagnosis. Foolish tampering with the governor mechanism or injector pump can cause serious problems with engine operation.

Due to the complex and sensitive mechanical nature of the governor mechanism, no procedures for adjustment or disassembly are given. It is recommended that only a qualified diesel engine mechanic with proper testing equipment perform any kind of adjustments or repairs to the governor mechanism.

Fig. 49 Adjustments to the governor are strongly discouraged by the manufacturer; foolish tampering can cause major engine problems

Fig. 50 Notice that the lever adjustment and governor adjustment mechanisms are wired together to prevent adjustment

Fuel Lines

The fuel lines are one of the most overlooked components of the fuel system. Amazingly, the high pressure metal fuel lines of a diesel engine contain pressure measured in **thousands** of pounds per square inch. Since the fuel lines are subjected to such high pressure, they should be inspected frequently.

INSPECTION

When inspecting the fuel lines, look for signs of leakage or cracking, usually identified with dirt or wetness surrounding a fuel line. If an area on a fuel line is suspect, wipe the line clean, operate the engine, and look for any signs of wetness. If the wetness returns, a leak is present, and the line should be replaced.

The most common problem with high pressure lines is cracking, caused by the constant vibration of the engine. Cracked fuel lines may occur on older engines with high hours.

If there is leakage from a line end, try loosening the nut, and retightening it. In most cases, this will stop a small leak from the line end. If performing this procedure does not help correct the leak, the mating surfaces of the line end and delivery valve retainer or fuel injector may be damaged, requiring replacement of the fuel line, or in extreme cases, the delivery valve retainer or injector. If the fuel injector uses a banjo type fitting, it is advisable to replace the copper gaskets on both sides of the fitting to prevent leakage.

REMOVAL & INSTALLATION

▶ **See Figures 51 thru 56**

Removal and installation of fuel lines varies slightly between models, but the general procedure is the same.

❊❊ CAUTION

Observe all applicable safety precautions when working around fuel. Whenever servicing the fuel system, always work in a well ventilated area. Do not allow fuel spray or vapors to come in contact with a spark or open flame. Keep a dry chemical fire extinguisher near the work area. Always keep fuel in a container specifically designed for fuel storage; also, always properly seal fuel containers to avoid the possibility of fire or explosion.

Fig. 51 It may be necessary to remove certain components, (like the air silencer) to allow the fuel lines to be removed

Fig. 52 This engine uses a spacer between the intake manifold and air silencer to allow for clearance of the spacers on the injectors

Fig. 53 Loosen the bolts that hold the fuel lines in position

❊❊ WARNING

When removing the fuel lines, EXTREME cleanliness is essential. Before removing the lines, thoroughly clean the lines, the fittings, all tools, and your hands. The internal components of the injector pump and injectors are very precise, and sensitive to contamination.

1. Before removing the fuel line, determine if it will be required to remove adjacent fuel lines or other items to allow for clearance.
2. Turn the petcock on the fuel tank to the **OFF** position.
3. Remove any rubber-lined brackets that secure the fuel lines.
4. Using an appropriately sized line wrench, loosen the nuts on both sides of the fuel line, at both the injector and the pump.

❊❊ WARNING

If the injector has a hex spacer on the inlet, make sure to hold it stationary with a wrench while loosening the line nut. Do NOT attempt to loosen the line by turning the spacer. Turning the spacer will twist the fuel line, requiring replacement of the line.

5. Remove the fuel line(s) from the engine.
6. Cap the end on the fuel injector(s) and the injector pump.

To install:

7. Remove the caps on the injector and pump.

8. Test fit the line to the engine. The bends of the line must match must match those of the old line. Most importantly, the ends of the line must align with the threads of the injector pump fitting and the injector.

9. If the line(s) fit properly, hand thread the line nuts onto the injector and the injector pump fitting(s). Leave the line nuts loose until all the lines are attached to allow easier threading. If the line(s) have banjo-type fittings, make sure to install new copper washers on both sides of each fitting.

10. Once all the line end nuts are threaded properly, they can be tightened.

11. If equipped, attach the rubber-lined brackets that secure the lines. It is very important that the lines are secured properly, since the constant vibration of the engine can cause a line to crack prematurely. Also, make sure that the metal fuel lines are not rubbing against each other; this can be another cause of breakage.

12. Once the line(s) are attached, the air will have to be bled from the line(s). Refer to the procedure in this section.

Fig. 54 Once the bolts are removed, remove the metal brackets, and the rubber insulators

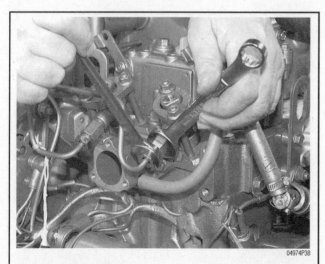

Fig. 55 Remove the line nuts from the injectors (note the backup wrench on the spacer) . . .

Fig. 56 . . . and the injector pump

Fuel injectors

The fuel injectors on a diesel engine are basically mechanical pop-off valves that spray fuel into the cylinder when the fuel pressure reaches the proper range. A spring inside the injector holds the nozzle valve tightly against the valve seat until the high pressure fuel from the injector pump overcomes the strength of the spring. This allows fuel past the nozzle valve, which sprays in a precise pattern into the combustion chamber. This cycle happens instantaneously, and is repeated on every combustion stroke for all the engine's cylinders.

Although fuel injectors are basic in design and construction, they are very precise, and can clog easily if contaminates are allowed into the system. Fuel injectors commonly fail when a small piece of debris becomes lodged in the injector nozzle, producing an insufficient spray pattern. When fuel is not properly sprayed from an injector, it will not burn properly, causing reduced engine power output.

TESTING & INSPECTION

▶ **See Figures 57, 58 and 59**

If the engine is not running correctly, you can check for operation of the injectors by loosening (but not removing) the nut that secures the high pressure line at the injector. The nut only needs to be loosened until fuel leaks from the threads. If the rpm drops slightly and the engine misfires, it can be assumed that the injector is functioning properly. If nothing happens when the line is loosened slightly, the injector is not firing, or there may be problems with the injector pump.

❊❊ CAUTION

Be sure to wear full eye and body protection when performing this procedure; fuel can spray from the line nut at high pressure, possibly into your eyes. Only LOOSEN the line nut to allow fuel to leak from the threads on the injector. If the line is REMOVED from the injector, and the engine turned over, fuel will be expelled from the line, possibly spraying into your eyes.

Another cause for improper injector function is carbon deposits. Over time, carbon deposits can accumulate on the injector nozzle, causing an improper spray pattern.

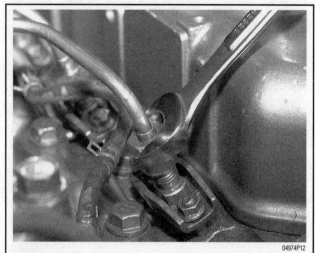

Fig. 57 The fuel injectors can be tested by slightly loosening the line while the engine is running

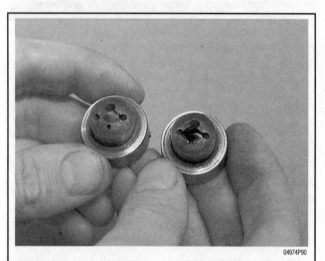

Fig. 58 Two examples of badly damaged precombustion chambers. Damaged chambers cause poor engine performance

Fig. 59 The injectors should be cleaned with a soft brush if there is carbon build-up forming around the tip of the injector

Checking for carbon deposits requires removal of the injectors from the engine. Once you have removed the injectors, inspect the nozzle of each injector for carbon deposits. A soft brass wire brush or carburetor cleaner can be used to clean for cleaning the nozzle. Use caution when cleaning the injector nozzles; the spray pattern can be altered if the nozzle is marred or distorted.

✳✳ WARNING

Do not use a steel brush on the injector nozzles. The steel can damage the hole(s) on the nozzle, causing the injector to spray improperly.

Other methods of testing injectors for proper operation require bench testing equipment and other special tools. If your engine's fuel injectors are not operating properly, they should be taken to a qualified diesel repair facility for testing.

REMOVAL AND INSTALLATION

QM and HM/GM Series Engines

▶ See Figures 60 thru 80

1. Turn the petcock on the fuel tank to the **OFF** position.
2. Remove any rubber-lined brackets that secure the fuel lines.
3. Disconnect the return lines on each injector.
4. Loosen the fuel lines at the injector pump.
5. Disconnect the fuel lines from the injectors. (If equipped, remember to use a wrench on the spacer to hold it stationary while loosening the line nut)

✳✳ WARNING

When removing the injectors, EXTREME cleanliness is essential. Before disconnecting the lines, thoroughly clean the injectors, fittings, fuel lines, all tools, and your hands. The internal components of the injector pump and injectors are very precise, and sensitive to contamination.

6. Using an appropriately sized wrench, remove the nuts and washers that hold the injector retainer(s) in place.
7. Remove the injector retainer(s) from the studs on the cylinder head.

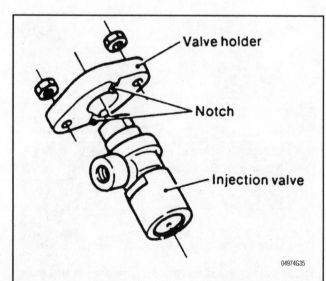

Fig. 60 Fuel injector mounting detail—QM and GM/HM engines

Fig. 61 Loosen the clamp that secures the return line . . .

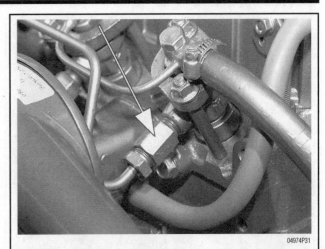

Fig. 64 Remove the high pressure line by loosening the line nut. If there are spacers on the injectors, be sure to use a backup wrench

Fig. 62 . . . and remove the line

Fig. 65 Once the line nut is loosened, unthread it from the injector (or spacer) and carefully move the line to one side

Fig. 63 Next, remove the hard return line that connects the injectors. Be careful not to lose the copper washers

Fig. 66 Remove the two nuts on the injector retainer, then remove the retainer

Fig. 67 The injector can now be lifted from the cylinder head

Fig. 70 The copper gasket may stick to the bottom of the heat shield; make sure it is removed before installation

Fig. 68 Once the injector is removed, the heat shield, prechamber and copper gaskets can be removed

Fig. 71 After the heat shield is removed, lift the precombustion chamber from the injector bore

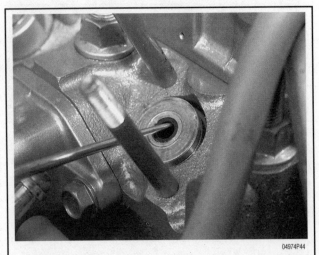

Fig. 69 Use a small pick or other similar tool to carefully lift the heat shield and gasket from the injector bore

Fig. 72 And finally, remove the last copper gasket that seals the precombustion chamber in the injector bore

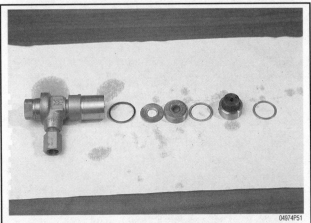

Fig. 73 The fuel injector, with the heat shield, precombustion chamber, and related gaskets

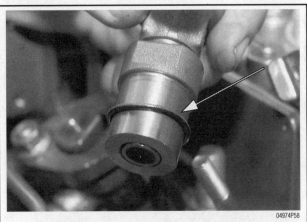

Fig. 75 This O-ring, found on later model GM engines, is used to keep rust from forming between the injector and bore

Fig. 74 Before installing a new gasket on the heat shield, make sure to scrape any old gasket material from the top of the shield

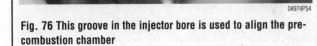

Fig. 76 This groove in the injector bore is used to align the precombustion chamber

8. Remove the injector(s) from the cylinder head.

9. Using a small pick or other similar tool, remove the heat shield, precombustion chamber, and copper gaskets from the injector bore.

➡Every time the injectors are removed, the copper gaskets that seal the precombustion chamber and heat shield must be replaced.

To install:

10. Install the heat shields and precombustion chambers as follows:

   a. Install a new copper gasket in the cylinder bore, followed by the precombustion chamber. Make sure that the pin on the chamber is fully aligned with the groove in the bore. Do not force the precombustion chamber in the bore!

   b. Install another copper gasket on the top of the precombustion chamber.

   c. Clean any gasket material from the top of the heat shield. Install the heat shield, and the gasket on the top of the heat shield.

   d. Repeat this procedure with each injector.

11. Install the injector(s) to the cylinder head. Align the line fittings on the injectors.

12. Place the injector retainer(s) onto the studs on the cylinder head. Make sure the retainer is mounted in the proper direction (see illustration).

13. Install and tighten the nuts and washers that hold the injector retainer(s). Be sure to use a torque wrench to tighten the nuts. Graduate the torque on the nuts, until the final torque of 14 ft. lbs. (19 Nm) is achieved.

14. Uncap the end(s) on the high pressure fuel line(s).

15. Install the fuel lines.

16. Connect the fuel return line(s) to the injectors.

17. If equipped, attach the rubber-lined brackets that secure the lines. It is very important that the lines are secured properly, since the constant vibration of the engine can cause a line to crack prematurely. Also, make sure that the metal fuel lines are not rubbing against each other; this can be another cause of breakage.

18. Once the line(s) are attached, the air will have to be bled from the line(s). Refer to the Fuel System Bleeding procedure in this section.

JH Series Engines

▶ See Figures 81 thru 91

1. Turn the petcock on the fuel tank to the **OFF** position.

2. Remove any rubber-lined brackets that secure the fuel lines.

3. Loosen the fuel lines from the injector pump.

4. Disconnect the return lines on each injector.

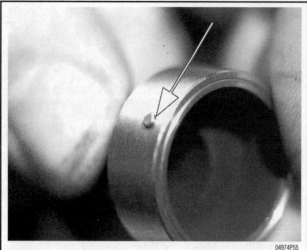

Fig. 77 Notice the small pin on the precombustion chamber; this pin fits into the groove in the injector bore

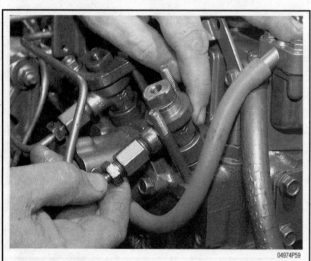

Fig. 78 After all the new gaskets are installed, carefully place the injector into the bore, and thread the line nut on by hand

Fig. 79 Make absolutely certain the injector retainer is installed with the protrusions facing down, on the injector

Fig. 80 ALWAYS use a torque wrench when installing the injectors

5. Remove the lines from the injectors. With the line brackets removed, and the lines loosened at the injector pump, there should be sufficient slack to place the lines to the side of the injector(s) to allow for removal. Do NOT bend the fuel lines!

✳✳ WARNING

When removing the injectors, EXTREME cleanliness is essential. Before disconnecting the lines, thoroughly clean the injectors, fittings, fuel lines, all tools, and your hands. The internal components of the injector pump and injectors are very precise, and sensitive to contamination.

6. Using an appropriately sized wrench, remove the nuts and washers that hold the injector retainers in place.
7. Remove the injector retainers from the studs on the cylinder head.
8. Remove the injectors from the cylinder head.
9. Using a small pick or other similar tool, remove the heat protectors from each injector bore if they do not come out with the injector.

➡Every time the injectors are removed, the heat protectors should be replaced.

To install:

10. Install a new gasket and heat protector on each injector in the proper direction.
11. Install the injectors to the cylinder head. Align the return line fittings on the injectors.
12. Place the injector retainers onto the studs on the cylinder head. Make sure the retainer is mounted in the proper direction (see illustration).
13. Install and tighten the nuts and washers that hold the injector retainers. Be sure to use a torque wrench to tighten the nuts. Graduate the torque on the nuts, until the final torque of 8–10 ft. lbs. (10.9–13.6 Nm) is achieved.
14. Uncap the ends on the high pressure fuel lines).
15. Connect the fuel return lines to the injectors.
16. Install the fuel lines.
17. Attach the rubber-lined brackets that secure the lines. It is very important that the lines are secured properly, since the constant vibration of the engine can cause a line to crack prematurely. Also,

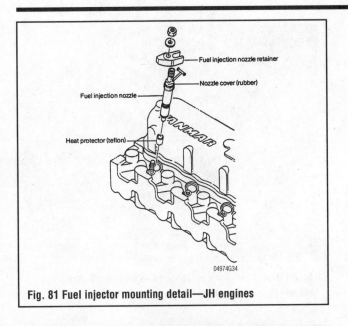

Fuel injection nozzle retainer

Nozzle cover (rubber)

Fuel injection nozzle

Heat protector (teflon)

04974G34

Fig. 81 Fuel injector mounting detail—JH engines

04974P15

Fig. 84 Remove the two nuts that secure the retainer . . .

04974P13

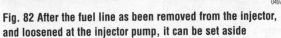

Fig. 82 After the fuel line as been removed from the injector, and loosened at the injector pump, it can be set aside

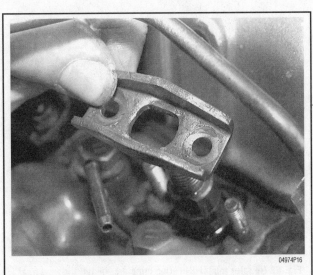

04974P16

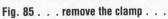

Fig. 85 . . . remove the clamp . . .

04974P14

Fig. 83 Remove the return line clamps with pliers, and disconnect the lines from the injector

04974P17

Fig. 86 . . . and pull the injector from the cylinder head

Fig. 87 This rubber gasket is used to keep water from getting between the injector and the bore in the cylinder head, which can cause rust

Fig. 88 On later model JH series injectors, install the new heat protector as shown

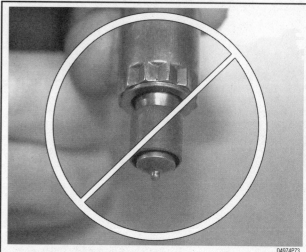

Fig. 90 If the heat protector is installed incorrectly, it will not properly fit the base of the injector

Fig. 91 Be sure to use a torque wrench when installing the injectors

make sure that the metal fuel lines are not rubbing against each other; this can be another cause of breakage.

18. Once the lines are attached, the air will have to be bled from the lines. Refer to the Fuel System Bleeding procedure in this section.

DISASSEMBLY

▶ **See Figures 92 and 93**

Because of the precise mechanical nature of diesel fuel injectors, no procedures for complete disassembly are given. Although a procedure for removal of the injectors is given, it is recommended that only a qualified diesel engine mechanic with proper testing equipment disassemble the injectors for adjustment, calibration, or repair.

✳✳ WARNING

Do NOT attempt to disassemble the fuel injectors. The components of the injectors are EXTREMELY close-fitting, precision parts, and can fail quickly if contaminated with dirt and dust. Injectors should be serviced only by a professional diesel service facility equipped with the proper tools and equipment.

Fig. 89 On earlier models, the heat protector is a sleeve, with an internal tapered end that fits around the base of the injector

Although the injectors on many Yanmar engines appear the same, they cannot be interchanged with those of other models. Each fuel injector is tested at the factory for flow and proper spray pattern, and are marked with identification numbers for each engine. For optimum engine performance, the injectors on a diesel engine should be "matched" so the same amount of fuel is injected into each cylinder. Only a diesel component repair and rebuilding facility with the proper testing and calibration equipment can rebuild and match a set of fuel injectors.

Fuel System Bleeding

▶ **See Figures 94 thru 99**

Any time the fuel system is exposed to air (e.g. changing a fuel filter, or removing the injectors) the accumulated air will have to be bled from the system to allow for proper delivery of fuel to the engine.

Besides changing the fuel filter, air can be drawn into the fuel system by several routes. Typically, air is sucked into the fuel system through a poor seal upstream of the lift pump. Air can also be drawn into the fuel system when the fuel tank is low, and the motor is being operated in rough seas. A small amount of air (about a teaspoon) in the fuel system is enough air to shut down the engine, and require fuel system bleeding.

To bleed air from the fuel system, proceed as follows:

1. Loosen the air bleed screw at the top of the secondary fuel filter (between the lift pump and injector pump).

➡ **When operating the priming lever on the lift pump, make sure to push down completely to ensure full operation of the diaphragm within the pump.**

2. Operate the priming lever on the lift pump. If the lever is only moving slightly, rotate the crankshaft by hand to reposition the lobe on the camshaft that operates the pump. This will allow for a full stroke of the priming lever.

3. Operate the priming lever until fuel free of air flows from the air bleed screw, then tighten the screw.

➡ **Air can be drawn into the fuel system by a faulty bleed screw washer. After the bleeder screw has been loosened a few times, it is a good idea to replace it to ensure a leak-free seal.**

4. If air remains in the system, check for leaks in the lift pump or primary filter.

5. If the engine has stalled from air in the system, (or if the high pressure side of the system has been opened) it will be necessary to bleed the injector pump, injector lines, and injectors.

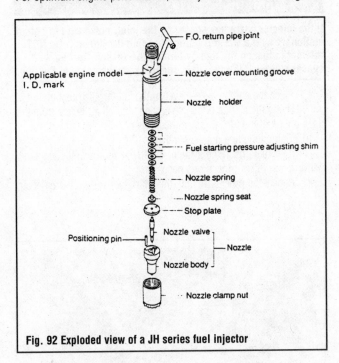

Fig. 92 Exploded view of a JH series fuel injector

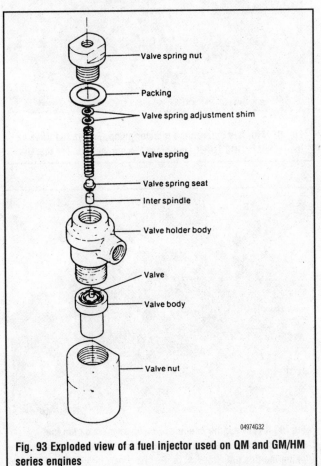

Fig. 93 Exploded view of a fuel injector used on QM and GM/HM series engines

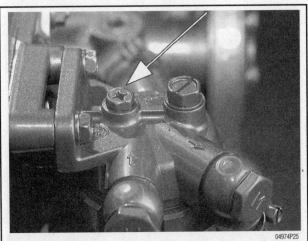

Fig. 94 Although the fuel filter is slightly different between models, the air bleed screw is always the 10mm bolt with the Phillips head

Fig. 95 Loosening the air bleed screw on a 3GM engine

Fig. 96 Loosening the air bleed screw on a JH series engine

Fig. 97 After the bleed screw has been loosened, operate the priming lever on the feed pump to push the air out of the system

6. Loosen the bleed screw on the injector pump and operate the priming lever on the lift pump until fuel free of air flows from the bleed screw, then tighten the screw.

7. Loosen the injector line nut(s) on the injector(s) **one** turn.

✳✳ WARNING

If the injector has a hex spacer on the inlet, make sure to hold it stationary with a wrench while loosening the line nut. Do NOT attempt to loosen the line by turning the spacer. Turning the spacer will twist the fuel line, requiring replacement of the line.

8. Place the engine speed lever to the full throttle position.

9. Using the starter, crank the engine until all air is free from the lines until fuel runs from the connections.

10. Tighten the line nuts securely.

11. If air remains in the system, locate the source of the air, and perform and repair procedures necessary.

12. Repeat the procedure until the entire fuel system is free of air.

Fig. 98 If air has entered the injector pump, loosen the lines on the injectors ONE TURN, and crank the engine to bleed the air

Fig. 99 If your engine is equipped with spacers on the fuel injectors, use a wrench to hold the spacer stationary while loosening the line nut

Troubleshooting Fuel Injection Systems

Condition	Cause	Correction
Difficulty starting	Injector pressure high	Incorrect pressure setting
	Cold starting aid not working	Check for electrical supply to aid / Check for fuel proper fuel supply / Check for proper grounding / Check aid for proper function
Difficulty starting, erratic rpm, engine dies	Air in fuel	Eliminate air leak at primary filter / Check for leaking pipes, hoses, and connections / Check for closed fuel valves / Check for defective lift pump / Check for sufficient fuel in tank / Check tank pickup
Difficulty starting, loss of power, black smoke	Clogged fuel filter	Maintain filter more frequently / Check for a dirty fuel tank
	Incorrect grade of fuel	Check for bad fuel / Use a fuel cetane booster
Difficulty starting, white smoke, loss of power	Injector pressure low	Check for worn injector / Incorrect pressure setting
Difficulty starting, low power, rough running	Injector spring broken	Check for spring fatigue
Difficulty starting, black smoke	Poor maintenance practices	Check for two sealing washers under injector
Difficulty starting, erratic rpm, engine dies, fuel in oil sump, engine oil level increase without adding oil	Lift pump not pumping	Check internal valve on lift pump for contamination / Check external non-return valve sticking closed / Check diaphragm for leakage
Engine will not run, difficulty starting, rough running, erratic rpm, engine dies	Injector pump not pumping	Check for air in fuel system / Check lift pump / Check for clogged fuel filters / Check injector pump
External fuel leaks, rough running, vibration, loss of power	Injector lines cracked	Check for severe vibration / Check for inadequate or missing pipe support / Check for incorrect pipe routing
	Injector lines not sealing	Check for loose injector nuts / Check for damaged sealing surfaces on injector
Filters clogged with black, stringy debris, difficulty starting, erratic engine rpm, engine dies	Algae or fungus growth in fuel	Eliminate algae with appropriate chemicals / Thoroughly clean fuel system
High rpm, erratic rpm	Governor spring defective	Check for spring fatigue / Check for loose governor components

04974C02

Troubleshooting Fuel Injection Systems

Condition	Cause	Correction
Knocking (pre-ignition), loss of power, white smoke, black smoke	Injector stuck open	Check for contamination in injector Check for worn injector
Knocking, rough idle, difficulty starting, black smoke	Injector nozzel eroded Timing advanced	Check for worn injector Check for water in fuel Check for loose injector pump Check for proper timing Check for a sticking governor
Loss of power, erratic rpm	Sticking governor	Check for governor wear Check for contamination of mechanism
Loss of power, overheating, white smoke	Timing retarded	Check for loose injector pump Check for proper timing Check for a sticking governor
Low maximum rpm	Weights defective	Check for loose governor components
Uneven injection, rough running, loss of power, vibration, difficulty starting	Injector stuck closed	Check for contamination in injector Check for worn injector
White Smoke, difficulty starting, loss of power	Water in fuel	Check for bad fuel Check deck plate filler for bad seal Eliminate condensation by keeping tank full

04974C03

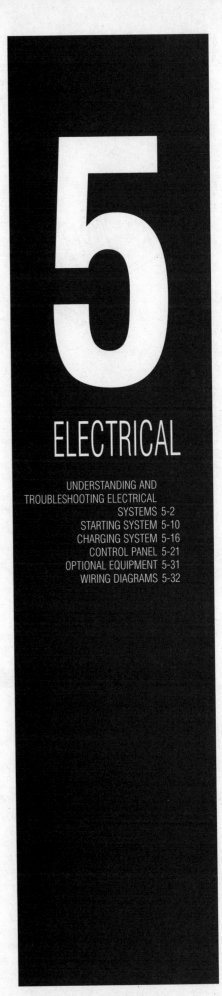

5

ELECTRICAL

UNDERSTANDING AND TROUBLESHOOTING ELECTRICAL SYSTEMS

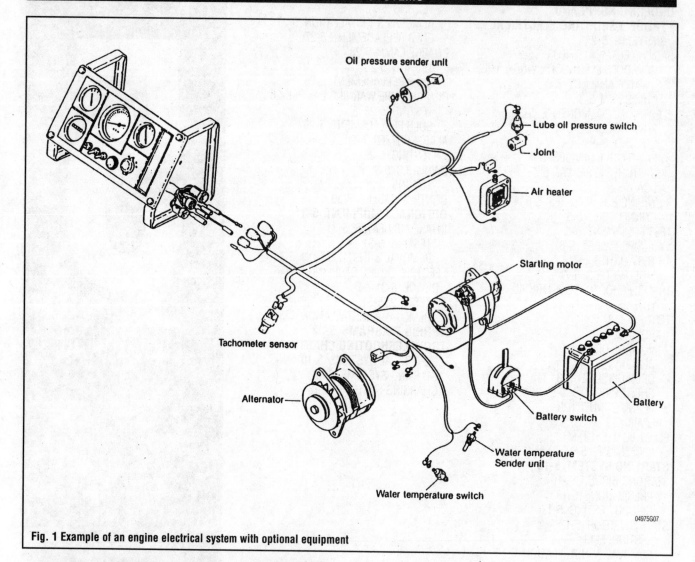

Fig. 1 Example of an engine electrical system with optional equipment

Basic Electrical Theory

▶ See Figure 2

For any 12 volt, negative ground, electrical system to operate, the electricity must travel in a complete circuit. This simply means that current (power) from the positive terminal (+) of the battery must eventually return to the negative terminal (−) of the battery. Along the way, this current will travel through wires, fuses, switches and components. If, for any reason, the flow of current through the circuit is interrupted, the component fed by that circuit will cease to function properly.

Perhaps the easiest way to visualize a circuit is to think of connecting a light bulb (with two wires attached to it) to the battery—one wire attached to the negative (−) terminal of the battery and the other wire to the positive (+) terminal. With the two wires touching the battery terminals, the circuit would be complete and the light bulb would illuminate. Electricity would follow a path from the battery to the bulb and back to the battery. It's easy to see that with longer wires on our light bulb, it could be mounted anywhere. Further, one wire could be fitted with a switch so that the light could be turned on and off.

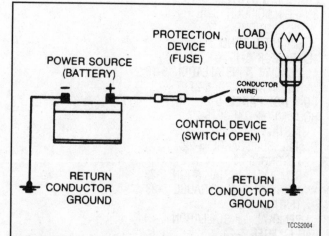

Fig. 2 This example illustrates a simple circuit. When the switch is closed, power from the positive (+) battery terminal flows through the fuse and the switch, and then to the light bulb. The light illuminates and the circuit is completed through the ground wire back to the negative (−) battery terminal.

The normal marine circuit differs from this simple example in two ways. First, instead of having a return wire from each bulb to the battery, the current travels through a single ground wire which handles all the grounds for a specific circuit. Secondly, most marine circuits contain multiple components which receive power from a single circuit. This lessens the amount of wire needed to power components.

HOW DOES ELECTRICITY WORK: THE WATER ANALOGY

Electricity is the flow of electrons—the sub-atomic particles that constitute the outer shell of an atom. Electrons spin in an orbit around the center core of an atom. The center core is comprised of protons (positive charge) and neutrons (neutral charge). Electrons have a negative charge and balance out the positive charge of the protons. When an outside force causes the number of electrons to unbalance the charge of the protons, the electrons will split off the atom and look for another atom to balance out. If this imbalance is kept up, electrons will continue to move and an electrical flow will exist.

Many people have been taught electrical theory using an analogy with water. In a comparison with water flowing through a pipe, the electrons would be the water and the wire is the pipe.

The flow of electricity can be measured much like the flow of water through a pipe. The unit of measurement used is amperes, frequently abbreviated as amps (a). You can compare amperage to the volume of water flowing through a pipe. When connected to a circuit, an ammeter will measure the actual amount of current flowing through the circuit. When relatively few electrons flow through a circuit, the amperage is low. When many electrons flow, the amperage is high.

Water pressure is measured in units such as pounds per square inch (psi); The electrical pressure is measured in units called volts (v). When a voltmeter is connected to a circuit, it is measuring the electrical pressure.

The actual flow of electricity depends not only on voltage and amperage, but also on the resistance of the circuit. The higher the resistance, the higher the force necessary to push the current through the circuit. The standard unit for measuring resistance is an ohm (Ω). Resistance in a circuit varies depending on the amount and type of components used in the circuit. The main factors which determine resistance are:

• Material—some materials have more resistance than others. Those with high resistance are said to be insulators. Rubber materials (or rubber-like plastics) are some of the most common insulators used, as they have a very high resistance to electricity. Very low resistance materials are said to be conductors. Copper wire is among the best conductors. Silver is actually a superior conductor to copper and is used in some relay contacts, but its high cost prohibits its use as common wiring. Most marine wiring is made of copper.

• Size—the larger the wire size being used, the less resistance the wire will have. This is why components which use large amounts of electricity usually have large wires supplying current to them.

• Length—for a given thickness of wire, the longer the wire, the greater the resistance. The shorter the wire, the less the resistance. When determining the proper wire for a circuit, both size and length must be considered to design a circuit that can handle the current needs of the component.

• Temperature—with many materials, the higher the temperature, the greater the resistance (positive temperature coefficient). Some materials exhibit the opposite trait of lower resistance with

higher temperatures (negative temperature coefficient). These principles are used in many of the sensors on the engine.

OHM'S LAW

There is a direct relationship between current, voltage and resistance. The relationship between current, voltage and resistance can be summed up by a statement known as Ohm's law.

Voltage (E) is equal to amperage (I) times resistance (R): $E=I \times R$
Other forms of the formula are $R=E/I$ and $I=E/R$

In each of these formulas, E is the voltage in volts, I is the current in amps and R is the resistance in ohms. The basic point to remember is that as the resistance of a circuit goes up, the amount of current that flows in the circuit will go down, if voltage remains the same.

The amount of work that the electricity can perform is expressed as power. The unit of power is the watt (w). The relationship between power, voltage and current is expressed as:

Power (W) is equal to amperage (I) times voltage (E): $W=I \times E$

This is only true for direct current (DC) circuits; The alternating current formula is a tad different, but since the electrical circuits in most vessels are DC type, we need not get into AC circuit theory.

Electrical Components

POWER SOURCE

Power is supplied to the vessel by two devices: The battery and the alternator. The battery supplies electrical power during starting or during periods when the current demand of the vessel's electrical system exceeds the output capacity of the alternator. The alternator supplies electrical current when the engine is running. The alternator does not just supply the current needs of the vessel, but it recharges the battery.

The Battery

In most modern vessels, the battery is a lead/acid electrochemical device consisting of six 2 volt subsections (cells) connected in series, so that the unit is capable of producing approximately 12 volts of electrical pressure. Each subsection consists of a series of positive and negative plates held a short distance apart in a solution of sulfuric acid and water.

The two types of plates are of dissimilar metals. This sets up a chemical reaction, and it is this reaction which produces current flow from the battery when its positive and negative terminals are connected to an electrical load. The power removed from the battery is replaced by the alternator, restoring the battery to its original chemical state.

The Alternator

On some vessels there isn't an alternator, but a generator. The difference is that an alternator supplies alternating current which is then changed to direct current for use on the vessel, while a generator produces direct current. Alternators tend to be more efficient and that is why they are used on almost all modern engines.

Alternators and generators are devices that consist of coils of wires wound together making big electromagnets. One group of coils spins within another set and the interaction of the magnetic fields causes a current to flow. This current is then drawn off the coils and fed into the vessel's electrical system.

GROUND

Two types of grounds are used in marine electric circuits. Direct ground components are grounded to the electrically conductive metal through their mounting points. All other components use some sort of ground wire which leads back to the battery. The electrical current runs through the ground wire and returns to the battery through the ground (–) cable; if you look, you'll see that the battery ground cable connects between the battery and a heavy gauge ground wire.

➡ It should be noted that a good percentage of electrical problems can be traced to bad grounds.

PROTECTIVE DEVICES

▶ See Figure 3

It is possible for large surges of current to pass through the electrical system of your vessel. If this surge of current were to reach the load in the circuit, the surge could burn it out or severely damage it. It can also overload the wiring, causing the harness to get hot and melt the insulation. To prevent this, fuses, circuit breakers and/or fusible links are connected into the supply wires of the electrical system. These items are nothing more than a built-in weak spot in the system. When an abnormal amount of current flows through the system, these protective devices work as follows to protect the circuit:

• Fuse—when an excessive electrical current passes through a fuse, the fuse "blows" (the conductor melts) and opens the circuit, preventing the passage of current.

• Circuit Breaker—a circuit breaker is basically a self-repairing fuse. It will open the circuit in the same fashion as a fuse, but when the surge subsides, the circuit breaker can be reset and does not need replacement.

• Fusible Link—a fusible link (fuse link or main link) is a short length of special, high temperature insulated wire that acts as a fuse. When an excessive electrical current passes through a fusible link, the thin gauge wire inside the link melts, creating an intentional open to protect the circuit. To repair the circuit, the link must be replaced. Some newer type fusible links are housed in plug-in modules, which are simply replaced like a fuse, while older type fusible links must be cut and spliced if they melt. Since this link is very early in the electrical path, it's the first place to look if nothing on the vessel works, yet the battery seems to be charged and is properly connected.

✳✳ CAUTION

Always replace fuses, circuit breakers and fusible links with identically rated components. Under no circumstances should a component of higher or lower amperage rating be substituted.

SWITCHES & RELAYS

▶ See Figure 4

Switches are used in electrical circuits to control the passage of current. The most common use is to open and close circuits between the battery and the various electric devices in the system. Switches are rated according to the amount of amperage they can handle. If a sufficient amperage rated switch is not used in a circuit, the switch could overload and cause damage.

Some electrical components which require a large amount of current to operate use a special switch called a relay. Since these circuits carry a large amount of current, the thickness of the wire in the circuit is also greater. If this large wire were connected from the load to the control switch, the switch would have to carry the high amperage load and the space needed for wiring in the vessel would be twice as big to accommodate the increased size of the wiring harness. To prevent these problems, a relay is used.

Relays are composed of a coil and a set of contacts. When the coil has a current passed though it, a magnetic field is formed and this field causes the contacts to move together, completing the circuit. Most relays are normally open, preventing current from passing through the circuit, but they can take any electrical form depending on the job they are intended to do. Relays can be considered "remote control switches." They allow a smaller current to operate devices that require higher amperages. When a small current operates the coil, a larger current is allowed to pass by the contacts. Some common circuits which may use relays are horns, lights, starter, electric fuel pumps and other high draw circuits.

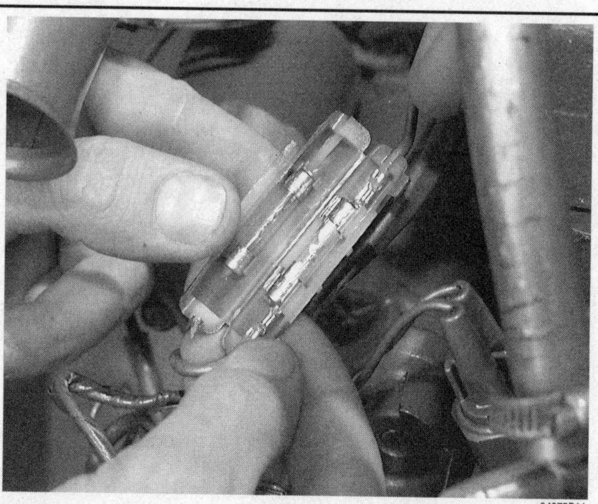

Fig. 3 Fuses protect the vessel's electrical system from abnormally high amounts of current flow

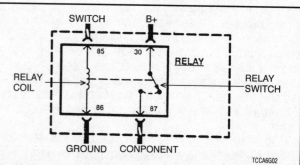

Fig. 4 Relays are composed of a coil and a switch. These two components are linked together so that when one operates, the other operates at the same time. The large wires in the circuit are connected from the battery to one side of the relay switch (B+) and from the opposite side of the relay switch to the load (component). Smaller wires are connected from the relay coil to the control switch for the circuit and from the opposite side of the relay coil to ground

LOAD

Every electrical circuit must include a "load" (something to use the electricity coming from the source). Without this load, the battery would attempt to deliver its entire power supply from one pole to another. This is called a "short circuit". All this electricity would take a short cut to ground and cause a great amount of damage to other components in the circuit by developing a tremendous amount of heat. This condition could develop sufficient heat to melt the insulation on all the surrounding wires and reduce a multiple wire cable to a lump of plastic and copper.

WIRING & HARNESSES

The average vessel contains miles of wiring, with hundreds of individual connections. To protect the many wires from damage and to keep them from becoming a confusing tangle, they are organized into bundles, enclosed in plastic or taped together and called wiring harnesses. Different harnesses serve different parts of the vessel. Individual wires are color coded to help trace them through a harness where sections are hidden from view.

Marine wiring or circuit conductors can be either single strand wire, multi-strand wire or printed circuitry. Single strand wire has a solid metal core and is usually used inside such components as alternators, motors, relays and other devices. Multi-strand wire has a core made of many small strands of wire twisted together into a single conductor. Most of the wiring in a marine electrical system is made up of multi-strand wire, either as a single conductor or grouped together in a harness. All wiring is color coded on the insulator, either as a solid color or as a colored wire with an identification stripe. A printed circuit is a thin film of copper or other conductor that is printed on an insulator backing. Occasionally, a printed circuit is sandwiched between two sheets of plastic for more protection and flexibility. A complete printed circuit, consisting of conductors, insulating material and connectors is called a printed circuit board. Printed circuitry is used in place of individual wires or harnesses in places where space is limited, such as behind instrument panels.

Since marine electrical systems are very sensitive to changes in resistance, the selection of properly sized wires is critical when systems are repaired. A loose or corroded connection or a replacement wire that is too small for the circuit will add extra resistance and an additional voltage drop to the circuit.

The wire gauge number is an expression of the cross-section area of the conductor. Vessels from countries that use the metric system will typically describe the wire size as its cross-sectional area in square millimeters. In this method, the larger the wire, the greater the number. Another common system for expressing wire size is the American Wire Gauge (AWG) system. As gauge number increases, area decreases and the wire becomes smaller. An 18 gauge wire is smaller than a 4 gauge wire. A wire with a higher gauge number will carry less current than a wire with a lower gauge number. Gauge wire size refers to the size of the strands of the conductor, not the size of the complete wire with insulator. It is possible, therefore, to have two wires of the same gauge with different diameters because one may have thicker insulation than the other.

It is essential to understand how a circuit works before trying to figure out why it doesn't. An electrical schematic shows the electrical current paths when a circuit is operating properly. Schematics break the entire electrical system down into individual circuits. In a schematic, usually no attempt is made to represent wiring and components as they physically appear on the vessel; switches and other

components are shown as simply as possible. Face views of harness connectors show the cavity or terminal locations in all multipin connectors to help locate test points.

CONNECTORS

▶ See Figures 5, 6 and 7

Three types of connectors are commonly used in marine applications—weatherproof, molded and hard shell.

• Weatherproof—these connectors are most commonly used where the connector is exposed to the elements. Terminals are protected against moisture and dirt by sealing rings which provide a weather tight seal. All repairs require the use of a special terminal and the tool required to service it. Unlike standard blade type terminals, these weatherproof terminals cannot be straightened once they are bent. Make certain that the connectors are properly seated and all of the sealing rings are in place when connecting leads.

• Molded—these connectors require complete replacement of the connector if found to be defective. This means splicing a new connector assembly into the harness. All splices should be soldered to insure proper contact. Use care when probing the connections or replacing terminals in them, as it is possible to create a short circuit between opposite terminals. If this happens to the wrong terminal pair, it is possible to damage certain components. Always use jumper wires between connectors for circuit checking and NEVER probe through weatherproof seals.

TCCA6P03

Fig. 5 Hard shell (left) and weatherproof (right) connectors have replaceable terminals

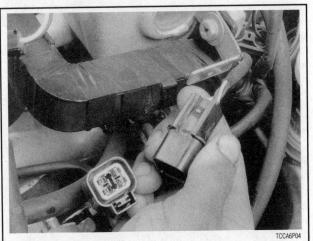

TCCA6P04

Fig. 6 Weatherproof connectors are most commonly used in the engine compartment or where the connector is exposed to the elements

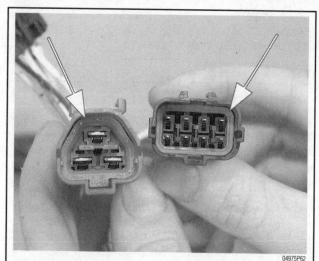

Fig. 7 The seals on weatherproof connectors must be kept in good condition to prevent the terminals from corroding

04975P62

• Hard Shell—unlike molded connectors, the terminal contacts in hard-shell connectors can be replaced. Replacement usually involves the use of a special terminal removal tool that depresses the locking tangs (barbs) on the connector terminal and allows the connector to be removed from the rear of the shell. The connector shell should be replaced if it shows any evidence of burning, melting, cracks, or breaks. Replace individual terminals that are burnt, corroded, distorted or loose.

Test Equipment

Pinpointing the exact cause of trouble in an electrical circuit is most times accomplished by the use of special test equipment. The following sections describe different types of commonly used test equipment and briefly explain how to use them in diagnosis. In addition to the information covered below, the tool manufacturer's instruction manual (provided with most tools) should be read and clearly understood before attempting any test procedures.

JUMPER WIRES

✳✳ CAUTION

Never use jumper wires made from a thinner gauge wire than the circuit being tested. If the jumper wire is of too small a gauge, it may overheat and possibly melt. Never use jumpers to bypass high resistance loads in a circuit. Bypassing resistances, in effect, creates a short circuit. This may, in turn, cause damage and fire. Jumper wires should only be used to bypass lengths of wire or to simulate switches.

Jumper wires are simple, yet extremely valuable, pieces of test equipment. They are basically test wires which are used to bypass sections of a circuit. Although jumper wires can be purchased, they are usually fabricated from lengths of standard marine wire and whatever type of connector (alligator clip, spade connector or pin connector) that is required for the particular application being tested. In cramped, hard-to-reach areas, it is advisable to have insulated boots over the jumper wire terminals in order to prevent accidental grounding. It is also advisable to include a standard marine fuse in any jumper wire. This is commonly referred to as a

"fused jumper". By inserting an in-line fuse holder between a set of test leads, a fused jumper wire can be used for bypassing open circuits. Use a 5 amp fuse to provide protection against voltage spikes.

Jumper wires are used primarily to locate open electrical circuits, on either the ground (–) side of the circuit or on the power (+) side. If an electrical component fails to operate, connect the jumper wire between the component and a good ground. If the component operates only with the jumper installed, the ground circuit is open. If the ground circuit is good, but the component does not operate, the circuit between the power feed and component may be open. By moving the jumper wire successively back from the component toward the power source, you can isolate the area of the circuit where the open is located. When the component stops functioning, or the power is cut off, the open is in the segment of wire between the jumper and the point previously tested.

You can sometimes connect the jumper wire directly from the battery to the "hot" terminal of the component, but first make sure the component uses 12 volts in operation. Some electrical components, such as fuel injectors or sensors, are designed to operate on about 4 to 5 volts, and running 12 volts directly to these components will cause damage.

TEST LIGHTS

♦ See Figure 8

The test light is used to check circuits and components while electrical current is flowing through them. It is used for voltage and ground tests. To use a 12 volt test light, connect the ground clip to a good ground and probe wherever necessary with the pick. The test light will illuminate when voltage is detected. This does not necessarily mean that 12 volts (or any particular amount of voltage) is present; it only means that some voltage is present. It is advisable before using the test light to touch its ground clip and probe across the battery posts or terminals to make sure the light is operating properly.

✳✳ WARNING

Do not use a test light to probe electronic ignition, spark plug or coil wires. Never use a pick-type test light to probe wiring

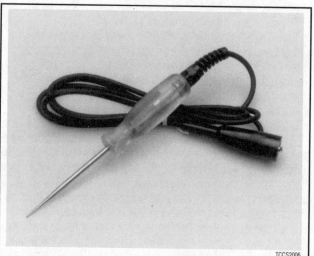

Fig. 8 A 12 volt test light is used to detect the presence of voltage in a circuit

TCCS2006

on electronically controlled systems unless specifically instructed to do so. Any wire insulation that is pierced by the test light probe should be taped and sealed with silicone after testing.

Like the jumper wire, the 12 volt test light is used to isolate opens in circuits. But, whereas the jumper wire is used to bypass the open to operate the load, the 12 volt test light is used to locate the presence of voltage in a circuit. If the test light illuminates, there is power up to that point in the circuit; if the test light does not illuminate, there is an open circuit (no power). Move the test light in successive steps back toward the power source until the light in the handle illuminates. The open is between the probe and a point which was previously probed.

The self-powered test light is similar in design to the 12 volt test light, but contains a 1.5 volt penlight battery in the handle. It is most often used in place of a multimeter to check for open or short circuits when power is isolated from the circuit (continuity test).

The battery in a self-powered test light does not provide much current. A weak battery may not provide enough power to illuminate the test light even when a complete circuit is made (especially if there is high resistance in the circuit). Always make sure that the test battery is strong. To check the battery, briefly touch the ground clip to the probe; if the light glows brightly, the battery is strong enough for testing.

➡A self-powered test light should not be used on any electronically controlled system or component. The small amount of electricity transmitted by the test light is enough to damage many electronic marine components.

MULTIMETERS

▸ See Figure 9

Multimeters are an extremely useful tool for troubleshooting electrical problems. They can be purchased in either analog or digital form and have a price range to suit any budget. A multimeter is a voltmeter, ammeter and ohmmeter (along with other features) com-

bined into one instrument. It is often used when testing solid state circuits because of its high input impedance (usually 10 megaohms or more). A brief description of the multimeter main test functions follows:

• Voltmeter—the voltmeter is used to measure voltage at any point in a circuit, or to measure the voltage drop across any part of a circuit. Voltmeters usually have various scales and a selector switch to allow the reading of different voltage ranges. The voltmeter has a positive and a negative lead. To avoid damage to the meter, always connect the negative lead to the negative (−) side of the circuit (to ground or nearest the ground side of the circuit) and connect the positive lead to the positive (+) side of the circuit (to the power source or the nearest power source). Note that the negative voltmeter lead will always be black and that the positive voltmeter will always be some color other than black (usually red).

• Ohmmeter—the ohmmeter is designed to read resistance (measured in ohms) in a circuit or component. Most ohmmeters will have a selector switch which permits the measurement of different ranges of resistance (usually the selector switch allows the multiplication of the meter reading by 10, 100, 1,000 and 10,000). Some ohmmeters are "auto-ranging" which means the meter itself will determine which scale to use. Since the meters are powered by an internal battery, the ohmmeter can be used like a self-powered test light. When the ohmmeter is connected, current from the ohmmeter flows through the circuit or component being tested. Since the ohmmeter's internal resistance and voltage are known values, the amount of current flow through the meter depends on the resistance of the circuit or component being tested. The ohmmeter can also be used to perform a continuity test for suspected open circuits. In using the meter for making continuity checks, do not be concerned with the actual resistance readings. Zero resistance, or any ohm reading, indicates continuity in the circuit. Infinite resistance indicates an opening in the circuit. A high resistance reading where there should be none indicates a problem in the circuit. Checks for short circuits are made in the same manner as checks for open circuits, except that the circuit must be isolated from both power and normal ground. Infinite resistance indicates no continuity, while zero resistance indicates a dead short.

✴✴ WARNING

Never use an ohmmeter to check the resistance of a component or wire while there is voltage applied to the circuit.

• Ammeter—an ammeter measures the amount of current flowing through a circuit in units called amperes or amps. At normal operating voltage, most circuits have a characteristic amount of amperes, called "current draw" which can be measured using an ammeter. By referring to a specified current draw rating, then measuring the amperes and comparing the two values, one can determine what is happening within the circuit to aid in diagnosis. An open circuit, for example, will not allow any current to flow, so the ammeter reading will be zero. A damaged component or circuit will have an increased current draw, so the reading will be high. The ammeter is always connected in series with the circuit being tested. All of the current that normally flows through the circuit must also flow through the ammeter; if there is any other path for the current to follow, the ammeter reading will not be accurate. The ammeter

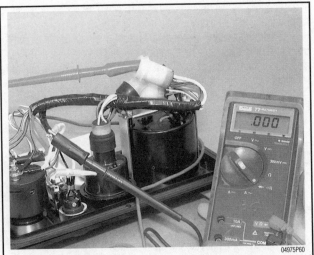

04975P60

Fig. 9 Multimeters are essential for diagnosing faulty wires, switches and other electrical components

itself has very little resistance to current flow and, therefore, will not affect the circuit, but it will measure current draw only when the circuit is closed and electricity is flowing. Excessive current draw can blow fuses and drain the battery, while a reduced current draw can cause motors to run slowly, lights to dim and other components to not operate properly.

Troubleshooting Electrical Systems

When diagnosing a specific problem, organized troubleshooting is a must. The complexity of a modern marine vessel demands that you approach any problem in a logical, organized manner. There are certain troubleshooting techniques, however, which are standard:

• **Establish when the problem occurs**. Does the problem appear only under certain conditions? Were there any noises, odors or other unusual symptoms? Isolate the problem area. To do this, make some simple tests and observations, then eliminate the systems that are working properly. Check for obvious problems, such as broken wires and loose or dirty connections. Always check the obvious before assuming something complicated is the cause.

• **Test for problems systematically to determine the cause once the problem area is isolated**. Are all the components functioning properly? Is there power going to electrical switches and motors. Performing careful, systematic checks will often turn up most causes on the first inspection, without wasting time checking components that have little or no relationship to the problem.

• **Test all repairs after the work is done to make sure that the problem is fixed**. Some causes can be traced to more than one component, so a careful verification of repair work is important in order to pick up additional malfunctions that may cause a problem to reappear or a different problem to arise. A blown fuse, for example, is a simple problem that may require more than another fuse to repair. If you don't look for a problem that caused a fuse to blow, a shorted wire (for example) may go undetected.

Experience has shown that most problems tend to be the result of a fairly simple and obvious cause, such as loose or corroded connectors, bad grounds or damaged wire insulation which causes a short. This makes careful visual inspection of components during testing essential to quick and accurate troubleshooting.

Testing

VOLTAGE

This test determines voltage available from the battery and should be the first step in any electrical troubleshooting procedure after visual inspection. Many electrical problems, especially on electronically controlled systems, can be caused by a low state of charge in the battery. Excessive corrosion at the battery cable terminals can cause poor contact that will prevent proper charging and full battery current flow.

1. Set the voltmeter selector switch to the 20V position.
2. Connect the multimeter negative lead to the battery's negative (−) post or terminal and the positive lead to the battery's positive (+) post or terminal.

3. Turn the ignition switch **ON** to provide a load.
4. A well charged battery should register over 12 volts. If the meter reads below 11.5 volts, the battery power may be insufficient to operate the electrical system properly.

VOLTAGE DROP

When current flows through a load, the voltage beyond the load drops. This voltage drop is due to the resistance created by the load and also by small resistances created by corrosion at the connectors and damaged insulation on the wires. The maximum allowable voltage drop under load is critical, especially if there is more than one load in the circuit, since all voltage drops are cumulative.

1. Set the voltmeter selector switch to the 20 volt position.
2. Connect the multimeter negative lead to a good ground.
3. Operate the circuit and check the voltage prior to the first component (load).
4. There should be little or no voltage drop in the circuit prior to the first component. If a voltage drop exists, the wire or connectors in the circuit are suspect.
5. While operating the first component in the circuit, probe the ground side of the component with the positive meter lead and observe the voltage readings. A small voltage drop should be noticed. This voltage drop is caused by the resistance of the component.
6. Repeat the test for each component (load) down the circuit.
7. If a large voltage drop is noticed, the preceding component, wire or connector is suspect.

RESISTANCE

✳✳ WARNING

Never use an ohmmeter with power applied to the circuit. The ohmmeter is designed to operate on its own power supply. The normal 12 volt electrical system voltage could damage the meter!

1. Isolate the circuit from the vessel's power source.
2. Ensure that the ignition key is **OFF** when disconnecting any components or the battery.
3. Where necessary, also isolate at least one side of the circuit to be checked, in order to avoid reading parallel resistances. Parallel circuit resistances will always give a lower reading than the actual resistance of either of the branches.
4. Connect the meter leads to both sides of the circuit (wire or component) and read the actual measured ohms on the meter scale. Make sure the selector switch is set to the proper ohm scale for the circuit being tested, to avoid misreading the ohmmeter test value.

OPEN CIRCUITS

▶ See Figure 10

This test already assumes the existence of an open in the circuit and it is used to help locate the open portion.

1. Isolate the circuit from power and ground.
2. Connect the self-powered test light or ohmmeter ground clip

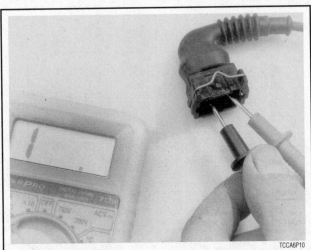

Fig. 10 The infinite reading on this multimeter (1 .) indicates that the circuit is open

to the ground side of the circuit and probe sections of the circuit sequentially.

3. If the light is out or there is infinite resistance, the open is between the probe and the circuit ground.

4. If the light is on or the meter shows continuity, the open is between the probe and the end of the circuit toward the power source.

SHORT CIRCUITS

➡**Never use a self-powered test light to perform checks for opens or shorts when power is applied to the circuit under test. The test light can be damaged by outside power.**

1. Isolate the circuit from power and ground.

2. Connect the self-powered test light or ohmmeter ground clip to a good ground and probe any easy-to-reach point in the circuit.

3. If the light comes on or there is continuity, there is a short somewhere in the circuit.

4. To isolate the short, probe a test point at either end of the isolated circuit (the light should be on or the meter should indicate continuity).

5. Leave the test light probe engaged and sequentially open connectors or switches, remove parts, etc. until the light goes out or continuity is broken.

6. When the light goes out, the short is between the last two circuit components which were opened.

Wire And Connector Repair

Almost anyone can replace damaged wires, as long as the proper tools and parts are available. Wire and terminals are available to fit almost any need. Even the specialized weatherproof, molded and hard shell connectors are now available from aftermarket suppliers.

Be sure the ends of all the wires are fitted with the proper terminal hardware and connectors. Wrapping a wire around a stud is never a permanent solution and will only cause trouble later. Replace wires one at a time to avoid confusion. Always route wires in the same manner of the manufacturer.

When replacing connections, make absolutely certain that the connectors are certified for marine use. Automotive wire connectors may not meet United States Coast Guard (USCG) specifications.

➡**If connector repair is necessary, only attempt it if you have the proper tools. Weatherproof and hard shell connectors require special tools to release the pins inside the connector. Attempting to repair these connectors with conventional hand tools will damage them.**

Electrical System Precautions

• Wear safety glasses when working on or near the battery.

• Don't wear a watch with a metal band when servicing the battery or starter. Serious burns can result if the band completes the circuit between the positive battery terminal and ground.

• Be absolutely sure of the polarity of a booster battery before making connections. Connect the cables positive-to-positive, and negative-to-negative. Connect positive cables first, and then make the last connection to ground on the body of the booster vessel so that arcing cannot ignite hydrogen gas that may have accumulated near the battery. Even momentary connection of a booster battery with the polarity reversed will damage alternator diodes.

• Disconnect both vessel battery cables before attempting to charge a battery.

• Never ground the alternator or generator output or battery terminal. Be cautious when using metal tools around a battery to avoid creating a short circuit between the terminals.

• When installing a battery, make sure that the positive and negative cables are not reversed.

• Always disconnect the battery (negative cable first) when charging.

• Never smoke or expose an open flame around the battery . Hydrogen gas accumulates near the battery and is highly explosive.

Troubleshooting Electrical Systems

Condition	Cause	Correction
Battery never reaches full charge	Undercharging	Allow more charging time Check alternator belt for slipping Battery defective Check for corroded or loose terminals
Electrolyte boils, smells of acid	Overcharging	Check voltage regulator output
Not increase in voltage when engine starts, charge light stays ON	Not charging	Check for broken alternator belt Check voltage regulator output Alternator worn out
Starter fails to operate, no voltage at starter solenoid	Open starter circuit	Check for defective switch Check for loose or corroded terminals
Starter engages as soon as battery isolator is turned on	Short in starter circuit	Check for defective switch Check for water contamination in switch
Starter fails to operate or turns slowly	Brushes worn	Normal wear condition, replace starter
Starter fails to operate or turns slowly, starter draws high current	Internal short circuit	Check for water in starter Check for loose or broken wiring
Starter fails to operate or has reduced power, starter draws low current	Internal open circuit	Check for corrosion Check for loose or broken wiring
	Internal contamination	Check for water in cylinders Check for flywheel sitting in high bilge water
Starter overspeeds and fails to start the engine	Drive gear defective	Starter drive defective Clutch mechanism jammed or broken
Starter will not operate	Starter solenoid jammed in OFF position	Check for contamination Check for a worn starter
Starter runs constantly, even when switch is released, starter may overspeed	Starter solenoid jammed in ON position	Check for contamination Check for a worn starter
Starter turns intermittently or slowly	Corroded or erroded contacts	Check for a worn starter Check for wet contacts
Starter fails to operate, starter turns engine over slowly, starter solenoid only clicks, lights dim excessively when starting	Low battery output	Check for a discharged battery Check electrolyte level Check for loose terminals Check for corroded terminals Battery worn out
	Battery will not hold a charge	Check electrolyte level Battery worn out

04975C01

STARTING SYSTEM

Starting Circuit

♦ See Figures 11 and 12

The starting system includes the battery, starter motor, solenoid, starter button, and ignition switch.

When the starter button on the instrument panel is depressed, current flows and energizes the starter's solenoid coil. The energized coil becomes an electromagnet, which pulls the plunger into the coil, and closes a set of contacts which allow high current to reach the starter motor. At the same time, the plunger also serves to push the starter pinion to mesh with the teeth on the flywheel.

To prevent damage to the starter motor when the engine starts, the pinion gear incorporates an over-running (one-way) clutch which is splined to the starter armature shaft. The rotation of the running engine may speed the rotation of the pinion but not the starter motor itself.

Once the starter button is released, the current flow ceases, stopping the activation of the solenoid. The plunger is pulled out of contact with the battery-to-starter cables by a coil spring, and the flow of electricity is interrupted to the starter. This weakens the magnetic fields and the starter ceases its rotation. As the solenoid plunger is released, its movement also pulls the starter drive gear from its engagement with the engine flywheel.

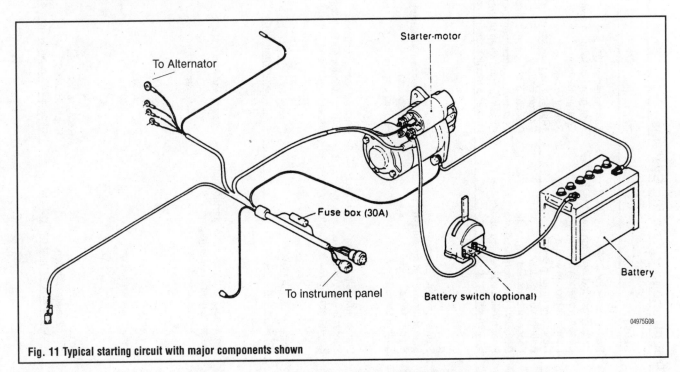

Fig. 11 Typical starting circuit with major components shown

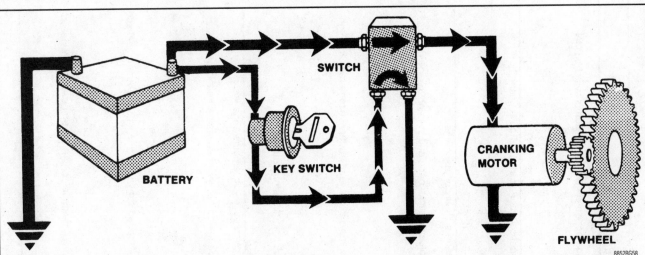

Fig. 12 A typical starting system converts electrical energy into mechanical energy to turn the engine. The components are: Battery, to provide electricity to operate the starter; Ignition switch, to control the energizing of the starter relay or solenoid; Starter relay or solenoid, to make and break the circuit between the battery and starter; Starter, to convert electrical energy into mechanical energy to rotate the engine; Starter drive gear, to transmit the starter rotation to the engine flywheel

Troubleshooting Starting Systems

Condition	Cause	Correction
Nothing happens when the ignition is switched ON	Battery selector OFF	Turn selector ON
	Battery ground isolation switch OFF	Turn ground isolation switch ON
	Battery voltage very low	Charge Battery
	Loose or broken wiring	Repair wiring
	Defective ignition switch	Test and as necessary replace ignition switch
No response when starter button is pushed	Battery voltage low	Charge battery
	Loose or corroded connections	Disconnect plugs and inspect for corrosion
	Defective ignition/start switch	Test and as necessary replace ignition/start switch
	Poor ground connection on engine	Check ground strap to engine
	Broken wiring	Find and repair broken wiring
Starter turns slowly or just clicks		
Solenoid clicks once and battery voltage drops	Starter shorting internally	Replace starter
	Starter jammed in flywheel ring gear	Free starter by manually turning over engine, then replace
	Starter seized	Replace starter
	Engine seized	Rebuild engine
Solenoid clicks and battery voltage remains high	Starter solenoid not making contact	Replace starter solenoid
	Starter B+ not connected or loose	Inspect and repair B+ connection
	Engine ground not connected or loose	Inspect and repair ground connection
Solenoid clicks repeatedly	Battery voltage low	Charge battery
	Loose or corroded connections	Inspect and repair connections
Starter overspeeds or remains engaged		
Overspeed occurs when starter button is pressed	Dirt, wear or corrosion on starter shaft prevents gear from engaging	Inspect and clean starter
	Drive gear defective	Replace drive gear
Overspeed continues when starter button released	Defective solenoid	Replace solenoid
	Defective start switch	Replace switch
Starter continues engaged after switch is released	Defective solenoid	Replace solenoid
	Defective start switch	Replace switch
	Damaged starter gear or flywheel	Inspect and replace gear or flywheel

0497SC04

PRECAUTIONS

To prevent damage to the starter, the following precautionary measures must be taken:
- Wear safety glasses when working on or near the battery.
- Don't wear a watch with a metal band when servicing the starter. Serious burns can result if the band completes the circuit between the positive battery terminal and ground.
- Be absolutely sure of the polarity of a booster battery before making connections. Connect the cables positive-to-positive, and negative-to-negative. Connect positive cables first, and then make the last connection to ground on the body of the booster vessel so that arcing cannot ignite hydrogen gas that may have accumulated near the battery. Even momentary connection of a booster battery with the polarity reversed will damage alternator diodes.
- Disconnect both battery cables before attempting to charge a battery.
- Be cautious when using metal tools around a the starter to avoid creating a short circuit between the terminals.
- When installing a battery, make sure that the positive and negative cables are not reversed.
- Always disconnect the battery ground cable before disconnecting the starter leads.

CIRCUIT TESTING

➡A good quality digital multimeter with at least 10 megaohm/volt impedance should be used when testing circuits. These meters can accurately detect very small amounts of voltage, current and resistance. This type of meter also has a high internal resistance that will not load the circuit being tested.

1. Check the battery and clean the connections as follows:
   a. If the battery cells have removable caps, check the water level, and add distilled water if low. Load test the battery and charge if necessary. See Battery Testing in this section for the procedure.
   b. Remove the cables and clean them with a wire brush. Reconnect the cables.
2. Check the starter motor ground circuit using a voltage drop test as follows:
   a. Set the meter to read DC voltage on the lowest possible scale.
   b. Connect the negative lead of your multimeter to the negative terminal of the battery.
   c. Connect the positive lead to the body of the starter. Make sure the starter mounting bolts are tight. The meter should read 0.2 volts or less. If the voltage reading is greater, remove and clean the negative battery connection on the engine block. The voltage reading should now be within specification: if not, replace the negative battery cable.
3. Check the motor feed circuit with a voltage drop test as follows:
   a. Move the engine speed lever to the OFF position to prevent the engine from starting.
   b. Connect the positive lead of your meter to the positive terminal of the battery.
   c. Connect the negative meter lead to the motor feed terminal.

The motor feed terminal comes out of the body of the starter motor and connects to the solenoid.
   d. Turn the ignition key to the **START** position. The meter should read 0.2 volts or less. If the voltage reading is greater, remove and clean the positive battery connection on the starter solenoid. The voltage reading should now be within specification; if not replace the positive battery cable.
4. Check for battery voltage at the **S** terminal on the starter solenoid as follows:
   a. Move the engine speed lever to the **OFF** position to prevent the engine from starting.
   b. Set the meter to read battery voltage. Move it to the next higher range if set on the 2 volt scale.
   c. Connect the positive lead to the **S** terminal on the starter solenoid and the negative lead to a good ground.
   d. Turn the ignition key to the **START** position, and push the starter button to crank the engine. The meter should read battery voltage. If battery voltage is not present, check the fuse(s) and wiring between the ignition switch and starter solenoid. If battery voltage is present at the **S** terminal on the solenoid and the starter does not operate, it can be assumed that the starter is faulty.

Starter Motor

♦ See Figure 13

TESTING

♦ See Figures 14, 15 and 16

Check the fuse attached to the wiring harness near the starter before starting any testing. If the fuse is good, this confirms the starter circuit is working correctly.

Although considered a unit, starters consist of two primary components—the starter motor and the starter solenoid. The solenoid can be tested independently to confirm if it is faulty.

Keep in mind that if you do find that one of the components of the starter is faulty, replacement as a unit is usually most practical. Most parts suppliers supply "rebuilt" starters, saving time and money on having to replace individual components.

Solenoid

Although removal of the starter solenoid (also known as the magnetic switch) requires disassembly of the starter, it can be tested for proper operation independently of the starter without removal.

SOLENOID PLUNGER TEST

1. Disconnect the negative battery cable.
2. Remove the starter motor power wire (the large braided wire) from the M terminal on the solenoid. Leave all other connections intact.
3. Reconnect the negative battery cable.
4. Turn the ignition switch to the **ON** position, and press the starter button. An audible "click" should be heard from the starter. If there is no noise from the starter when the button is pushed, the solenoid is faulty.

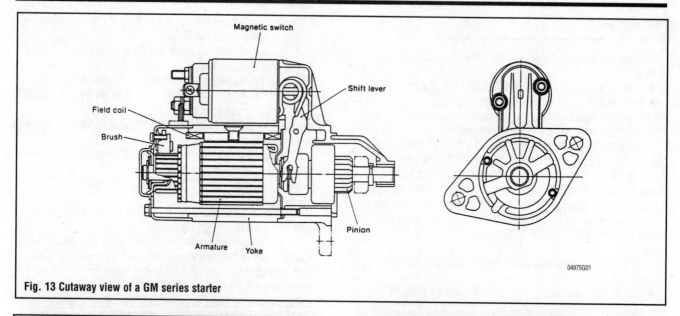

Fig. 13 Cutaway view of a GM series starter

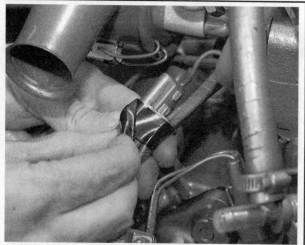

Fig. 14 An inline fuse is located on the wiring harness near the starter. To check the fuse, remove the tape . . .

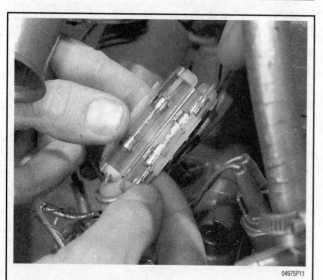

Fig. 15 . . . open the fuse box . . .

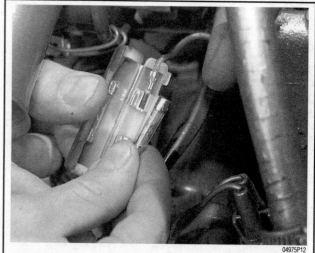

Fig. 16 . . . and unclip the fuse from the holder. Note the spare fuse inside the box

SOLENOID CONTACT TEST

♦ See Figures 17 and 18

1. Disconnect the negative battery cable.
2. Remove the starter motor power wire (the large braided wire) and the power (+12v) wire from the solenoid. Leave the solenoid coil connector attached.

➥Make sure that the power wire is not touching anything on the engine.

3. Reconnect the negative battery cable.
4. Check for continuity between the "B" and "M" terminals on the solenoid. Continuity should not exist when the starter button is not pressed (solenoid not energized). If continuity exists, the solenoid is faulty and should be replaced.
5. Turn the ignition switch to the **ON** position, and press the starter button. After the starter makes a "click" sound, check for continuity between the "B" and "M" terminals on the solenoid while continuing to hold down the starter button (this may require an

Fig. 17 To test the solenoid contact, remove the wires from the "B" and "M" terminals . . .

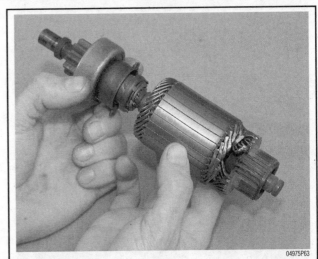

Fig. 19 This starter motor armature was destroyed from excessive engine cranking

Fig. 18 . . . and connect a multimeter between them to check for continuity

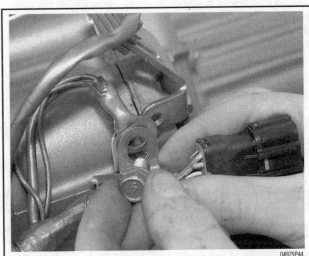

Fig. 20 It may be necessary to remove brackets or other components to make starter removal easier

assistant). If continuity does not exist, the solenoid is faulty and should be replaced.

Starter Motor

▶ See Figure 19

If the tests for the starter solenoid indicate proper function and the starter motor itself is not functioning, it can be assumed the starter is faulty. This is because power is being supplied to the motor, through the braided cable which connects to the M terminal on the starter solenoid.

REMOVAL & INSTALLATION

▶ See Figures 20, 21, 22 and 23

➡If you decide to use a generic brand of starter on your engine, make absolutely certain that it is certified for marine use. Automotive starters do not meet United States Coast Guard (USCG) standards because they do not have shielding for the contacts to prevent sparks. Under no circumstances should an automotive-type starter

Fig. 21 Remove the power connections from the solenoid . . .

Fig. 22 . . . loosen the starter mounting bolts . . .

Fig. 23 . . . and lift the starter from the engine

be used on a marine engine. If an automotive starter (or any other non-USCG approved electrical device) is used on your marine engine, and it catches fire, your insurance company will **NOT** cover your loss, and you will be issued a heavy fine by the USCG.

1. Disconnect the negative battery cable.
2. If necessary, remove any components to gain access to the starter motor, such as the air filter or wiring brackets.
3. Label and disconnect the wiring from the starter.
4. Remove the starter mounting bolts.
5. Remove the starter assembly from the engine. In some cases, the starter will have to be turned to a different angle to clear obstructions.

To install:

6. If necessary, measure and adjust the pinion-to-ring gear clearance. This should be done by the manufacturer or rebuilder. If in doubt, ask the parts distributor if this will be necessary.
7. Position the starter motor on the mounting boss. Tighten the mounting bolts securely.
8. Connect the wiring to the starter.
9. Install any components that were removed to gain access to the starter.
10. Connect the negative battery cable.
11. Test the starter for proper operation.

CHARGING SYSTEM

Charging Circuit

▶ See Figures 24 and 25

A typical charging system contains an alternator, drive belt, battery, voltage regulator (integrated into alternator housing) and the associated wiring. The charging system, like the starting system is a series circuit with the battery wired in parallel. After the engine is started and running, the alternator takes over as the source of power and the battery then becomes part of the load on the charging system.

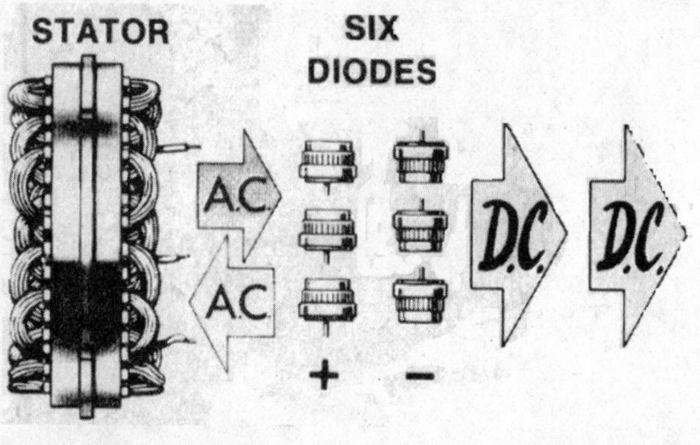

The alternator, which is driven by the drive belts, consists of a rotating coil of laminated wire called the rotor. Surrounding the rotor are more coils of laminated wire that remain stationary (which is how we get the term stator) just inside the alternator case. When current is passed through the rotor via the slip rings and brushes, the rotor becomes a rotating magnet with, of course, a magnetic field. When a magnetic field passes through a conductor (the stator), alternating current (A/C) is generated. This A/C current is rectified, turned into direct current (D/C), by the diodes located within the alternator.

The voltage regulator controls the alternator's field voltage by grounding one end of the field windings very rapidly. The frequency varies according to current demand. The more the field is grounded, the more voltage and current the alternator produces. Voltage is maintained at about 13.5–15 volts. During high engine speeds and low current demands, the regulator will adjust the voltage of the alternator field to lower the alternator output voltage. Conversely, when the engine is idling and the current demands may be high, the regulator will increase the field voltage, increasing the output of the alternator. Depending on the manufacturer, voltage regulators can be found in different locations, including inside or on the alternator.

➡ Drive belts are often overlooked when diagnosing a charging system failure. Check the belt tension on the alternator pulley and

Fig. 24 Negative and positive diodes convert AC current into DC current—note that the AC current reverses direction while the DC current flows in one direction

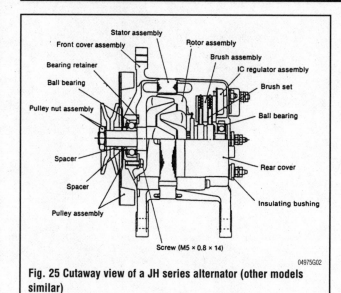

Fig. 25 Cutaway view of a JH series alternator (other models similar)

replace/adjust the belt. A loose belt will result in an undercharged battery and a no-start condition. This is especially true in wet weather conditions when the moisture causes the belt to become even more slippery.

PRECAUTIONS

To prevent damage to the alternator, the following precautionary measures must be taken:
• Wear safety glasses when working on or near the battery.
• Don't wear a watch with a metal band when servicing the battery or alternator. Serious burns can result if the band completes the circuit between the positive battery terminal and ground.
• Be absolutely sure of the polarity of a booster battery before making connections. Connect the cables positive-to-positive, and negative-to-negative. Connect positive cables first, and then make the last connection to ground on the body of the booster vessel so that arcing cannot ignite hydrogen gas that may have accumulated near the battery. Even momentary connection of a booster battery with the polarity reversed will damage alternator diodes.
• Disconnect both battery cables before attempting to charge a battery.
• Never ground the alternator or generator output or battery terminal. Be cautious when using metal tools around a battery to avoid creating a short circuit between the terminals.
• Never ground the field circuit between the alternator and regulator.
• Never run an alternator without an electrical load unless the field circuit is disconnected.
• Never attempt to "polarize" an alternator.
• When installing a battery, make sure that the positive and negative cables are not reversed.
• Never operate the alternator with the battery disconnected or on an otherwise uncontrolled open circuit.
• Do not short across or ground any alternator or regulator terminals.
• Always disconnect the battery ground cable before disconnecting the alternator lead.
• Always disconnect the battery (negative cable first) when charging.

CIRCUIT TESTING

The charging circuit should be inspected if:
• The charging system warning light on the control panel is illuminated.
• The voltmeter on the instrument panel indicates improper charging (either high or low) voltage.
• The battery is overcharged (electrolyte level is low and/or boiling out).
• The battery is undercharged (insufficient power to crank the starter).
The starting point for all charging system problems begins with the inspection of the battery, related wiring and the alternator drive belt. The battery must be in good condition and fully charged before system testing.
The charging system warning light will illuminate if the charging voltage is either too high or too low. The warning light should light when the key is turned to the **ON** position as a bulb check. When the alternator starts producing voltage due to the engine starting, the light should go out.
A good sign of voltage that is too high, are lights that burn out and/or burn very brightly. Over-charging can also cause damage to the battery and electronic circuits.

➡**Before testing, make sure all connections and mounting bolts are clean and tight. Many charging system problems are related to loose and corroded terminals or bad grounds. Don't overlook the engine ground connection to the battery, or the tension of the alternator drive belt.**

Voltage Drop Test

POSITIVE SIDE OF THE CIRCUIT

1. Make sure the battery is in good condition and fully charged.
2. Start the engine and allow it to reach normal operating temperature.
3. Turn on all the electrical accessories on the vessel.
4. Bring the engine to a high idle, of approximately 2,000 rpm.
5. Connect the negative (-) voltmeter lead directly to the battery positive (+) terminal.
6. Touch the positive voltmeter lead directly to the alternator B+ output stud, not the nut. The meter should read no higher than about 0.5 volts. If it does, then there is higher than normal resistance between the positive side of the battery and the B+ output at the alternator.
7. Move the positive (+) meter lead to the nut and compare the voltage reading with the previous measurement. If the voltage reading drops substantially, then there is resistance between the stud and the nut.
8. The theory is to keep moving closer to the battery terminal one connection at a time in order to find the area of high resistance (bad connection).

NEGATIVE SIDE OF THE CIRCUIT

1. Start the engine and allow it to reach normal operating temperature.
2. Turn on all the electrical accessories on the vessel.
3. Bring the engine to a high idle, of approximately 2,000 rpm.
4. Connect the negative (-) voltmeter lead directly to the negative battery terminal.
5. Touch the positive (+) voltmeter lead directly to the alternator case or ground connection. The meter should read no higher than about 0.3 volts. If it does, then there is higher than normal resistance between the battery ground terminal and the alternator ground.

Troubleshooting Charging Systems

Condition	Cause	Correction
Ignition warning light fails to extinguish No voltage increase at the alternator output terminal Alternator does not get warm/hot No indication on the ammeter	Alternator not charging	Check for broken or slipping alternator belt Check for disconnected, broken, loose or corroded output wire Check for poor ground connection Replace worn out alternator
Minimal voltage increase on the engine panel voltmeter Batteries fail to fully charge	Alternator undercharging	Charge batteries more Check for broken or slipping alternator belt Check for loose or corroded output connection Check for loose or corroded ground connection Replace faulty voltage regulator Replace worn out alternator
Battery very hot Electrolyte bolting violently Strong acid smell Voltmeter reads over 14.5 volts Low electrolyte level	Alternator overcharging	Check for damaged or disconnected battery voltage sensing wire Replace faulty voltage regulator Replace defective battery

04975C03

6. Move the positive (+) meter lead to the alternator mounting bracket, if the voltage reading drops substantially then you know that there is a bad electrical connection between the alternator and the mounting bracket.

➡The theory is to keep moving closer to the battery terminal one connection at a time in order to find the area of high resistance (bad connection).

Current Output Test

▶ **See Figure 26**

➡The current output test requires the use of a volt/amp tester with battery load control and an inductive amperage pick-up. Follow the manufacturer's instructions on the use of the equipment.

1. Start the engine and allow it to reach normal operating temperature.
2. Turn OFF all electrical accessories.
3. Connect the tester to the battery terminals and cable according to the instructions.
4. Bring the engine to a high idle, of approximately 2,000 rpm.
5. Apply a load to the charging system with the rheostat on the tester. Do not let the voltage drop below 12 volts.
6. The alternator should deliver to within 10 percent of the rated output. If the amperage is not within 10 percent and all other components test properly, replace the alternator.

Alternator

Yanmar offers alternators of various outputs, each designed for providing power to the electrical applications on your particular vessel. Upgrading your alternator to one of higher output may be as simple as requesting a higher output unit for replacement from your local Yanmar parts dealer.

REMOVAL & INSTALLATION

▶ **See Figures 27 thru 36**

➡If you decide to use a generic brand of alternator on your engine, make absolutely certain that the alternator is certified for marine use.

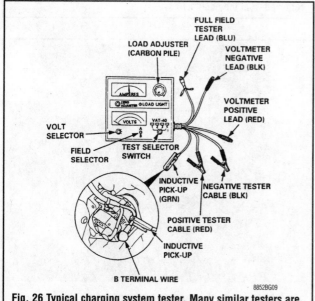

Fig. 26 Typical charging system tester. Many similar testers are available that perform equally as well

Automotive alternators do not meet United States Coast Guard (USCG) standards because they do not have shielding for the contacts to prevent sparks. Under no circumstances should an automotive-type alternator be used on a marine engine. If an automotive alternator (or any other non USCG approved electrical device) is used on your marine engine, and it catches fire, your insurance company will NOT cover your loss, and you may be issued a heavy fine by the USCG.

1. Remove any components necessary for access to the alternator.
2. Disconnect the negative battery cable.
3. Loosen the alternator pivot nut, followed by the adjustment nut.
4. Pivot the alternator so that the belt can be removed.
5. Remove the drive belt from the alternator pulley.

➡In some cases, it may be easier to disconnect the wiring after the alternator has been removed. Be sure to support the alternator by hand while removing the wiring.

6. Label and disconnect the wiring from the alternator. If necessary, mark the wires with paint or tape to identify them for proper installation.
7. Remove the pivot and adjustment bolts, and remove the alternator.

Fig. 27 Before removing the alternator from the engine, remove the wires and connectors from the back of the alternator

Fig. 28 When loosening the wire connections, use a backup wrench on the terminal to keep it from spinning. Spinning a terminal can damage the alternator internally

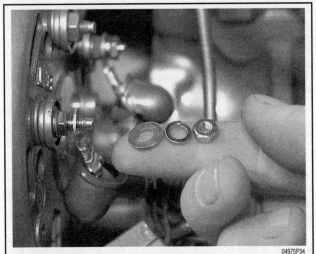

Fig. 29 Note that the terminals use two washers; be sure to install them in the proper order

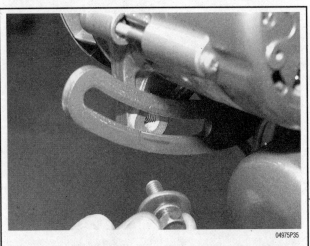

Fig. 30 Once the wire terminals are disconnected, The alternator can be unbolted from the engine. First, remove the adjustment bolt . . .

Fig. 31 . . . then loosen the alternator pivot bolt . . .

Fig. 32 . . . and slide it out of the bracket

Fig. 33 Once the bolts are removed, the alternator can be removed from the engine

Fig. 34 When installing the alternator, the rubber terminal covers should not just be pushed over the terminal . . .

Fig. 35 . . . but should be completely sealed against the terminal base. This prevents the entry of moisture

Fig. 36 Inspect the drive belt and replace it with a new one if it is found to be worn

To install:

→If necessary, attach the wiring to the alternator before installation.

8. Install the alternator and attach the wiring using the markings made during disassembly. Be sure to attach the wires to their proper locations.
9. Install the drive belt on the alternator pulley and properly tension.
10. Connect the negative battery cable.

CONTROL PANEL

♦ See Figures 37, 38, 39 and 40

Since an engine is so far below deck, it can be difficult to hear problems when they first start. A problem may go unnoticed for a considerable amount of time, causing a considerable amount of damage to the engine. The best way to keep track of your engine's condition is to use the gauges and warning circuits.

BELT TENSIONING

Drive belts are often overlooked when diagnosing a charging system failure. Check the belt tension on the alternator pulley and replace/adjust the belt. A loose belt will result in an undercharged battery and a no-start condition. Refer to the "Maintenance Section" for information on proper belt tensioning.

Control panels usually contain a number of indicating devices (gauges and warning lights). These devices are composed of two separate components. One is the sending unit, mounted on the engine and the other is the actual gauge or light in the instrument panel.

Several types of sending units exist, however most can be characterized as being either a pressure-type or a resistance-type. Pressure-type sending units convert liquid pressure into an electrical signal

Fig. 37 The typical control panel contains several gauges and warning lights in addition to the key switch—Type "B" control panel shown

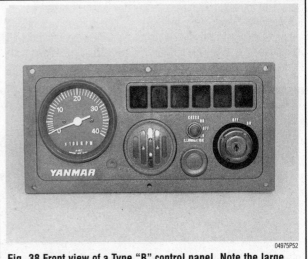

Fig. 38 Front view of a Type "B" control panel. Note the large warning buzzer at the center of the panel

Fig. 39 This Type "C" control panel has enough gauges to keep any helmsman aware of the engines condition

Fig. 40 Some control panel circuits are protected by fuses. These fuses are accessed by removing the cap on the control panel

which is sent to the gauge. Resistance-type sending units are most often used to measure temperature and use variable resistance to control the current flow back to the indicating device. Both types of sending units are connected in series by a wire to the battery (through the ignition switch). When the ignition is turned **ON**, current flows from the battery through the indicating device and on to the sending unit.

Hour Meter Gauge

▶ See Figure 41

An hour meter is used to help keep track of maintenance intervals, which ensures a long life for the engine. Basically, an hour meter is operated by a voltage signal when the engine is in use, and runs like a clock. When the engine is turned off, the voltage is cut to the hour meter, and it simply stops until the next time it is supplied voltage.

TESTING

Testing of an hour meter is simply a matter of checking if voltage is being supplied to the meter when the engine is operating.

Fig. 41 Hour meters are essential for keeping track of maintenance intervals

1. Remove the control panel from the housing, it is not necessary to disconnect any wires.
2. Place the positive lead of the multimeter to the positive terminal on the back of the hour meter. Place the negative terminal of the multimeter to the ground terminal on the back of the hour meter.
3. Turn the key switch to the **ON** position.
4. There should be battery voltage between the two terminals.
5. If there is no voltage, check the ignition switch for proper operation.
6. If voltage is being supplied to the meter, and the meter is not functioning, the meter is faulty and should be replaced.

Oil Pressure Gauge

▶ See Figure 42

The oil pressure gauge indicates the pressure of the lubricating system. The pressure is usually measured just after the oil pump.

Depending on the rpm and condition of the engine, oil pressure should be constant, somewhere around 30–40 psi (207–275 kPa)

Fig. 42 Oil pressure gauges offer accurate monitoring of oil pressure

at cruising speed, and less at idle. Under load, you can expect the oil pressure to rise slightly and to fall off with deceleration.

At fast idle, when the engine (and oil) are cold, the pressure will probably be at maximum on the gauge. As the engine warms, it is not uncommon to see a slight decrease in oil pressure.

Low oil pressure can warn of low oil level, wrong viscosity oil, overheating, clogged oil filter or worn engine (high hours).

TESTING

Most diesel engines today use an electric gauge with a sending unit mounted on the engine. The following procedure tests the gauge sending unit, which is the most likely item to fail.

➡On JH series engines, if the oil pressure gauge reads "0" and the oil pressure warning lamp and buzzer are not functioning, the oil pressure sending unit adapter is clogged, and must be cleaned. Typically, this occurs every 500–600 hours of engine operation.

1. The sending unit is most efficiently diagnosed by removing it from the engine and replacing it with a mechanical type oil pressure gauge.
2. Label and disconnect the sending unit wiring.
3. Place a catch pan under the sending unit to catch any oil that may leak from the engine when the unit is removed.
4. Using an appropriately sized wrench, loosen the unit using the fitting at the base of the gauge near the engine.

➡Do not turn the gauge using the sending unit housing. Doing so will damage the unit.

5. Replace the sending unit with an mechanical oil pressure gauge. Follow the manufacturer's instructions for installation.
6. Start the engine and note the pressure reading on the gauge. If correct oil pressure is noted, the sending unit may be faulty.
7. Prior to purchasing a new sending unit, carefully inspect all connections for corrosion. Gauge testers are also available to properly test the control panel gauge.

REMOVAL & INSTALLATION

♦ See Figures 43 and 44

1. Label and disconnect the sending unit wiring.
2. Place a catch pan under the sending unit to catch any oil that may leak from the engine when the unit is removed.

Fig. 43 The oil pressure sending unit converts mechanical pressure to an electric signal. It can easily be recognized by its can-like appearance

Fig. 44 After installing the wiring on the unit, replace the protecting cover to prevent corrosion from forming on the terminal

3. Using an appropriately sized wrench, loosen the unit using the fitting at its base near the engine.

➡Do not turn the sending unit using the housing, doing so will damage the unit.

To install:
4. Remove the sending unit from the engine.
5. Prepare the new sending unit by applying Teflon® sealant to the threads to prevent the unit from leaking.

➡On some replacement sending units this may have already been done at the factory. If this is the case, do not apply additional sealant.

6. Carefully thread the sending unit into the engine, taking care not to crossthread.
7. Tighten the sending unit securely.
8. Connect the wiring and test the gauge for proper operation.

Water Temperature Gauge

♦ See Figure 45

The water temperature gauge is used to monitor the temperature of the fresh water cooling circuit. As the engine warms, the temperature will should rise to approximately 160°F–200°F (83°C–94°C).

If you're under full load in rough seas, the temperature will rise slightly. It will also rise slightly immediately after shutting the engine off, because the coolant is not being cooled, but will return to normal when the engine is started.

Once the engine reaches operating temperature, the gauge should not deviate from its position. Cool temperatures indicate a faulty thermostat. Too hot readings indicate low coolant level, worn hoses, defective radiator cap, incorrect ignition timing or slipping belts.

If the normal operating temperature rises over the course of time, and the above factors are OK, suspect a worn water pump or a clogged system.

TESTING

♦ See Figure 46

Most diesel engines today use an electric gauge with a sending unit mounted on the engine. The following procedure tests the gauge sending unit, which is the most likely item to fail.

Fig. 45 The water temperature gauge will warn you of problem with the cooling system before the engine overheats and is seriously damaged

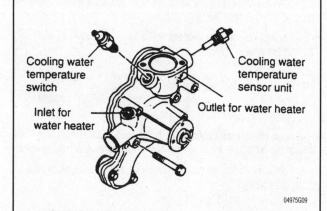

Fig. 46 The cooling temperature sensor (on the right) operates the gauge, while the switch (on the left) operates the warning lamp

1. Two types of coolant temperature sending units are commonly used in marine engines. The first is the Negative Temperature Coefficient (NTC) unit. In this type of sending unit the resistance of the unit will rise as the temperature of the engine drops. In the second type, a Positive Temperature Coefficient (PTC) unit, resistance will rise as engine temperature rises.

2. To test the sending units, label and disconnect the wiring.

3. Using a multimeter set to the ohms setting, probe the sending unit terminals (on two terminal models) or probe between the sending unit and a good ground. Note the reading.

4. Start the engine and allow it to warm to operating temperature.

5. Monitor the resistance as the engine warms and ensure that the resistance changes smoothly. Remember, if the resistance was high with the engine cold it should gradually lower. If resistance was low with the engine cold, it should gradually rise.

6. If resistance is does not change smoothly or in the proper direction, the sending unit is faulty and should be replaced.

REMOVAL & INSTALLATION

1. Label and disconnect the sending unit wiring.
2. Drain the coolant below the level of the sending unit.
3. Place a catch pan under the sending unit to catch any coolant that may leak from the engine when the unit is removed.
4. Using an appropriately sized wrench, loosen the unit using the fitting at its base near the engine.

➡ Do not turn the sending unit using the housing, doing so will damage the unit.

5. Remove the sending unit from the engine.

To install:

6. Prepare the new sending unit by applying Teflon® sealant to the threads to prevent the unit from leaking.

➡ On some replacement sending units this may have already been done at the factory. If this is the case, do not apply additional sealant.

7. Carefully thread the sending unit into the engine, taking care not to crossthread.
8. Tighten the sending unit securely.
9. Connect the wiring and test the gauge for proper operation.

Tachometer

◆ See Figures 47 and 48

The tachometer indicates the number of revolutions per minute of the engine by means of a pulse signal generated by the teeth on the

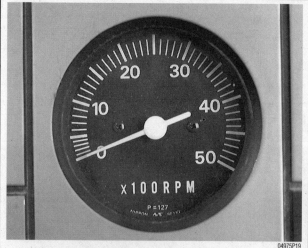

Fig. 47 The tachometer indicates the number of revolutions per minute the engine is rotating

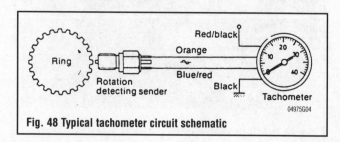

Fig. 48 Typical tachometer circuit schematic

flywheel. A magnetic pickup sender mounted on the flywheel housing converts the rotary motion of the flywheel into an electrical signal, and sends the signal to the tachometer gauge on the control panel.

TESTING

▶ **See Figures 49, 50 and 51**

Most diesel engines today use a tachometer with a sending unit mounted on the engine. The following procedure tests the sending unit, which is the most likely item to fail.

1. Label and disconnect the electrical wires from both terminals from the sender on the flywheel housing.

2. Using a multimeter, measure the resistance of the sender. The resistance should be 1.6 kilohms.

3. If the resistance is not correct, the sender is faulty and should be replaced.

4. If the resistance is correct, and the tachometer fails to function, check the wires leading to the tachometer, making sure they are not broken or disconnected.

Fig. 51 . . . and attach a multimeter to the sender. The reading should be approximately 1.6 kilohms; this sender is functioning properly

5. If the sender wires are intact and connected properly, the tachometer gauge may be faulty.

REMOVAL & INSTALLATION

▶ **See Figure 52**

The magnetic pickup sender is located on the flywheel housing of the engine.

1. Label and disconnect the electrical wires from both terminals from the sender.

2. Using an appropriately sized wrench, loosen the sender unit by turning it at the fitting near the base. Attempting to turn the sender near the top may damage the sender.

3. Once loose, the sender should unscrew from the block easily.

To install:

4. Install the sender and screw it into the engine block. Tighten the sender securely.

5. Connect the electrical wires to their proper terminals on the sender.

6. Start the engine and check the tachometer for proper operation.

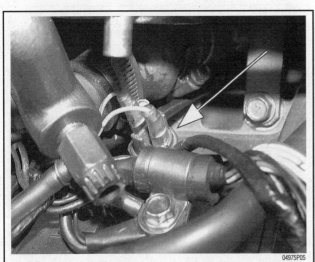

Fig. 49 Though not easily seen, the tachometer sender on this JH engine is mounted on the flywheel housing

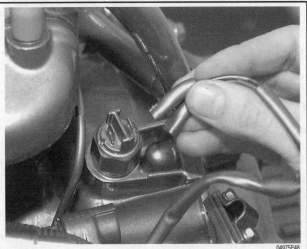

Fig. 50 To test the tachometer sender, disconnect the wire terminals . . .

Fig. 52 The sender is replaced by simply unscrewing it from the engine block

Water Temperature Warning Lamp

♦ See Figure 53

The water temperature warning lamp is connected to a switch that is activated when the coolant temperature within the engine reaches a predetermined critical temperature. When the coolant rises to the predetermined temperature (well above normal operating temperature), the switch is activated, and completes a circuit, causing the warning light on the control panel to illuminate and the alarm buzzer to sound.

✳✳ WARNING

If the water temperature warning lamp and/or alarm buzzer are activated when the engine is operating, TURN THE ENGINE OFF IMMEDIATELY! The engine is overheating, and will be severely damaged if the engine continues to operate.

TESTING

♦ See Figures 54, 55 and 56

If the water temperature warning lamp and/or alarm buzzer fail to operate when the water temperature gauge indicates that the

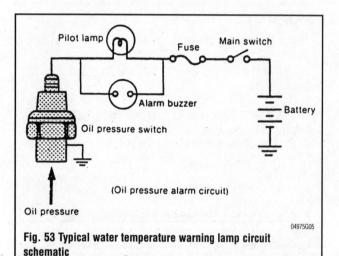
Fig. 53 Typical water temperature warning lamp circuit schematic

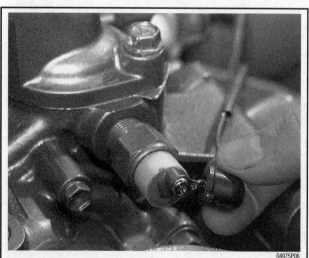
Fig. 54 To test the water temperature warning lamp, remove the connector boot . . .

Fig. 55 . . . and disconnect the wire terminal from the switch. With the key on the control panel in the ON position . . .

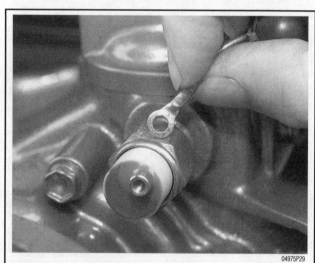
Fig. 56 . . . touch the terminal to the side of the switch; the light on the control panel should illuminate

engine is overheating, the fuse on the control panel should be checked first. Secondly, the warning light bulb on the control panel should be checked. If the system still fails to operate properly, the switch can be checked by performing the following procedure.

1. Turn the key switch on the control panel to the **ON** position.
2. Disconnect the wire lead from the cooling water temperature switch on the engine.
3. Ground the cooling water temperature switch wire by touching the terminal on the side of the switch. Make sure that there is not paint on the side of the switch, or the circuit will not ground.
4. The warning light on the control panel and the alarm buzzer (if equipped) should be working.
5. If the warning light and/or alarm buzzer are working, and the engine has NOT been overheating, the cooling water switch can be considered faulty and should be replaced.
6. If the warning light and/or alarm buzzer are not activated when the water temperature switch is grounded, check the supply of voltage to the light, and to the alarm buzzer.

REMOVAL & INSTALLATION

▶ **See Figures 57 and 58**

1. Label and disconnect the wiring at the switch on the engine.
2. Drain the coolant below the level of the switch.
3. Place a catch pan under the switch to catch any coolant that may leak from the engine when the unit is removed.
4. Using an appropriately sized wrench, loosen the unit using the fitting at its base near the engine.

➡ **Do not turn the switch using the housing, doing so will damage the unit.**

5. Remove the switch from the engine.

To install:

6. Prepare the new switch by applying Teflon® sealant to the threads to prevent the unit from leaking.

➡ **On some replacement sending units this may have already been done at the factory. If this is the case, do not apply additional sealant.**

Fig. 57 Water temperature warning switches are usually located near the thermostat housing . . .

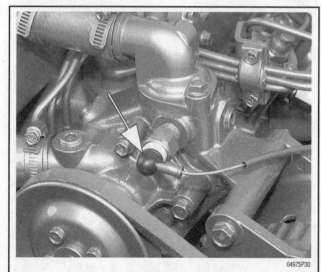

Fig. 58 . . . or on the thermostat housing itself

7. Carefully thread the switch into the engine, taking care not to crossthread.
8. Tighten the switch securely.
9. Connect the wiring and test the gauge for proper operation.

BULB REPLACEMENT

▶ **See Figures 59 and 60**

It is common for warning lamp bulbs to burn out. This happens primarily due to the harsh conditions which the lamps encounter out on the water. Saltwater, vibration and the heat of the sun baking on a black control panel all contribute to their early failure.

It is important for the helmsman to visually inspect the bulbs during the "lamp check" each time the ignition switch is turned to the **ON** position. If it is noted that a bulb is burned out, it should be replaced immediately. It is always a good idea to carry spare bulbs.

Replacing the bulbs is as simple as pulling the socket from the housing on the rear of the control panel and pull the bulb straight out of the socket. Prior to installing the new bulb, take the time to clean the socket if corrosion is noted.

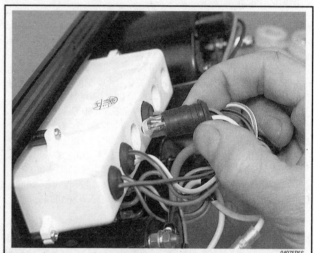

Fig. 59 To replace a warning lamp bulb, pull the socket from the housing on the rear of the control panel . . .

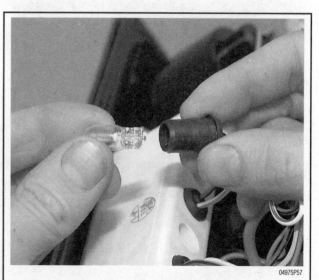

Fig. 60 . . . and pull the bulb straight out of the socket

Charge Lamp

The charge lamp on the instrument panel is used to indicate when the alternator is charging properly. The signal from the charge lamp light comes directly from the alternator.

TESTING

1. When the charge lamp is operating properly, it will illuminate when the key switch on the control panel is in the **ON** position.
2. If the charge lamp does not illuminate when the key switch is turned to the **ON** position, the bulb should be checked, and replaced if necessary.
3. Once the engine is started and brought to a moderate idle, the light should not illuminate.
4. If the light continues to be illuminated after the engine is running, the alternator is not charging properly.
5. Check the drive belt for proper tension.
6. If drive belt tension is within specification, perform the charging system test procedures.

BULB REPLACEMENT

▶ **See Figures 59 and 60**

It is common for warning lamp bulbs to burn out. This happens primarily due to the harsh conditions which the lamps encounter out on the water. Saltwater, vibration and the heat of the sun baking on a black control panel all contribute to their early failure.

It is important for the helmsman to visually inspect the bulbs during the "lamp check" each time the ignition switch is turned to the **ON** position. If it is noted that a bulb is burned out, it should be replaced immediately. It is always a good idea to carry spare bulbs.

Replacing the bulbs is as simple as pulling the socket from the housing on the rear of the control panel and pull the bulb straight out of the socket. Prior to installing the new bulb, take the time to clean the socket if corrosion is noted.

Oil Pressure Warning Lamp

▶ **See Figure 61**

The oil pressure warning lamp is connected to an engine mounted switch that is activated when the oil pressure drops below a predetermined limit. When the key switch on the control panel is turned to the **ON** position, the oil pressure warning lamp (and alarm buzzer, if equipped) will be activated. When the engine is started,

the oil pressure increases, opening the circuit which illuminates the oil pressure warning lamp. If the engine oil pressure drops to a low enough pressure while the engine is operating, the warning lamp will illuminate, and the buzzer will sound (if equipped).

If the oil pressure lamp (or buzzer) is activated when the engine is operating, TURN THE ENGINE OFF IMMEDIATELY! The oil pressure within the engine has dropped below tolerable limits, and may not be able to supply adequate lubrication to the internal components. If the warning light is illuminated and ignored, the engine may be completely destroyed by lack of lubrication.

TESTING

▶ **See Figures 62, 63 and 64**

1. Turn the key switch on the control panel to the **ON** position.
2. Disconnect the wire lead from the oil pressure switch on the engine.
3. Ground the cooling water temperature switch wire by touching the terminal on the side of the switch. Make sure that there is not paint on the side of the switch, or the circuit will not ground.

Fig. 62 The oil pressure switch is usually located near the oil filter. To test the switch . . .

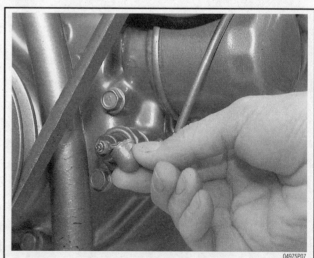
Fig. 63 . . . remove the rubber boot, disconnect the terminal . . .

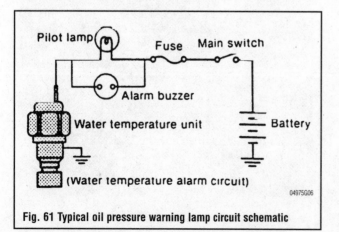
Fig. 61 Typical oil pressure warning lamp circuit schematic

Fig. 64 . . . and touch it against the side of the switch; the light on the control panel should illuminate

Fig. 65 To remove the alarm buzzer from the control panel, unscrew the ring on the rear . . .

4. With the wire grounded, and the key switch in the **ON** position, the warning light on the control panel and the alarm buzzer (if equipped) should be working.

5. If the warning light and/or alarm buzzer are not activated when the water temperature switch is grounded, check the supply of voltage to the light, and to the alarm buzzer.

6. If the warning circuit is operating properly, and the warning light and buzzer continue to operate when the engine is running, immediately turn the engine off, and have a qualified marine technician attach a mechanical oil pressure gauge to the engine to verify the oil pressure is inadequate.

BULB REPLACEMENT

▶ See Figures 59 and 60

It is common for warning lamp bulbs to burn out. This happens primarily due to the harsh conditions which the lamps encounter out on the water. Saltwater, vibration and the heat of the sun baking on a black control panel all contribute to their early failure.

It is important for the helmsman to visually inspect the bulbs during the "lamp check" each time the ignition switch is turned to the **ON** position. If it is noted that a bulb is burned out, it should be replaced immediately. It is always a good idea to carry spare bulbs.

Replacing the bulbs is as simple as pulling the socket from the housing on the rear of the control panel and pull the bulb straight out of the socket. Prior to installing the new bulb, take the time to clean the socket if corrosion is noted.

Alarm Buzzer

The alarm buzzer is integrated into the control panel warning lamp circuits. The alarm buzzer is essentially a secondary warning device to indicate abnormal engine operating conditions in case the vessel operator is not within view of the control panel.

TESTING

▶ See Figures 65, 66 and 67

1. Remove the control panel from its housing to access the rear of the buzzer. In some cases, it may be easier to remove the buzzer from the panel by unscrewing the retaining ring.

Fig. 66 . . . and pull the buzzer from the front of the panel

Fig. 67 Using a DC 12 volt power source, apply positive voltage to the top terminal, while touching the negative on the adjacent terminals

2. Unplug the connector from the rear of the buzzer.
3. Connect the positive terminal of a 12 volt DC power source to the **top** terminal of the buzzer The connector of the top terminal should have a red wire with a black stripe attached to the lead.
4. With the negative side of the 12 volt DC power source, touch the remaining terminals; the buzzer should sound.
5. If the buzzer does not sound, it may be faulty.

Switches

▶ **See Figures 68 and 69**

The switches on the control panel can be broken down into two types—toggle switches and momentary contact switches.

A toggle switch is usually has a two position lever which.is supplies current to the accessory when in one position and cuts current in the other. Some toggle switches transfer current between two accessories with the center position being off.

A momentary contact switch only supplies current when the switch is held in position. Once the button or lever is released, the switch contacts are disconnected cutting the flow of electrical cur-

rent. On a Yanmar control panel, momentary switches include the engine start switch, the warning light check switch, and on some installations, an electric engine stop switch.

TESTING

▶ **See Figures 70 and 71**

1. Label and disconnect the wires from the switch.
2. Connect a multimeter between the terminals on the switch.
3. With the switch in the **OFF** position (released), there should be no continuity.
4. Operate the switch while paying attention to the multimeter. There should now be continuity.
5. If continuity does not exist as stated, the switch is faulty.

Control Cables

Mechanical control cables are also a part of the control panel on your vessel. These cables are covered in a separate section of this manual called "Remote Control". Please refer to this section for more information.

Fig. 68 Moving this switch upward allows you to perform a bulb check, while moving it down turns the instrument panel lights on

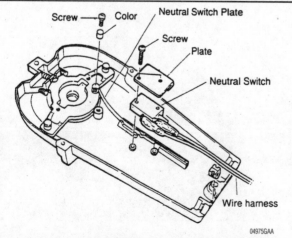

Fig. 69 Some remote control heads contain a neutral safety switch that prevents the engine from being started with the transmission in gear

Fig. 70 The continuity setting on a multimeter can be used to check for proper starter button operation

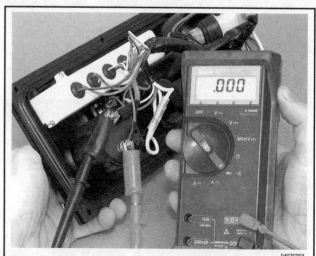

Fig. 71 Connect the multimeter between the two terminals on the switch to check for continuity

OPTIONAL EQUIPMENT

Intake Air Heater

An air heater is available for warming intake air during starting in cold areas during winter. The air heater is mounted between the intake manifold and manifold coupling. A switch on the control panel is used to supply current to the heater.

TESTING

1. Label and disconnect the wires from the heater.
2. Connect a multimeter, set to the ohms range, between the terminals on the heater.
3. Continuity should exist. If continuity does not exist, there is a break in the heating element and the unit is faulty.
4. If continuity exists, check the circuit back to the switch for proper power and ground.
5. Repair the circuit as necessary.

REMOVAL & INSTALLATION

1. Label and disconnect the wires from the heater unit.
2. Remove the air intake silencer and all other components necessary to gain access to the intake manifold coupling.
3. Note the position of the heating element in the intake tract and matchmark it for installation reference.
4. Remove the through bolts attaching the intake joint and heater element to the intake manifold.
5. Carefully remove the intake joint and heater element from the engine. Take care not to damage the delicate heating elements in the unit.
6. Separate the heater from the intake joint and discard the joint-to-heater and heater-to-intake manifold gaskets.
   To install:
7. Using new gaskets, assemble the heater and joint unit on the intake manifold.
8. Insert the through bolts and tighten securely using a criss-cross pattern to compress the gasket properly.
9. Install the air intake silencer and all other components previously removed.
10. Connect the wires to the heater unit.
11. Test the intake heater for proper operation.

Electric Engine Stopping Device

▶ **See Figures 72 and 73**

Some engines are fitted with an electric stop device that is used to cut fuel flow to the engine by means of an engine mounted solenoid. On these engines, a mechanical engine stop cable is not used.

The electric engine stopping system consists of an engine stop solenoid mounted on the injection pump, a stop button mounted on the control panel, a relay and associated wiring.

TESTING

1. Label and disconnect the wires from the solenoid.
2. Connect a multimeter, set to the ohms range, between the terminals on the solenoid.
3. Continuity should exist. If continuity does not exist, the solenoid is faulty.
4. If continuity exists, check the circuit back to the switch for

Fig. 72 The electric engine stopping device on this JH series engine is actually a solenoid that operates the engine stop lever

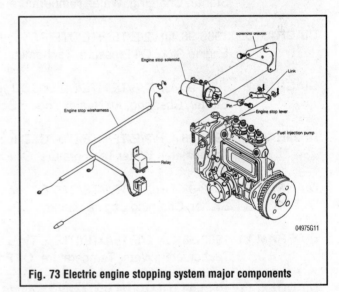

Fig. 73 Electric engine stopping system major components

proper power and ground. This includes inspecting and as necessary replacing the circuit relay.
5. Repair the circuit as necessary.

REMOVAL & INSTALLATION

1. Label and disconnect the wires from the stopping device.
2. Remove the pin which connects the stopping solenoid to the engine stop lever.
3. Remove the bolts attaching the stopping solenoid to the solenoid bracket.
4. Remove the solenoid from the injection pump.
   To install:
5. Install the solenoid on the injection pump and tighten the solenoid to bracket bolts securely.
6. Install the pin which connects the solenoid to the engine stop lever.
7. Connect the wires to the stopping device.
8. Test the electric stopping system for proper operation.

WIRING DIAGRAMS

→The following wiring diagrams cover circuits for Yanmar engines and control panels. However, due to the many different combinations of boats and engines, the wiring diagrams shown here are representative and my not illustrate all circuits on your particular vessel.

INDEX OF WIRING DIAGRAMS

04975W01

SAMPLE DIAGRAM: HOW TO READ & INTERPRET WIRING DIAGRAMS

WIRE COLOR ABBREVIATIONS

BLACK	B	PINK	PK
BROWN	BR	PURPLE	P
RED	R	GREEN	G
ORANGE	O	WHITE	W
YELLOW	Y	LIGHT BLUE	LBL
GRAY	GY	LIGHT GREEN	LG
BLUE	BL	DARK GREEN	DG
VIOLET	V	DARK BLUE	DBL
TAN	T	NO COLOR AVAILABLE-	NCA

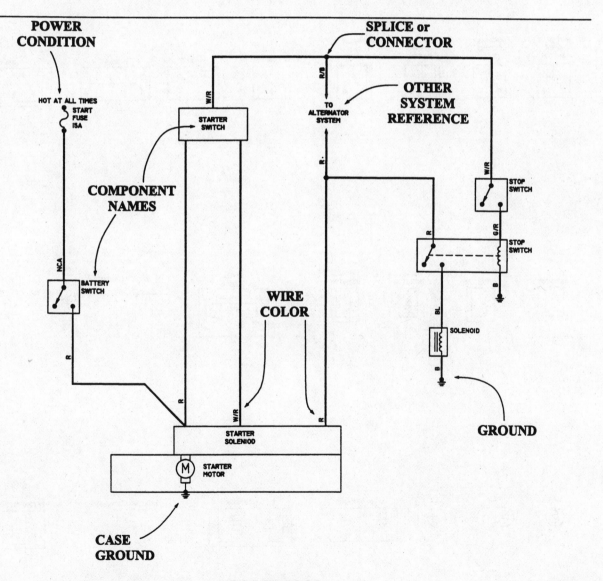

DIAGRAM 1

04975W02

WIRING DIAGRAM SYMBOLS

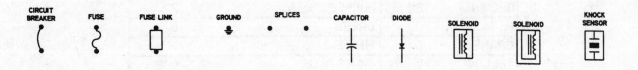

CIRCUIT BREAKER   FUSE   FUSE LINK   GROUND   SPLICES   CAPACITOR   DIODE   SOLENOID   SOLENOID   KNOCK SENSOR

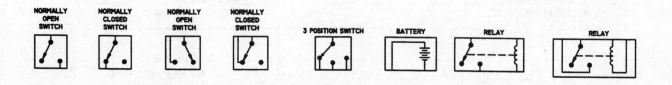

NORMALLY OPEN SWITCH   NORMALLY CLOSED SWITCH   NORMALLY OPEN SWITCH   NORMALLY CLOSED SWITCH   3 POSITION SWITCH   BATTERY   RELAY   RELAY

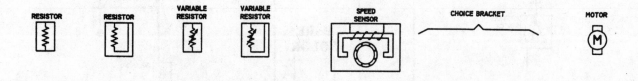

RESISTOR   RESISTOR   VARIABLE RESISTOR   VARIABLE RESISTOR   SPEED SENSOR   CHOICE BRACKET   MOTOR

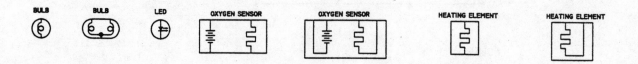

BULB   BULB   LED   OXYGEN SENSOR   OXYGEN SENSOR   HEATING ELEMENT   HEATING ELEMENT

DIAGRAM 2

04975W03

1975-80 2QM20/3QM30 ENGINE SCHEMATICS

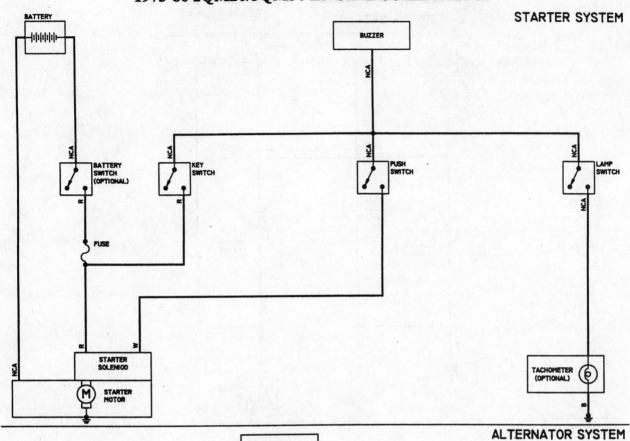

STARTER SYSTEM

ALTERNATOR SYSTEM

DIAGRAM 3

04975W04

1975-80 2QM20/3QM30 ENGINE SCHEMATICS

WATER TEMPERATURE/OIL PRESSURE SYSTEMS

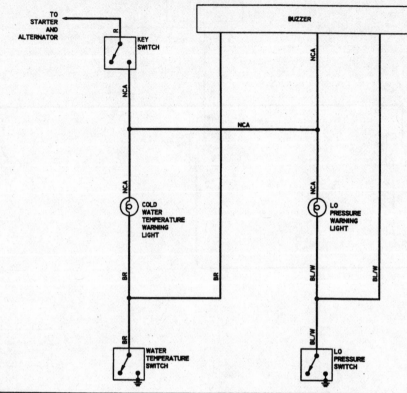

TACHOMETER SYSTEM

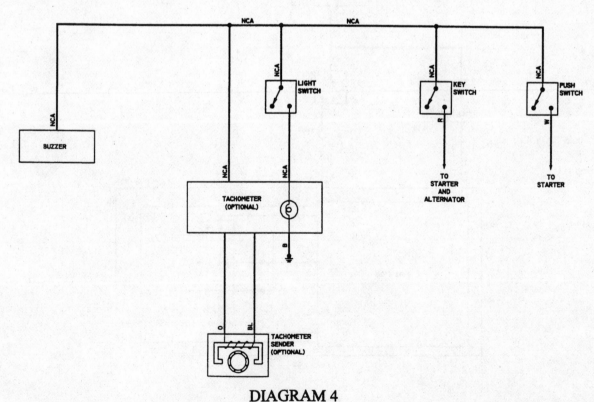

DIAGRAM 4

04975W05

1983-98 4JH/2/E/TE/HTE/DTE ENGINE SCHEMATICS
B TYPE INSTUMENT PANEL

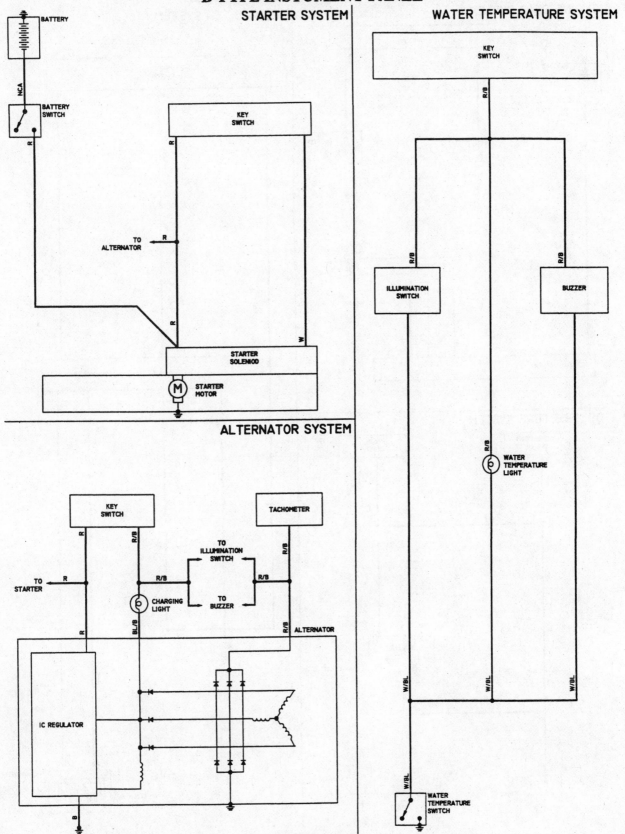

DIAGRAM 5

04975W06

1983-98 4JH/2/E/TE/HTE/DTE ENGINE SCHEMATICS
B TYPE INSTRUMENT PANEL

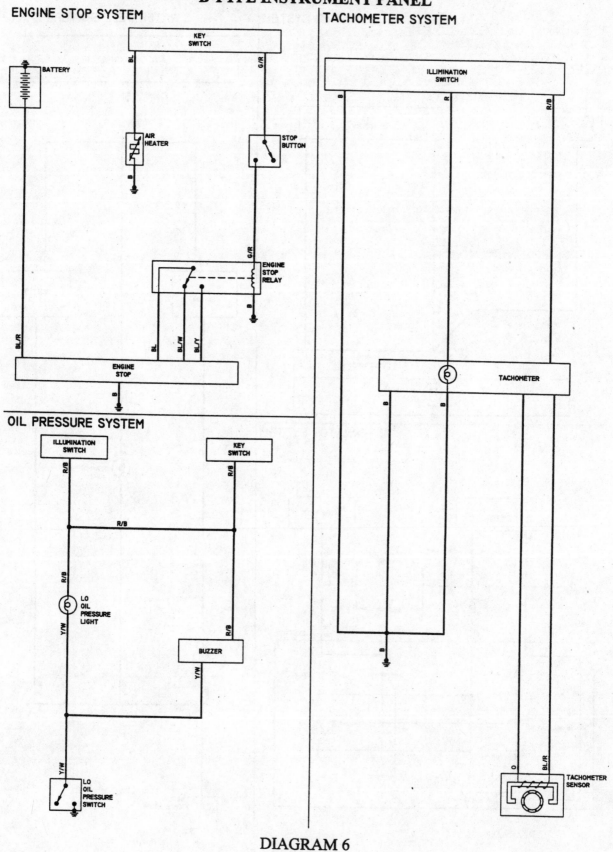

DIAGRAM 6

04975W07

1983-98 4JH/2/E/TE/HTE/DTE ENGINE SCHEMATICS
C&D TYPE INSTRUMENT PANELS

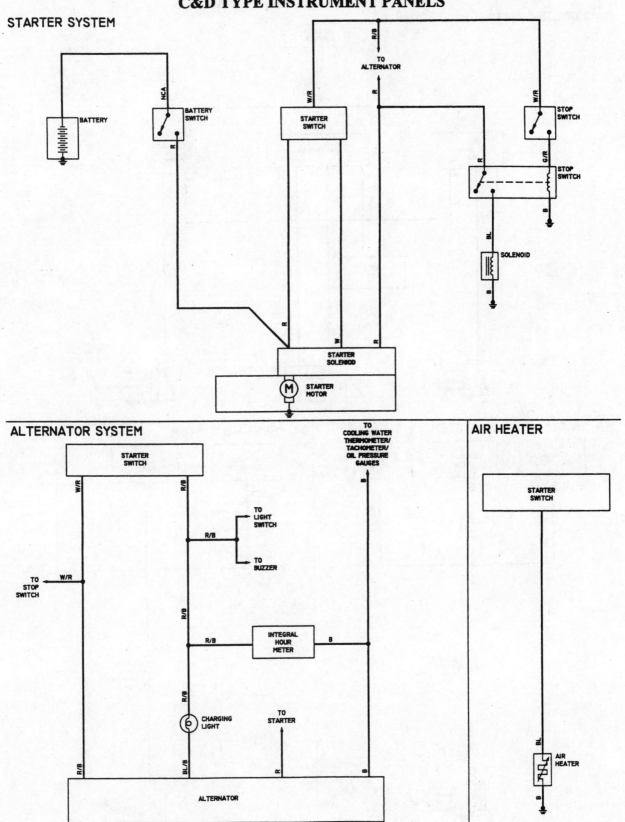

DIAGRAM 7

04975W08

1983-98 4JH/2/E/TE/HTE/DTE ENGINE SCHEMATICS
C&D TYPE INSTRUMENT PANELS

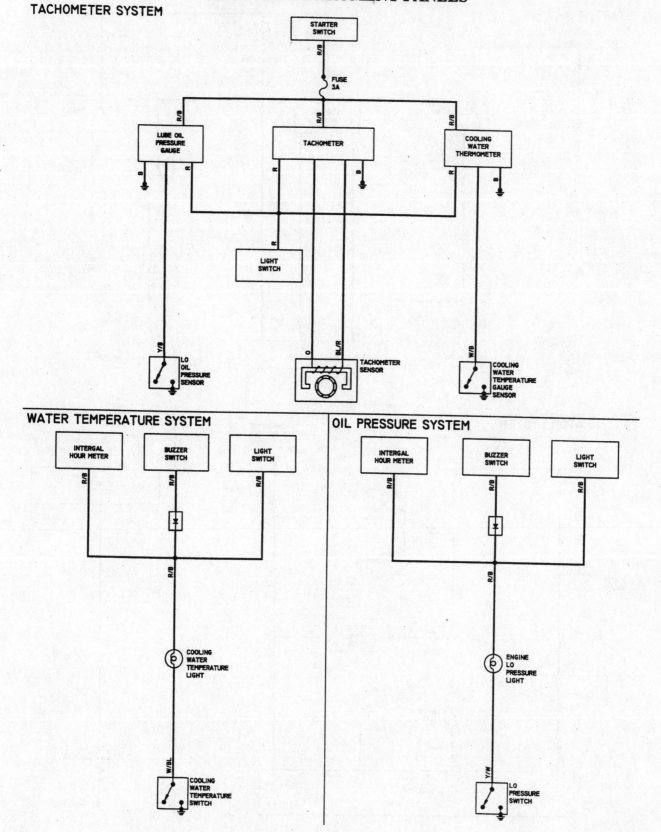

DIAGRAM 8

04975W09

1983-98 4JH/2/E/TE/HTE/DTE ENGINE SCHEMATICS
E TYPE INSTRUMENT PANEL

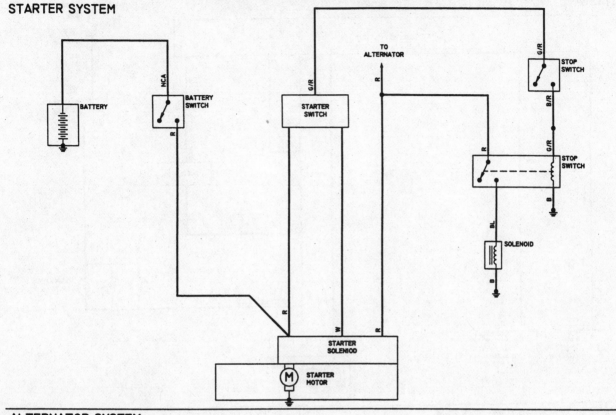

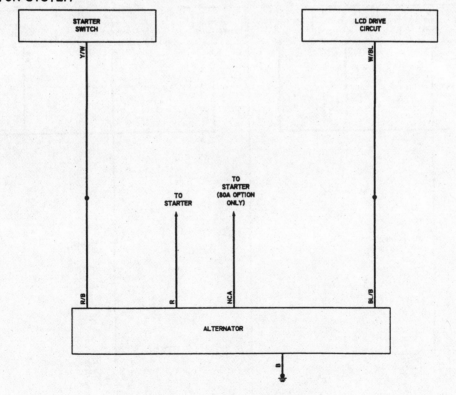

DIAGRAM 9

04975W10

1983-98 4JH/2/E/TE/HTE/DTE ENGINE SCHEMATICS
E TYPE INSTRUMENT PANEL

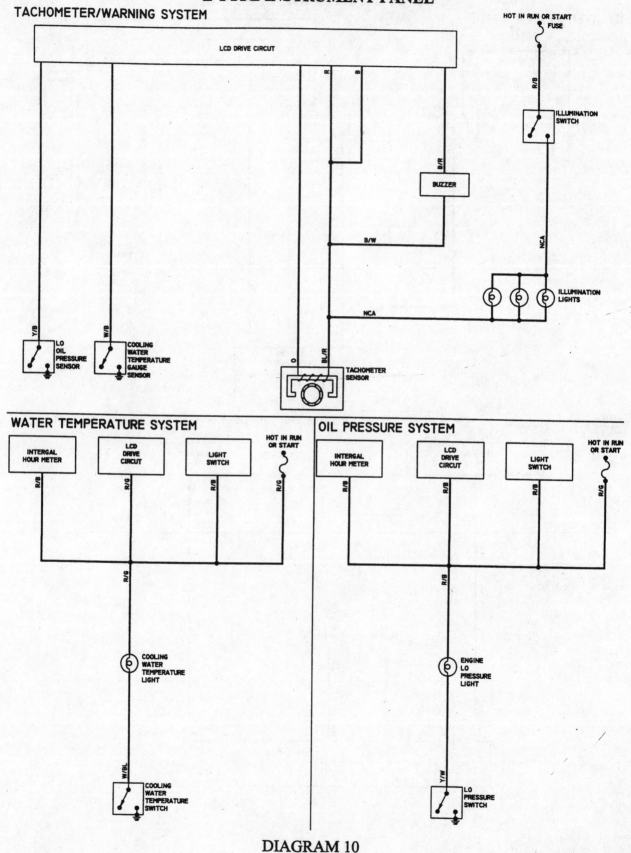

DIAGRAM 10

04975W11

1980-98 1GM10(C/L)/2GM20(C/F/L)/3GM30(C/D/F/L)3HM35(C/F/L) A&B-TYPE INSTRUMENT BOARD ENGINE SCHEMATICS

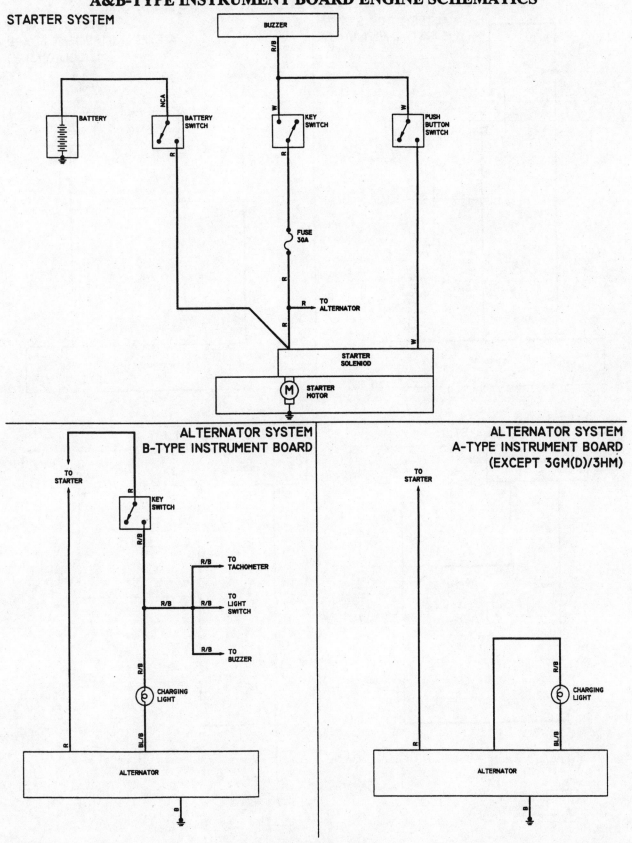

DIAGRAM 11

04975W12

1980-98 1GM10(C/L)/2GM20(C/F/L)/3GM30(C/D/F/L)/3HM35(C/F/L)
A&B-TYPE INSTRUMENT BOARD ENGINE SCHEMATICS

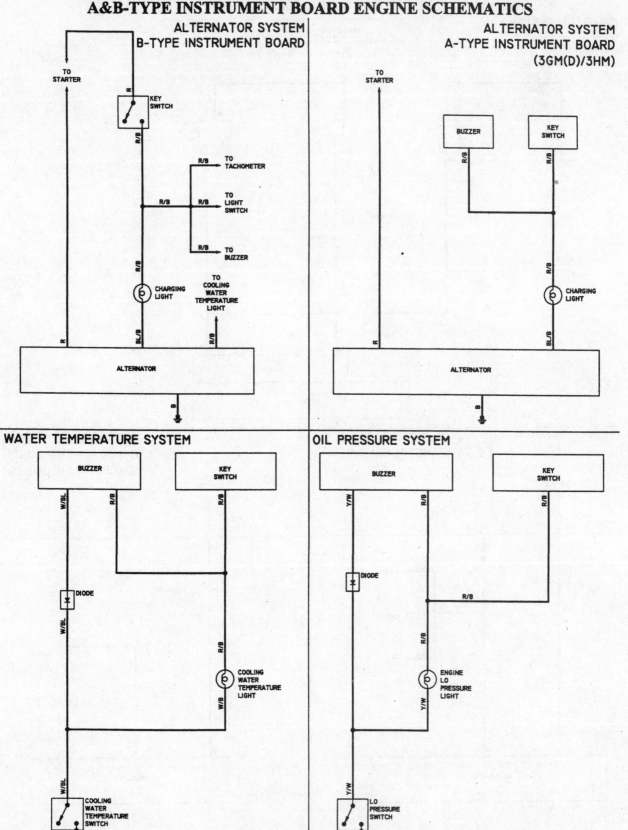

DIAGRAM 12

04975W13

1980-98 1GM10(C/L)/2GM20(C/F/L)/3GM30(C/D/F/L)/3HM35(C/F/L) B-TYPE INSTRUMENT BOARD ENGINE SCHEMATICS

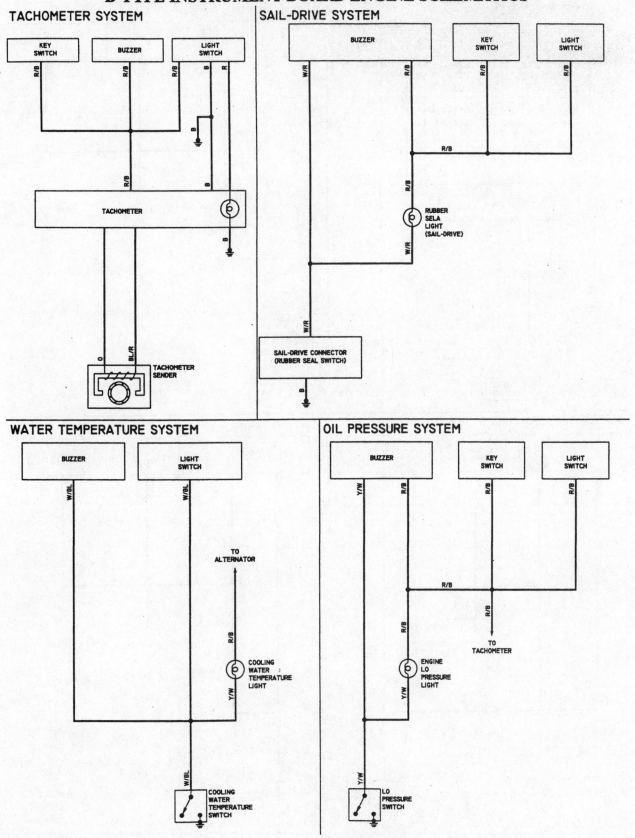

DIAGRAM 13

04975W14

1980-98 1GM10(L)/2GM(F/L)/3GM(D/F/L)/3HM(F/L)
B-TYPE INSTRUMENT BOARD ENGINE SCHEMATICS

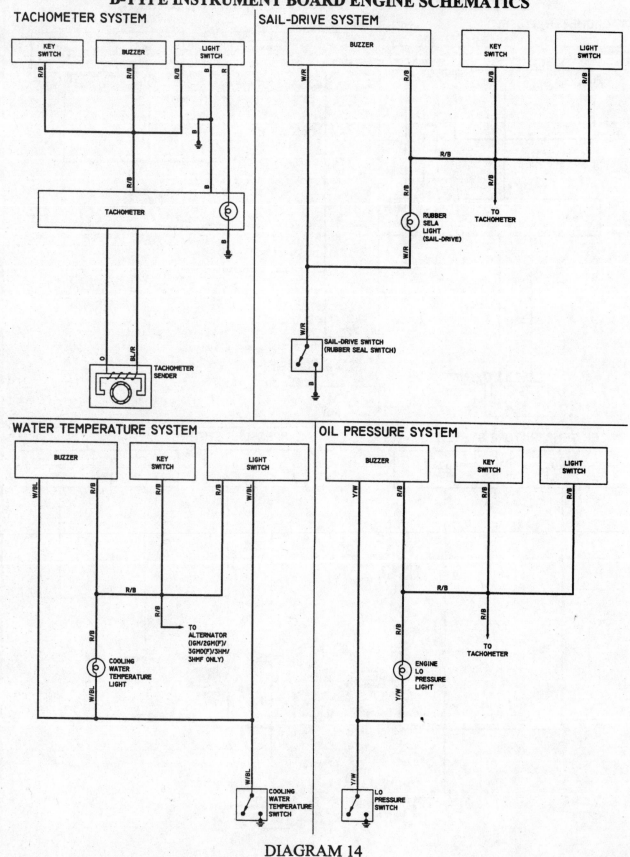

DIAGRAM 14

04975W15

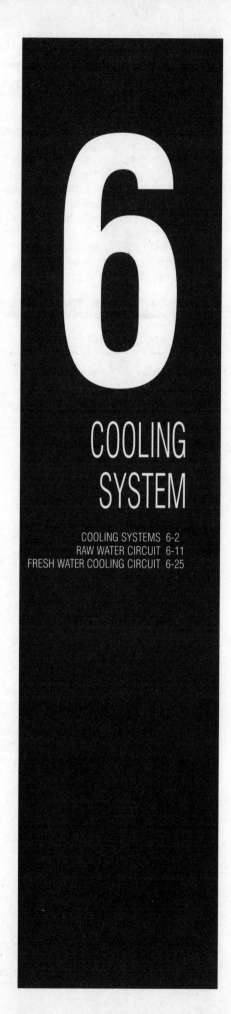

6

COOLING
SYSTEM

COOLING SYSTEMS

Diesel engines generate a lot of heat. Only 33 percent of which is converted to useful work. The remaining percentage must somehow be released to the environment so that temperatures in the engine compartment do not become dangerously high. Excessively high temperatures can break do much harm to an engine. Lubricating oils break down, causing the engine to seize and excessive coolant temperature can cause the cylinder head to crack.

About 50 percent of the excess heat the engine produces goes out with the exhaust gases. A similar amount is carried away by the cooling system. Any remainder radiates from the engine's hot surfaces.

Three principal types of cooling systems are used in boats: raw water, heat exchangers, and keel coolers.

Raw-Water Cooling

DESCRIPTION & OPERATION

▶ **See Figures 1, 2, 3, 4 and 5**

Raw-water cooling systems draw directly from the body of water on which the boat floats. With raw-water cooling, water enters through a seacock, passes through a strainer, circulates through the engine and oil coolers , and finally discharges overboard. On most 4-cycle diesels, after the water circulates through the engine, it enters the exhaust pipe and discharges with the exhaust gases. This is called a wet exhaust.

A raw-water cooling system is simple and economical to install, but it has a number of drawbacks. Over time a certain amount of silt inevitably finds its way into the engine and coolers and begins to plug cooling passages.

Regulating the engine's temperature in a raw-water system is difficult. The temperature of the raw-water may vary from the freezing point in northern climates in the winter to 90°F (32°C) in tropical climates in the summer. Raw-water systems sometimes do not use a thermostat because silt can clog it and cause it to malfunction over time the. This complicates engine temperature regulation even more.

In salt water, salts crystallize in the hottest parts of the cooling system, notably around cylinder walls and in cylinder heads. This leads to a reduction in cooling efficiency and localized hot spots. The rate of scale formation is related to the temperature of the water, and accelerates when coolant temperatures are above 160°F (71°C).

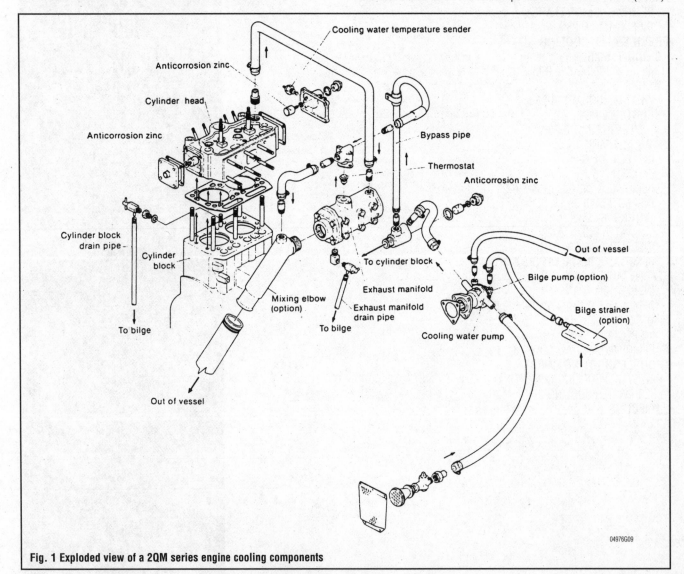

04976G09

Fig. 1 Exploded view of a 2QM series engine cooling components

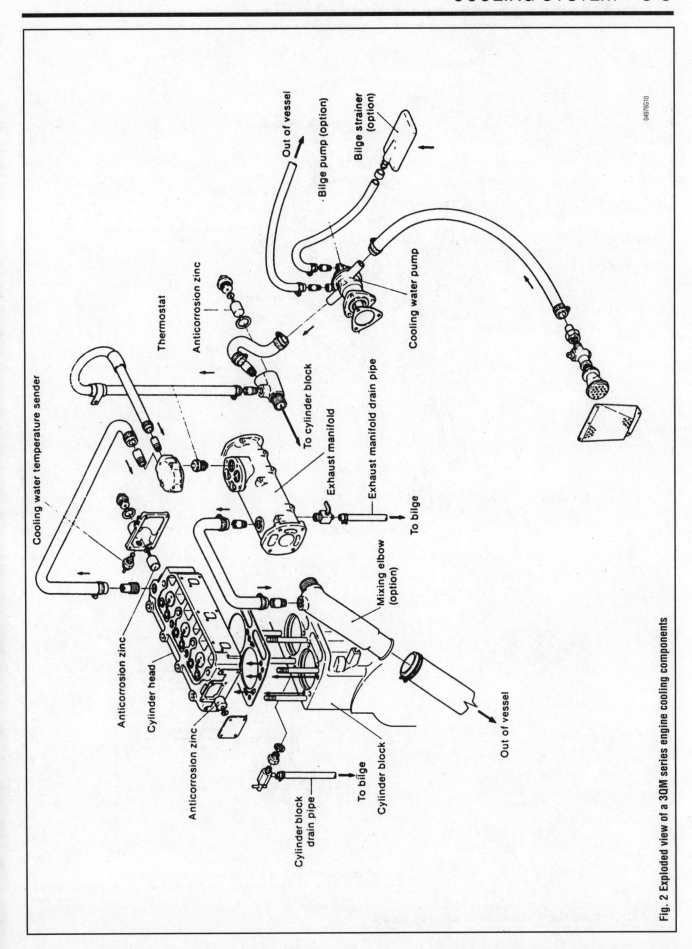

Out of vessel

Bilge pump (option)

Bilge strainer (option)

Thermostat

Anticorrosion zinc

Cooling water pump

Cooling water temperature sender

To cylinder block

Exhaust manifold

Exhaust manifold drain pipe

To bilge

Mixing elbow (option)

Anticorrosion zinc

Cylinder head

Anticorrosion zinc

Cylinder block drain pipe

To bilge

Cylinder block

Out of vessel

Fig. 2 Exploded view of a 3QM series engine cooling components

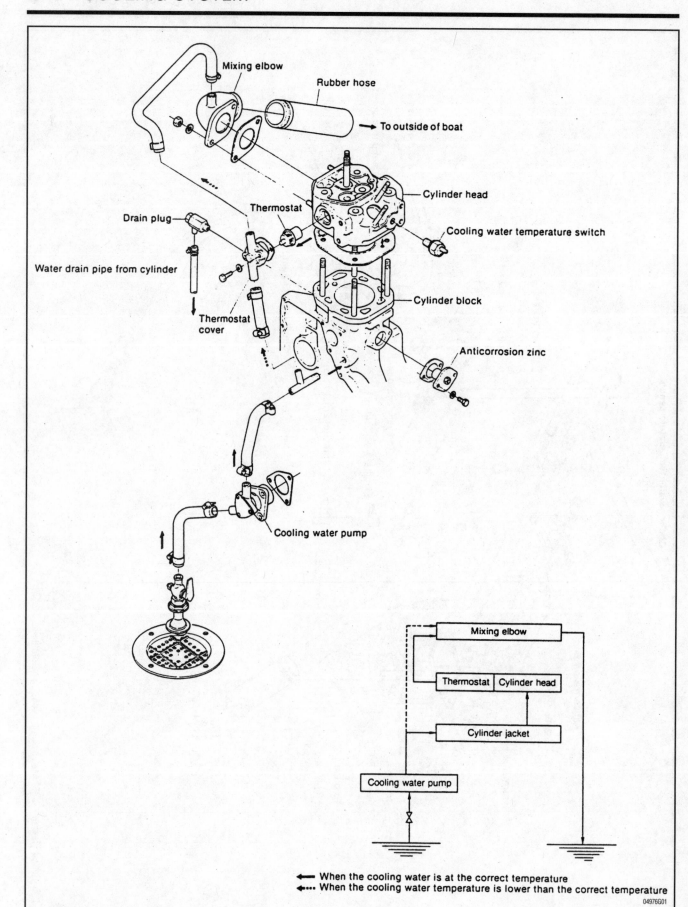

Mixing elbow

Rubber hose

→ To outside of boat

Cylinder head

Drain plug

Thermostat

Cooling water temperature switch

Water drain pipe from cylinder

Thermostat cover

Cylinder block

Anticorrosion zinc

Cooling water pump

Mixing elbow

Thermostat | Cylinder head

Cylinder jacket

Cooling water pump

◄— When the cooling water is at the correct temperature
◄···· When the cooling water temperature is lower than the correct temperature

04976G01

Fig. 3 Exploded view of a 1GM series engine cooling components

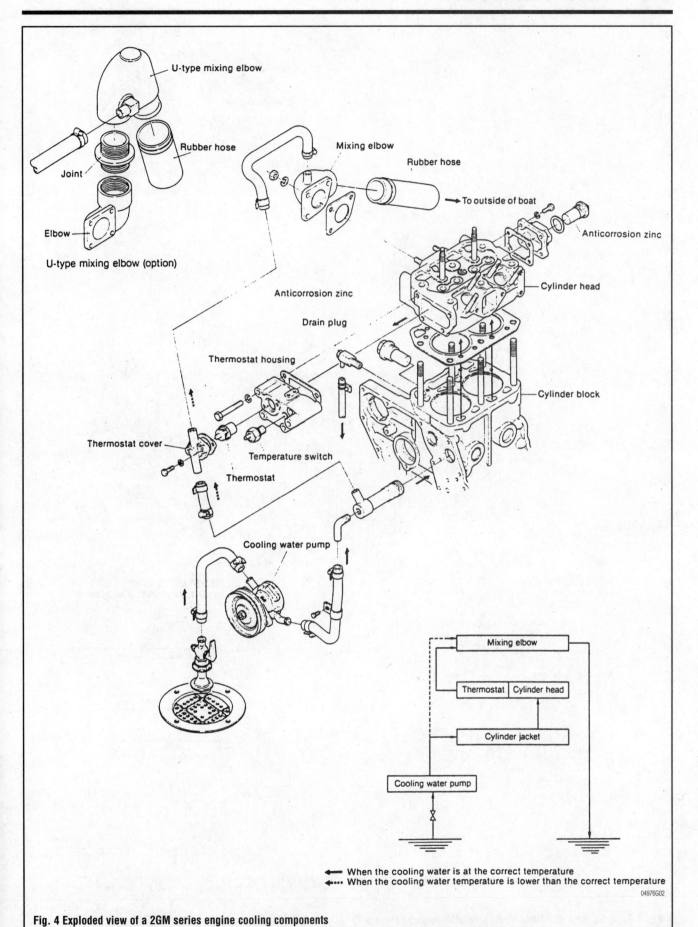

U-type mixing elbow

Rubber hose

Joint

Elbow

U-type mixing elbow (option)

Mixing elbow

Rubber hose

→ To outside of boat

Anticorrosion zinc

Cylinder head

Anticorrosion zinc

Drain plug

Thermostat housing

Cylinder block

Thermostat cover

Temperature switch

Thermostat

Cooling water pump

Mixing elbow		
Thermostat	Cylinder head	
Cylinder jacket		

Cooling water pump

◄━ When the cooling water is at the correct temperature
◄┄┄ When the cooling water temperature is lower than the correct temperature

04976G02

Fig. 4 Exploded view of a 2GM series engine cooling components

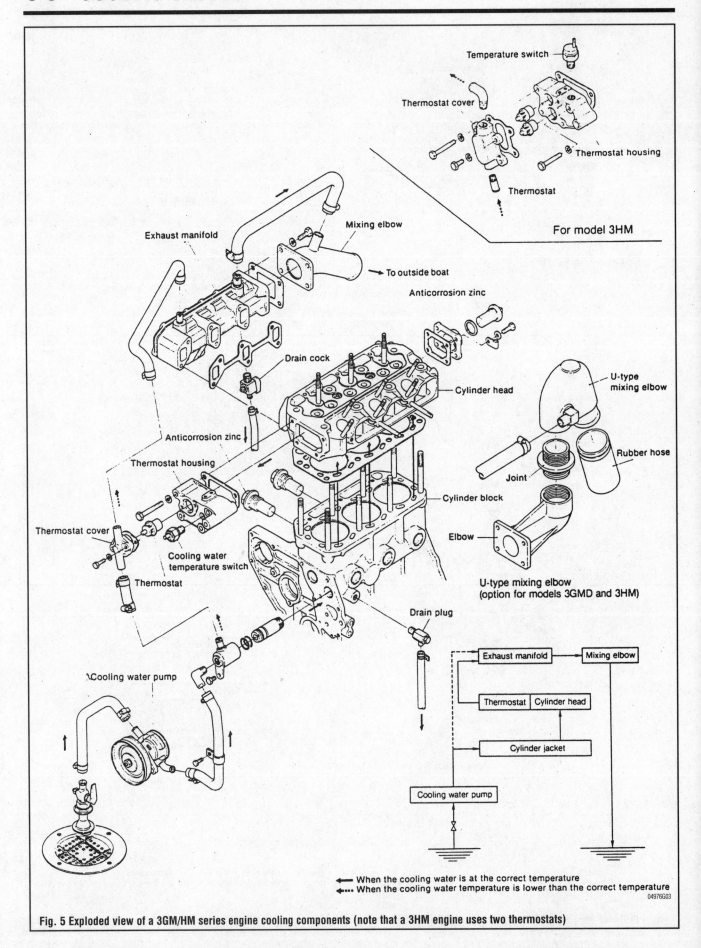

Temperature switch

Thermostat cover

Thermostat housing

Thermostat

For model 3HM

Exhaust manifold

Mixing elbow

To outside boat

Anticorrosion zinc

Drain cock

Cylinder head

U-type mixing elbow

Anticorrosion zinc

Rubber hose

Thermostat housing

Joint

Cylinder block

Thermostat cover

Elbow

Cooling water temperature switch

U-type mixing elbow (option for models 3GMD and 3HM)

Thermostat

Drain plug

Cooling water pump

Exhaust manifold	Mixing elbow
Thermostat	Cylinder head
Cylinder jacket	

Cooling water pump

◄— When the cooling water is at the correct temperature
◄•••• When the cooling water temperature is lower than the correct temperature

04976G03

Fig. 5 Exploded view of a 3GM/HM series engine cooling components (note that a 3HM engine uses two thermostats)

As a result, raw-water-cooled engines generally run cooler than other engines. This lowers the thermal efficiency of the engine and can also cause prevent harmful acids from being burned off in the engine oil.

The combination of heat, salt water, and dissimilar metals creates an ideal environment for galvanic corrosion. Raw-water cooled engines should be made of galvanically compatible materials. The best example is that of a cast-iron block which should have a cast-iron cylinder head, not an aluminum one. Sacrificial zinc anodes are a must in the cooling circuit. Since they have to be inspected and changed regularly, this creates a maintenance problem.

In most cases raw-water cooled engines that are subject to freezing temperatures cannot be protected with antifreeze. Since the water is drawn from the bottom of the boat, the boat must be dry docked in order to drain the system. All piping runs and pumps need to be installed without low spots in order to facilitate draining. Today's powerful, high-speed, lightweight diesels demand a greater degree of cooling efficiency than that provided by raw-water cooling. Almost all of these power plants use a fresh-water cooled heat exchanger or keel cooler.

Fresh-Water Cooling

♦ See Figures 6, 7, 8 and 9

DESCRIPTION & OPERATION

Heat Exchanger

Almost all modern marine engines use a fresh-water enclosed cooling system with a heat exchanger. The engine cooling pump

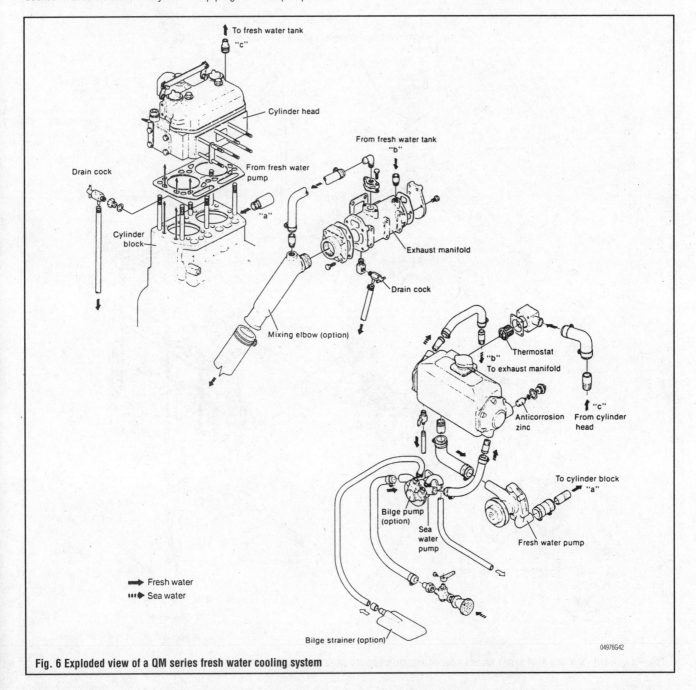

Fig. 6 Exploded view of a QM series fresh water cooling system

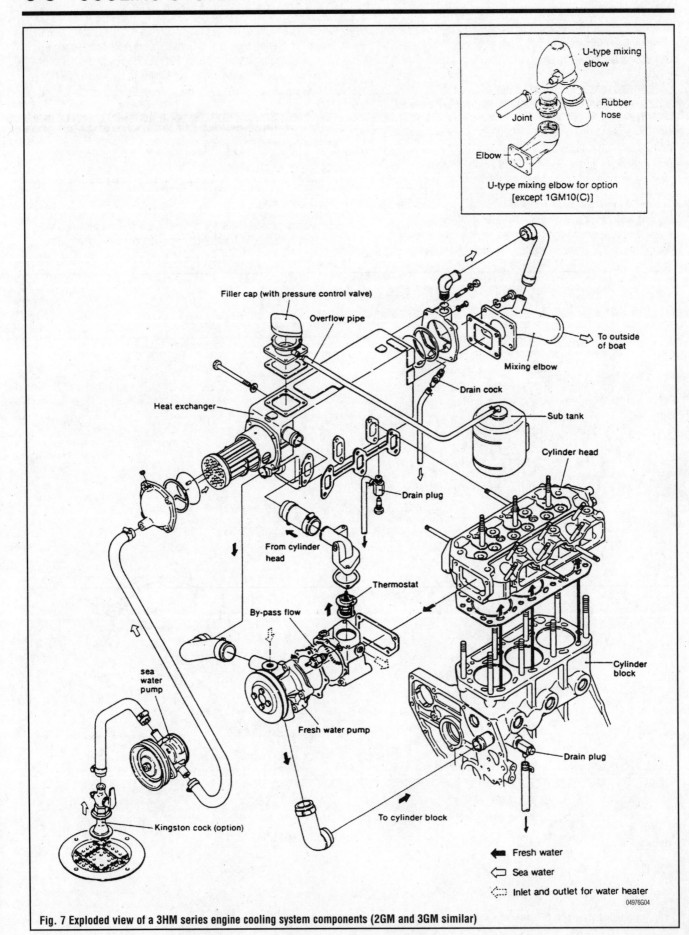

U-type mixing elbow

Rubber hose

Joint

Elbow

U-type mixing elbow for option
[except 1GM10(C)]

Filler cap (with pressure control valve)

Overflow pipe

To outside of boat

Mixing elbow

Drain cock

Heat exchanger

Sub tank

Cylinder head

Drain plug

From cylinder head

Thermostat

By-pass flow

sea water pump

Fresh water pump

Cylinder block

Kingston cock (option)

To cylinder block

Drain plug

← Fresh water

⇦ Sea water

⇚ Inlet and outlet for water heater

04976G04

Fig. 7 Exploded view of a 3HM series engine cooling system components (2GM and 3GM similar)

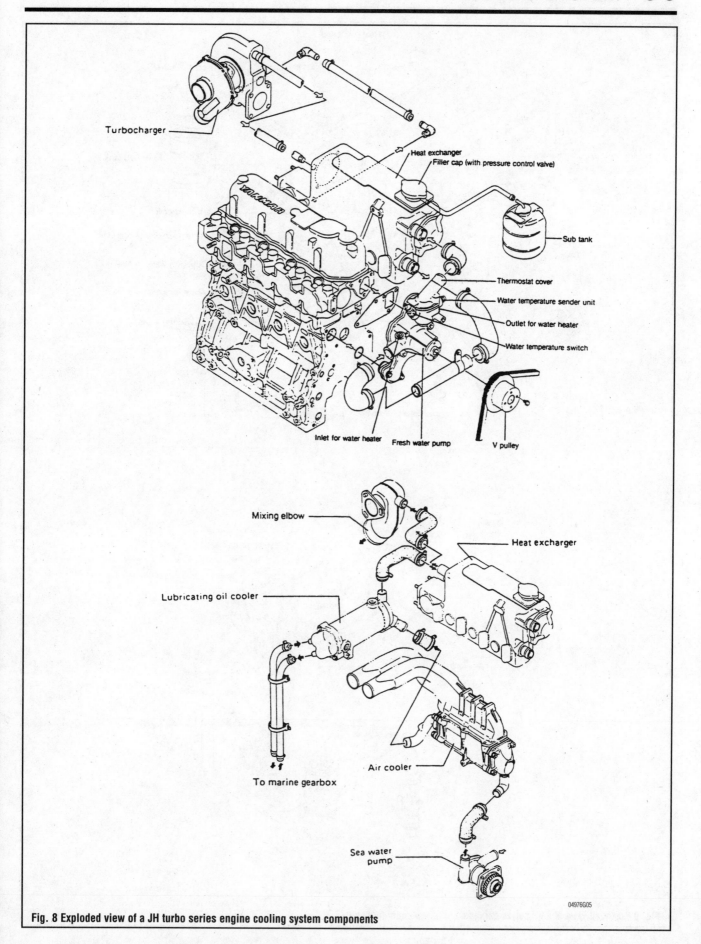

Turbocharger

Heat exchanger

Filler cap (with pressure control valve)

Sub tank

Thermostat cover

Water temperature sender unit

Outlet for water heater

Water temperature switch

Inlet for water heater

Fresh water pump

V pulley

Mixing elbow

Heat exchanger

Lubricating oil cooler

Air cooler

To marine gearbox

Sea water pump

04976G05

Fig. 8 Exploded view of a JH turbo series engine cooling system components

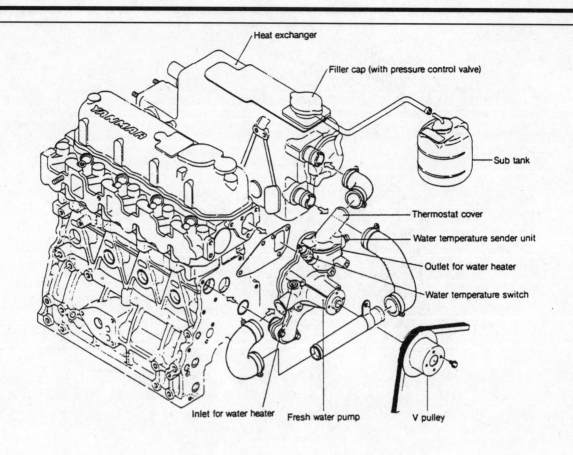

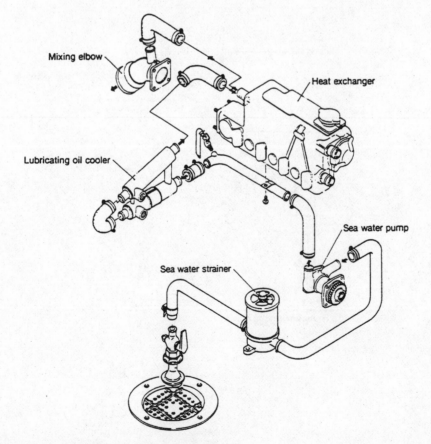

Fig. 9 Exploded view of a JH series engine cooling system components

04976G06

circulates the coolant from a header tank through oil cooler and then the engine. The coolant then passes through a heat exchanger where heat is extracted or radiated to lower its temperature. The coolant is then circulated once again.

A heat exchanger consists of a cylinder with a number of small copper-nickel tubes running through it. The hot coolant passes through the exchanger (located within the header tank) while cold raw-water is pumped through the tubes. The cold water carries off the heat from coolant then discharges directly overboard or via a wet exhaust.

Heat exchangers are expensive, require more piping and an extra pump. However, antifreeze and corrosion inhibitors can be added to the coolant to protect the engine against freezing and corrosion. Another advantage is that no silt finds its way into the engine and no salts crystallize around cylinder liners and in the cylinder head. The engine can also be operated at higher temperatures, which are more thermally efficient.

As with most automotive systems, the heat exchanger system is a closed loop with an expansion tank and a pressurized cap. This is an advantage because as the pressure of the coolant is increased, so too is the boiling point. Allowing the pressure to rise in a closed cooling system greatly reduces the risk of localized pockets of steam forming and causing damage to the engine.

The raw-water side of a heat-exchanger circuit will still suffer from the problems associated with raw-water cooling, particularly silting up of the heat exchanger tubes and the potential for damage from corrosion and freezing. Sacrificial zincs are essential once again, and the piping must have no low spots so that it can be easily drained in cold weather.

Keel Cooler

Another type of fresh water cooling system is keel cooling. Keel cooling does not require a raw-water circuit. Instead of placing a heat exchanger in the boat and bringing raw-water to it, a heat exchanger is placed outside the boat and is immersed directly in the sea water. This is normally accomplished by running a pipe around the keel of the boat and circulating the engine coolant through it or by installing an arrangement of cooling pipes on the outside of the hull.

A keel cooler has all the advantages of a heat exchanger plus a few more. It does not suffer any problems with silting, corrosion or freezing on any part of the cooling circuit. If the engine has a dry exhaust, it does not need any kind of a raw-water pump or circuit. If the engine has a wet exhaust, a separate raw-water pump is necessary to supply the water injected into the exhaust. This pump often also circulates water through an aftercooler and perhaps the oil cooler for the engine or transmission. The best keel coolers are made of bronze with a copper-nickel tubing.

RAW WATER CIRCUIT

Raw Water Strainer

▶ **See Figure 10**

A filter or strainer is used to prevent large particles of sand, mud, and other debris from entering the cooling circuit and damaging the raw water pump. Depending on water conditions, the strainer may become clogged with debris and should be cleaned. If engine temperature rises, water flow decreases and no other fault can be found in the cooling system, it can be assumed the intake screen is blocked.

INSPECTION & CLEANING

Most strainers have sight glasses or other means of visual inspection to determine when maintenance is necessary. When debris has accumulated within the strainer, it should be promptly cleaned to avoid clogging.

1. Close the cooling water seacock.
2. Using an appropriately sized wrench, loosen the nut on the center of the strainer.
3. Remove the cap nut and washer, followed by the top of the strainer.
4. The sight glass, element, and gaskets can be removed from the strainer body.
5. Clean the strainer element, and any dirt and sediment accumulated in the bottom of the strainer body.
6. Install the element, gaskets, and sight glass to the strainer body.
7. Install the top on the strainer, followed by the washer and cap nut.
8. Carefully tighten the nut on the strainer.
9. Open the cooling water seacock. Make sure the air in the strainer is bled through the breather plug. If air remains in the strainer, loosen the cap nut and allow water to flow from the strainer until it is free of air.
10. Tighten the nut on the strainer.

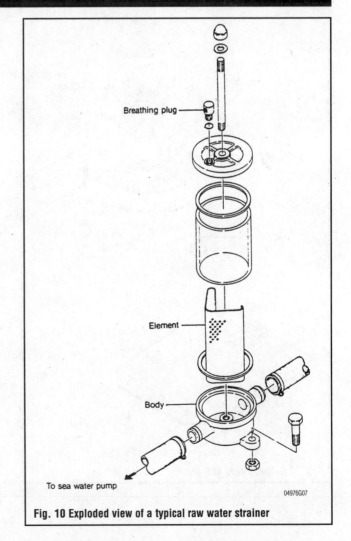

Fig. 10 Exploded view of a typical raw water strainer

Breathing plug

Element

Body

To sea water pump

04976G07

Seacock

▶ **See Figure 11**

The intake petcock, also known as a seacock, is the entry point for sea water in a raw-water cooling system. Each time the engine is operated, the seacock should be opened, and closed immediately after the engine has been shut down.

If the engine is operated without the seacock in the open position, the water pump will be starved of water causing the impeller to be destroyed. In worse cases, damage to the engine can result from overheating.

If the seacock is left open after the engine is shut down, water may be forced into the exhaust system through the mixing elbow, filling the cylinders with water. If the engine is cranked over with water in the cylinders, serious engine damage will occur.

INSPECTION

Little maintenance is necessary on the seacock other than to lubricate it with light oil and check it for proper operation approximately once a month. Make sure the handle operates smoothly, and there is no leakage. The main danger is if the seacock becomes

frozen. If this condition occurs, the valve must be replaced. This requires the boat to be removed from the water, which can be costly and time consuming.

If the screen underneath the hull becomes clogged, it MUST be cleaned before the engine is operated. As discussed earlier, the engine will overheat without adequate cooling.

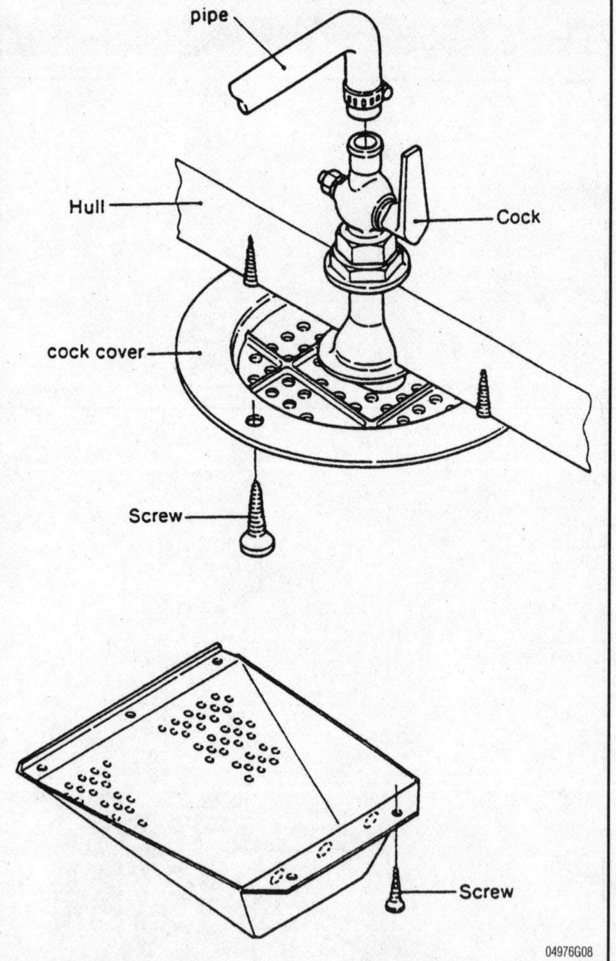

Antisiphon Valve

The antisiphon valve is used to prevent water from siphoning backwards into the exhaust manifold. If the exhaust manifold becomes filled with water, the cylinders can also become filled with water, which can cause serious damage.

The antisiphon valve is typically fitted between the raw water pump and the heat exchanger, or between the heat exchanger and exhaust elbow. In certain installations, it may be excluded if the exhaust elbow is above the waterline. Positioning is all dependent on the way the boat builder fitted the engine during installation.

INSPECTION

To inspect the antisiphon valve, simply remove the valve and attempt to blow air though one end. Air should pass through one end and not the other.

Raw Water Pump

▶ **See Figures 12 thru 18**

A raw water pump is used to circulate water from outside the vessel through the engine to provide cooling to the engine. The pump consists of an offset flexible impeller that provides a constant circulation of water with the impeller vanes.

Typically, raw water pumps last for around two seasons. Of course, if the vessel is operated in sandy or dirty water, and there is not a strainer installed to clean the incoming water, the impeller can have a short life.

➡**Never rotate the engine backwards; the water pump impeller can be damaged.**

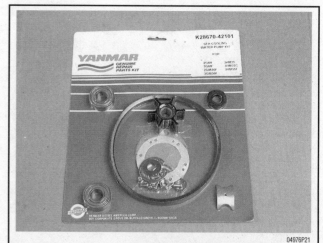

Fig. 11 Cutaway view of a seacock installation

04976G08

Fig. 12 If you're away from your home port for any length of time, items like this rebuild kit are essential when parts are not available

04976P21

Fig. 13 An impeller replacement kit should always be aboard your vessel for emergency repairs

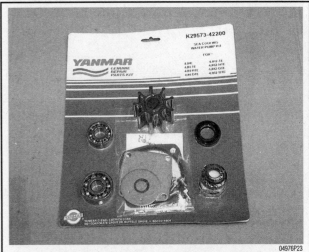

Fig. 14 A rebuild kit for a JH series raw water pump can save you money by allowing you to rebuild the pump yourself

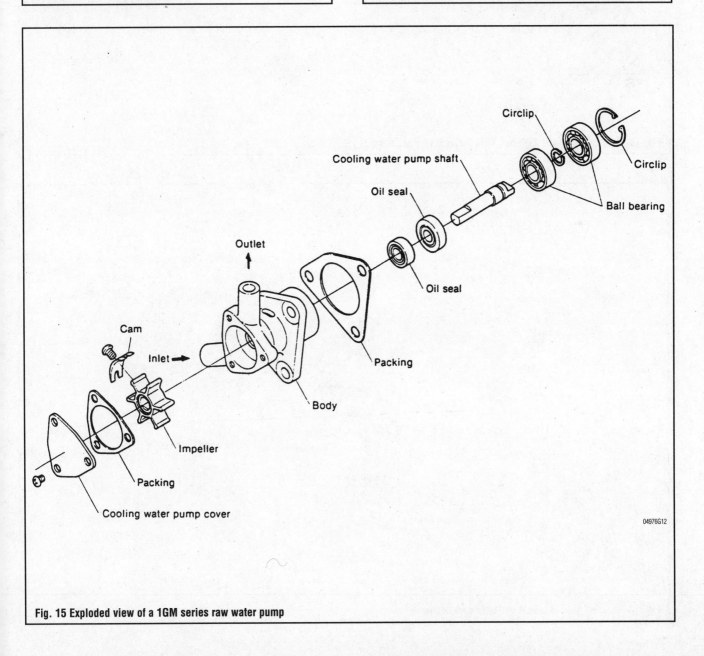

Fig. 15 Exploded view of a 1GM series raw water pump

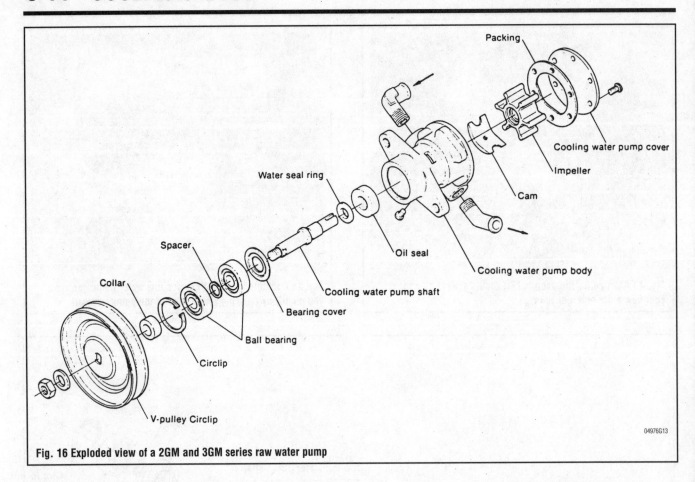

Fig. 16 Exploded view of a 2GM and 3GM series raw water pump

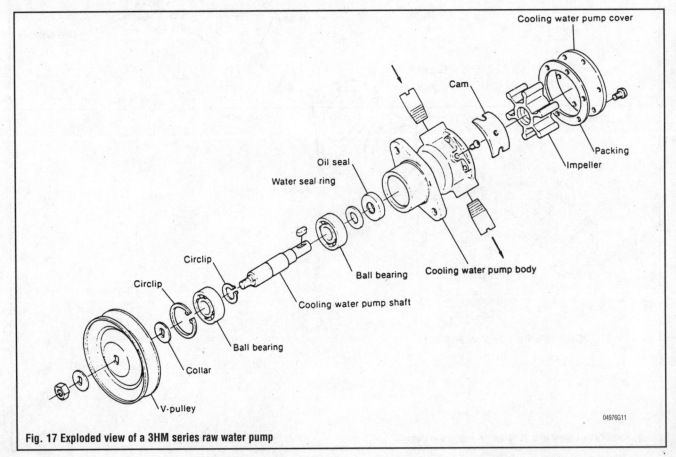

Fig. 17 Exploded view of a 3HM series raw water pump

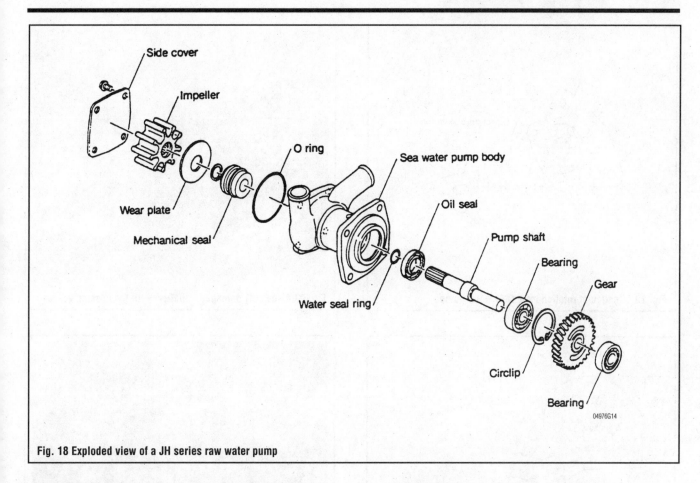

Fig. 18 Exploded view of a JH series raw water pump

The soft rubber vanes of the water pump will be quickly destroyed if the pump operates without a continuous flow water. If the seacock is accidentally left closed (or the inlet is clogged) when the engine is started, the water pump impeller can be destroyed in a matter of seconds due to lack of lubrication.

Typically, the raw water pump requires total replacement (or rebuilding) when the seal fails, or the bearing(s) fail. If a bearing fails, the water pump will produce a grinding metallic sound. A bearing failure can be a cause of a overtightened belt, or quite simply, age. If the seal fails, the water pump will leak from the drain hole in the bottom of the pump.

IMPELLER INSPECTION & REPLACEMENT

♦ **See Figures 19 thru 32**

It is not necessary to replace the entire water pump if the impeller is damaged. Usually, impeller replacement kits are available at a fraction of the cost of a new pump. The impeller of a fresh-water pump should be inspected regularly, and replaced if needed

1. Turn the seacock to the **OFF** position.
2. Remove the screws that retain the cover to the pump housing.
3. Remove the impeller cover, and the gasket or O-ring that seals the cover.
4. Using angled needlenose pliers, gently grasp an impeller vane, and pull outward. Grab an opposing vane, and pull outward; repeat until the impeller is free of the housing. Be careful not to pull too hard on the impeller vanes or they may tear.

5. Inspect the impeller for damage. Cracked, melted, or permanently arched vanes indicate replacement.
6. Inspect the impeller housing for wear and/or damage. If the housing is damaged, the entire water pump may need to be replaced.

To install:

7. Using an approved grease, or olive oil, lubricate the impeller and the inner pump housing surface. This will help protect the

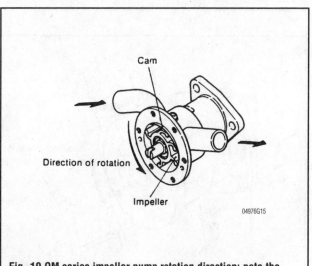

Fig. 19 QM series impeller pump rotation direction; note the direction in which the impeller vanes are curved

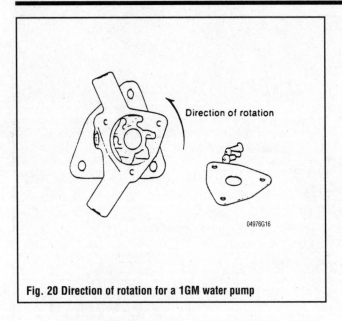

Fig. 20 Direction of rotation for a 1GM water pump

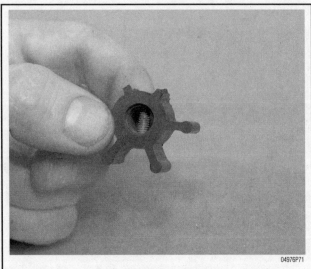

Fig. 23 A severely damaged impeller. Note the missing vanes

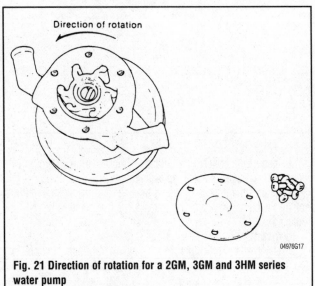

Fig. 21 Direction of rotation for a 2GM, 3GM and 3HM series water pump

Fig. 24 To remove the impeller on this JH series raw water pump, remove the 4 bolts

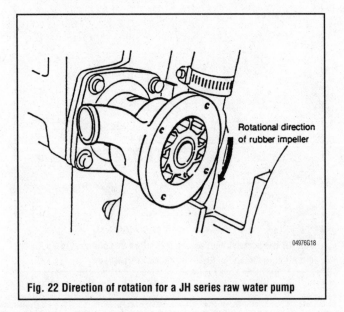

Fig. 22 Direction of rotation for a JH series raw water pump

Fig. 25 Although the raw water pump on this GM series engine is different, removing the 6 screws also allows access to the impeller

Fig. 26 On this 1GM engine, the impeller can be accessed by removing the three bolts that secure the cover

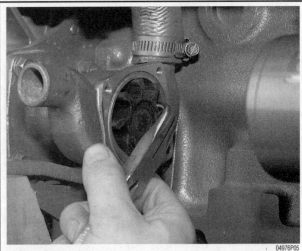

Fig. 29 Using angled needlenose pliers, gently grasp a vane, and pull outward

Fig. 27 After the bolts are removed, remove the impeller cover

Fig. 30 When the impeller is about ⅔ out of the housing, remove it by hand

Fig. 28 Always replace the O-ring or gasket that seals the impeller cover to the pump

Fig. 31 This impeller has been overheated. This was caused by operating the engine without opening the seacock

Fig. 32 Before installing a new impeller, inspect the inside of the pump housing for scoring

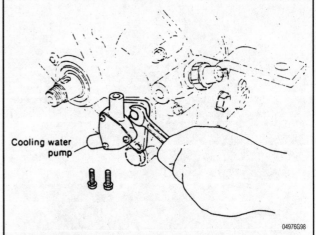

Fig. 33 After the hoses have been disconnected, remove the three bolts that attach the raw water pump to the engine block (1GM)

impeller from damage upon startup, since there will not be any water in the pump to provide lubrication.

8. Install the impeller into the pump housing. Make sure that the vanes are curved in the proper direction. Use the illustrations as a reference.

✸✸ WARNING

If the impeller vanes are incorrectly positioned, the pump will not function properly, and the engine may overheat.

9. Lightly coat the sliding surface of the impeller cover with an approved grease or olive oil.
10. Install the impeller cover, using a new O-ring or gasket.
11. Install the cover screws, and tighten them evenly.
12. Turn the seacock to the **ON** position.
13. Start the engine, and check for leaks.

PUMP REMOVAL & INSTALLATION

QM & 1 GM series engines

◆ See Figures 33 and 34

1. Turn the seacock to the **OFF** position.
2. Drain the water from the engine block.
3. Disconnect the inlet and outlet hoses from the engine.
4. Remove the three bolts that attach the water pump assembly to the engine.
   To install:
5. Using a new gasket, install the pump assembly to the engine. Make sure to align the impeller shaft.
6. Install and tighten the three screws that attach the pump assembly and tighten them securely.
7. Attach the inlet and outlet hoses to the water pump. Tighten the clamps securely.
8. Turn the seacock to the **ON** position.
9. Operate the engine and check for proper operation.

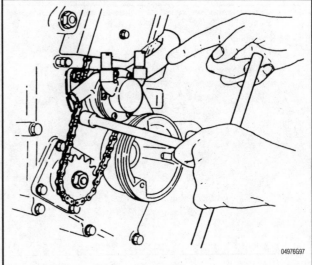

Fig. 34 Removing the raw water pump from a QM series engine

GM/HM series engines

◆ See Figures 35, 36, 37 and 38

1. Turn the seacock to the **OFF** position.
2. Drain the water from the engine block. After the water is drained, tighten the draincocks securely.
3. Disconnect the inlet and outlet hoses from the pump.
4. Loosen the water pump adjustment bracket bolts, and slip the drive belt from the water pump pulley.
5. Remove the two bolts that attach the water pump and bracket to the engine.
6. Remove the water pump from the engine.
7. If necessary, the pump can be unbolted from the bracket.
   To install:
8. If removed, install the water pump to the adjustment bracket with the two mounting bolts.

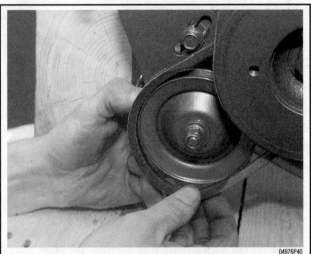

Fig. 35 After the pump bracket is loosened, slip the belt from the drive pulley

Fig. 36 Loosen the clamps on the inlet and outlet hoses, and remove them from the pump

Fig. 37 Then remove the two bolts that secure the bracket to the engine

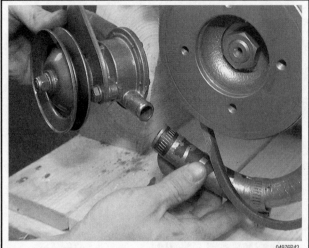

Fig. 38 In some cases, it may be easier to remove the hoses after the pump has been removed from the engine

9. Inspect the drive belt, and replace it if necessary.
10. Install the pump and bracket to the engine with the two bolts.
11. Place the drive belt on the pulley, and properly adjust the tension on the belt. Refer to the maintenance section for more information.
12. Attach the inlet and outlet hoses to the water pump. Tighten the clamps securely.
13. Turn the seacock to the **ON** position.
14. Operate the engine and check for proper operation.

JH series engines

▶ See Figure 39

1. Turn the seacock to the **OFF** position.
2. Drain the water from the engine block.

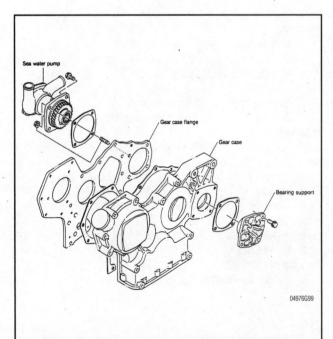

Fig. 39 On JH series engines, the water pump is attached to the engine by 4 bolts

➡On some installations, removal of the engine mount may be required to allow adequate clearance of the water pump.

3. Disconnect the inlet and outlet hoses from the water pump.

4. Remove the four nuts on the studs that secure the pump to the engine case.

5. The water pump assembly can be removed from the engine case.

To install:

6. Clean the gasket surfaces of the water pump and mating surface on the engine case.

7. Using a new gasket, install the pump to the engine. If necessary, turn the gear slightly to facilitate installation.

8. Tighten the four water pump mounting nuts securely.

9. If removed, install the engine mount bracket.

10. Attach the inlet and outlet hoses to the water pump. Tighten the hose clamps securely.

11. Turn the seacock to the **ON** position.

12. Operate the engine and check for proper operation.

Raw Water Thermostat

The raw water thermostat is a simple temperature sensitive valve that opens and closes to control cooling water flow through the engine. In operation the thermostat hovers somewhere between open and closed. As engine load and temperature increase, the thermostat opens to allow more cooling water into the engine. As temperature and load decrease the thermostat closes.

A sticking thermostat will either allow the temperature to rise well above the normal operating temperature before it opens, or, if stuck in the open position, will never allow the engine to reach operating temperature.

REMOVAL & INSTALLATION

QM series engines

▶ **See Figure 40**

On QM series engines, the thermostat is located on the exhaust manifold.

1. Turn the seacock to the **OFF** position.

2. Drain the water from the exhaust manifold by using the draincock. After the water is drained, make sure it is in the closed position.

3. Disconnect the hoses from the thermostat cover.

4. Remove the bolts that secure the thermostat cover to the exhaust manifold.

5. Remove the thermostat cover. If necessary, tap it lightly with a mallet to break the gasket.

6. Remove the thermostat(s) from the exhaust manifold.

To install:

7. Clean the gasket surfaces of the exhaust manifold and thermostat cover.

8. Install the new thermostat(s) into the exhaust manifold.

9. Using a new gasket or O-ring, install the thermostat cover to the exhaust manifold.

10. Tighten the thermostat cover bolts securely.

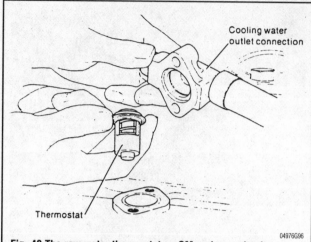

Fig. 40 The raw water thermostat on QM series engine is located under the cooling water outlet connection on the exhaust manifold

11. Attach the hoses to the thermostat cover. Tighten the hose clamps securely.

12. Turn the seacock to the **ON** position.

GM/HM series engines

▶ **See Figures 41, 42, 43, 44 and 45**

On GM/HM series engines, the thermostat is located on the front of the cylinder head. Note that 3HM engines use two thermostats.

➡Before removing the thermostat, note the direction of the cover. It MUST be mounted in the proper direction, or the engine may overheat.

1. Turn the seacock to the **OFF** position.

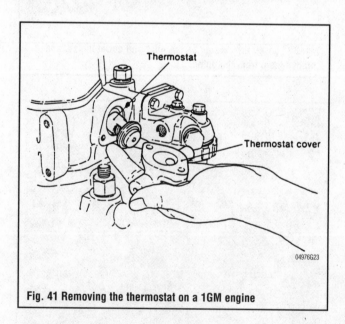

Fig. 41 Removing the thermostat on a 1GM engine

2. Drain the water from the cylinder head by using the drain-cock(s) on the engine. After the water is drained, make sure the draincock(s) is in the closed position.

3. Disconnect the hoses from the thermostat cover.

4. Remove the bolts that secure the thermostat cover to the cylinder head.

5. Remove the thermostat cover. If necessary, tap it lightly with a mallet to break the gasket.

6. Remove the thermostat(s) from the cylinder head.

To install:

7. Clean the gasket surfaces of the cylinder head and thermostat cover.

8. Install the new thermostat(s) to the engine in the proper direction.

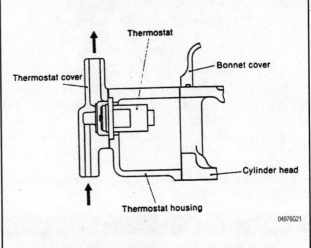

Fig. 44 Cutaway view of the cylinder head on 2/3GM engines showing mounting of the thermostat

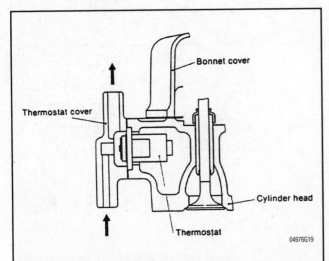

Fig. 42 Cutaway view of the 1GM cylinder head showing mounting of the thermostat (note the direction of the thermostat cover)

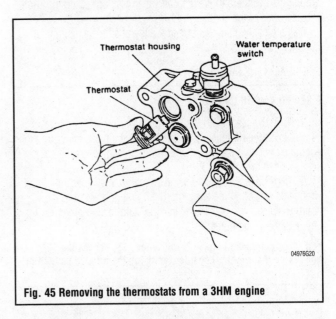

Fig. 45 Removing the thermostats from a 3HM engine

9. Using a new gasket, install the thermostat cover to the cylinder head, making sure it is in the proper direction.

10. Tighten the thermostat cover bolts securely.

11. Attach the hoses to the thermostat cover. Tighten the hose clamps securely.

12. Turn the seacock to the **ON** position.

13. Operate the engine and check for signs of leakage.

INSPECTION

The best way to inspect the thermostat is to remove it from the engine and visually inspect the mechanism. If the thermostat shows signs of contamination, corrosion, or wear marks, it is time to install a new thermostat.

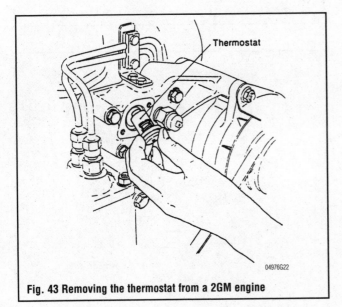

Fig. 43 Removing the thermostat from a 2GM engine

For some people the cost involved in removing the thermostat is enough to warrant installing a new one each time it is serviced.

TESTING

Check the rating stamped on the flange or body of the thermostat. This will tell you its rated temperature (the temperature at which it should be fully open.

Test the thermostat by suspending it in container of water. Slowly bring the water to a boil and using a thermometer, not when the thermostat opens.

If the thermostat does not open near the rated temperature, it is faulty and should be replaced.

Anodes

To reduce corrosion within the engine, anodes, (also known as zincs) are fitted to raw-water cooling systems. Anodes are mounted directly to the engine block, and protrude into the cooling jackets around the cylinders to prevent salt water from corroding the cooling passages of the engine.

Well maintained anodes extend the life of all metal components in contact with the sea water.

REMOVAL & INSTALLATION

▶ **See Figures 46 thru 52**

1. Turn the seacock to the **OFF** position.
2. Using the draincocks on the engine, drain the water from the engine by opening the draincocks. After the engine is drained, be sure to close the draincocks.
3. Remove any components necessary to allow placement of a wrench or socket on the anode mounting plug(s). Refer to the appropriate section(s) of this manual for information on removing any necessary components.

➡**If other components have to be removed to access the anodes on the engine, consider replacing the anodes instead of cleaning them. The amount of work necessary to access and remove the anodes from the engine may outweigh the cost of new anodes.**

4. Unscrew the anode mounting plugs from the engine. If necessary, use a penetrating oil to loosen the threads.
5. While holding a mounting plug stationary with a wrench, remove the anode by turning it counterclockwise using pliers. If the anode will not break free, spray it with penetrating oil and allow it to seep into the threads.
6. Repeat this step with all of the anodes removed from the engine.

To install:

7. Thread the new anodes onto the mounting plugs. Tighten them securely.
8. Coat the threads of the anode mounting plugs with sealant. (on 1GM engines, a new gasket is needed)

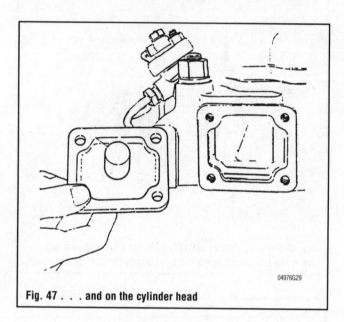

Fig. 47 . . . and on the cylinder head

04976G29

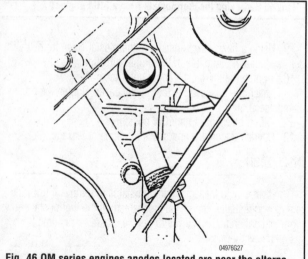

Fig. 46 QM series engines anodes located are near the alternator . . .

04976G27

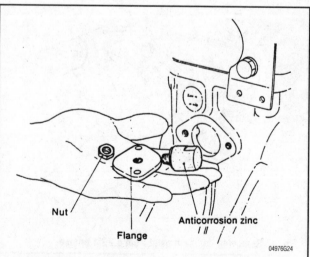

Nut

Flange

Anticorrosion zinc

Fig. 48 On 1GM engines, an anode is bolted to a plate on the engine

04976G24

Fig. 49 To replace an anode on a 1GM engine, remove the two bolts that attach the plate to the engine block

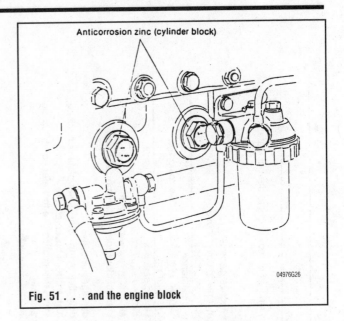
Anticorrosion zinc (cylinder block)

Fig. 51 . . . and the engine block

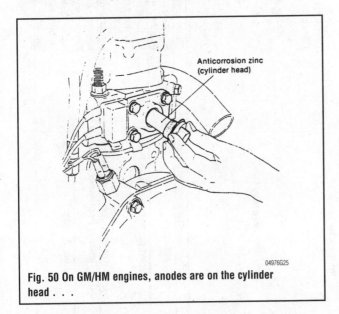
Anticorrosion zinc (cylinder head)

Fig. 50 On GM/HM engines, anodes are on the cylinder head . . .

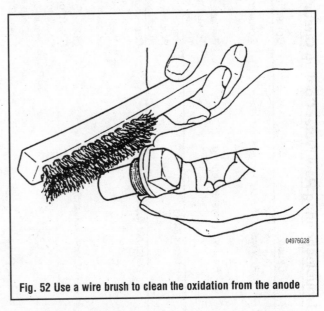
Fig. 52 Use a wire brush to clean the oxidation from the anode

9. Install the anode mounting plugs with the new anodes to the engine. Tighten them securely.
10. Install any components that may have been removed to access the anodes.
11. Turn the seacock to the **ON** position.
12. Operate the engine and check for leaks.

INSPECTION

The frequency of anode inspection depends where your boat is sailed and many other factors, so check your anodes frequently until you have learned how quickly they dissolve.

Anode inspection involves removing the anodes from the engine, heat exchanger, etc. and visually inspecting them for size. A worn anode is approximately half its original size.

Once removed, knock off the crusty oxides with a hammer or scrape them off with a wire brush. If the anode is less than half its original size, replace it. When replacing anodes, always use flexible sealer on the plug threads to prevent leakage and make removal at a later time easier.

➡Teflon® tape may prevent the anode from functioning properly so it is important to use some other type of sealant.

Troubleshooting Raw Water Cooling Circuits

Condition	Cause	Correction
External raw water leaks, engine overheats	Raw water strainer above water line	Reposition strainer or move boat to deeper water
	Raw water system leaking	Replace defective gaskets
		Tighten water pump cover loose
		Tighten hose clamps loose
Heat exchanger or header tank has significant gain or loss of coolant	Internal leaks	Replace heat exchanger
		Properly maintain system to avoid corrosion
		Repair solder failure
High operating temperature, engine overheats	Water not flowing	Check for closed seacock
		Check for blocked strainer
		Check for air leak in raw water system
		Check for internal blockage in raw water system
	Water flow insufficient	Check for damaged impeller
		Check for defective pump body, cover plate or backing plate
		Check for high exhaust backpressure
		Check for excessive suction due to water pump too high above water line
		Check for inadequate water supply
	Impeller torn	Check for debris in pump
		Check for overloaded pump
		Check for high exhaust backpressure
		Check for excessive suction due to water pump too high above water line
	Impeller frequently fails	Check for defective pump body, cover plate or backing plate
		Correct source of poor water supply
		Not properly lubricated when installed
		Check for high exhaust backpressure
		Check for excessive suction due to water pump too high above water line
Insufficient raw water flow, engine overheats	Raw water strainer restricted	Clean clogged raw water strainer
Metal in heat exchanger looks pink especially after cleaning	Corrosion	Replace or install anodes in raw water circuit
		Properly maintain existing anodes
Visible water leaks, corrosion and salt crystal trails on engine or boat	Water pump cover plate leaking	Replace defective cover plate gasket
		Tighten cover plate screws
Visible oil leak from raw water pump	Water leak from water pump body	Replace defective water pump
	Oil leaking from water pump body	Replace defective water pump
		Correct excessive crankcase pressure problem
Visible water leaks in heat exchanger	External leaks	Replace defective gaskets or seals
		Tighten loose hose clamps
		Replace corroded housing or end caps

04976C01

FRESH WATER COOLING CIRCUIT

How the Cooling System Works

The main parts of the engine cooling system are the pressure cap, hoses, thermostat and water pump. The system is filled with coolant, which should be a 50–50 mixture of antifreeze and water. No matter where you live or how hot or cold the weather becomes, the mixture should be maintained year round.

The water pump is usually driven by a belt connected to the engine. The pump draws coolant from heat exchanger and forces it through passages surrounding the hot area—the cylinders, combustion chambers and valves. From there the coolant flows through a hose back to the heat exchanger to give off its heat to the sea water.

WHAT IS COOLANT

✳✳ CAUTION

When draining the coolant, keep in mind that cats and dogs are attracted by ethylene glycol antifreeze, and are quite likely to drink any that is left in an uncovered container or in puddles on the ground. This will prove fatal in sufficient quantity. Always drain the coolant into a leak-proof container. To avoid injuries from scalding fluid and steam, DO NOT remove the radiator cap while the engine and radiator are still HOT. The best way to dispose of coolant is through an approved recycling center.

Ethylene Glycol

Coolant is at least 50–50 mixture of ethylene glycol and water. This mixture in older engines was required not only in the winter to prevent freezing, but also to prevent corrosion in cooling system components, and to provide lubricants to the water pump.

Good quality antifreezes also contain water pump lubricants, rust inhibitors, and other corrosion inhibitors along with acid neutralizers.

Propylene Glycol

Appearing in the early 90's a new, less-toxic antifreeze/coolant emerged. This is a propylene glycol base. As compared to ethylene glycol, propylene glycol is less toxic and safer for humans, pets, and wildlife in the environment. Its coolant and engine protection properties are similar to the ethylene glycol coolant listed above. Most of the coolant providers now offer a choice between Ethylene Glycol or Propylene Glycol based products.

CONTROLLING THE TEMPERATURE

It's important to get the coolant up to normal operating temperature as quickly as possible to ensure smooth engine operation and free flow of oil. When the engine is cold, the thermostat blocks the passage from the cylinder head to the heat exchanger and sends coolant on a shortcut to the water pump. The cooling fluid is not exposed to the cooling effects of the sea water, so it warms up rapidly. As temperature increases, the thermostat gradually opens and allows coolant to flow.

Expansion tank

♦ See Figures 53 and 54

Most engines use a coolant recovery system. The recover system allows coolant that would normally overflow to be caught in an

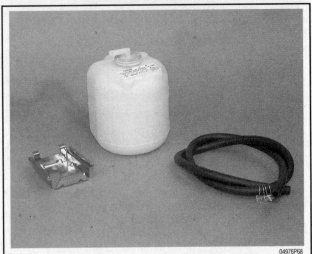

Fig. 53 If your engine is not equipped with an expansion tank, a kit is available from your local marine dealer

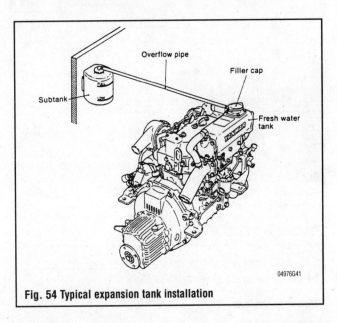

Fig. 54 Typical expansion tank installation

expansion tank; it will automatically be drawn back into the system when the coolant cools down.

INSPECTION

To inspect the coolant level, simply check the level of coolant in the plastic tank. If the coolant level is low, remove the pressure cap and add coolant directly to the header tank. If the coolant level is constantly low, check for leaks in the system.

Thermostat

♦ See Figure 55

The thermostat is a simple temperature sensitive valve that opens and closes to control cooling water flow through the engine. In operation the thermostat hovers somewhere between open and closed. As engine load and temperature increase, the thermostat

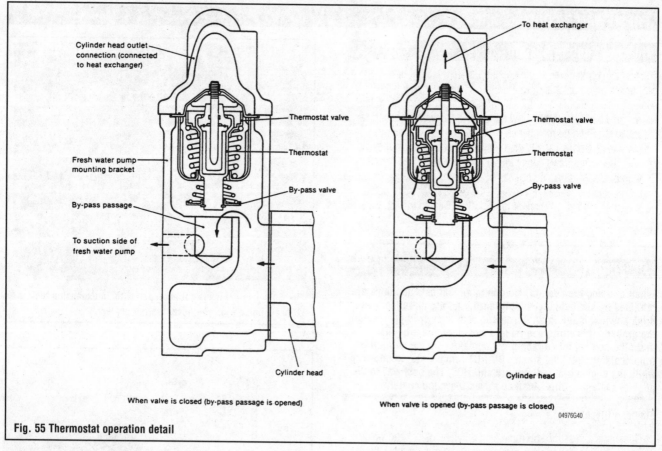

Cylinder head outlet connection (connected to heat exchanger)

Thermostat valve

Thermostat

Fresh water pump mounting bracket

By-pass valve

By-pass passage

To suction side of fresh water pump

Cylinder head

When valve is closed (by-pass passage is opened)

To heat exchanger

Thermostat valve

Thermostat

By-pass valve

Cylinder head

When valve is opened (by-pass passage is closed)

04976G40

Fig. 55 Thermostat operation detail

opens to allow more cooling water into the engine. As temperature and load decrease the thermostat closes.

A sticking thermostat will either allow the temperature to rise well above the normal operating temperature before it opens, or, if stuck in the open position, will never allow the engine to reach operating temperature.

REMOVAL & INSTALLATION

On fresh water cooled QM engines, the thermostat is located on the side of the header tank.

QM series engines

♦ See Figure 56

⁕⁕ CAUTION

Never open, service or drain the fresh water cooling system when the engine is hot; serious burns can occur from the steam and hot coolant. Also, when draining engine coolant, keep in mind that cats and dogs are attracted to ethylene glycol antifreeze and could drink any that is left in an uncovered container. This will prove fatal in sufficient quantities. Always drain coolant into a sealable container.

1. Drain the coolant from header tank.
2. Disconnect the hose from the fresh water outlet.
3. Remove the bolts that secure the water outlet to the header tank.
4. Remove the water outlet. If necessary, tap it lightly with a mallet to break the gasket.
5. Remove the thermostat from the header tank.

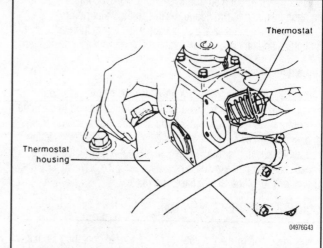

Thermostat

Thermostat housing

04976G43

Fig. 56 When installing the thermostat, make sure it is positioned in the right direction

To install:
6. Clean the mating surfaces of the header tank and water outlet.
7. Install the new thermostat in the header tank, in the proper direction.
8. Using a new gasket, install the water outlet to the header tank.
9. Install the four bolts and washers, and tighten them securely.
10. Attach the hose to the water outlet. Tighten the hose clamp securely.
11. Add coolant to the header tank.
12. Operate the engine and check for signs of leakage.

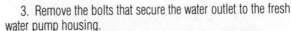

GM/HM series engines

▶ **See Figures 57, 58, 59, 60 and 61**

On fresh water cooled GM/HM engines, the thermostat is located on the top of the fresh water pump housing, underneath the fresh water outlet connection.

✳✳ CAUTION

Never open, service or drain the fresh water cooling system when the engine is hot; serious burns can occur from the steam and hot coolant. Also, when draining engine coolant, keep in mind that cats and dogs are attracted to ethylene glycol antifreeze and could drink any that is left in an uncovered container. This will prove fatal in sufficient quantities. Always drain coolant into a sealable container.

1. Drain the coolant from the engine to a point where it is below the thermostat housing.
2. Loosen the hose clamp the hose that attaches to the fresh water outlet.

3. Remove the bolts that secure the water outlet to the fresh water pump housing.
4. Remove the water outlet. If necessary, tap it lightly with a mallet to break the gasket.
5. Remove the thermostat(s) from the water pump housing.

To install:

6. Clean the mating surfaces of the fresh water pump housing and water outlet.
7. Install the new thermostat in the fresh water pump housing, in the proper direction.

➡ **When installing the thermostat, make sure the vent hole aligns with the bolt hole as shown in the illustration. If the thermostat is not aligned properly, it will not seat in the housing properly, causing the water outlet to leak.**

8. Using a new gasket, install the water outlet to the fresh water pump housing.
9. Install the two bolts and washers, and tighten them securely.
10. Attach the hose to the water outlet. Tighten the hose clamp securely.

Fig. 57 After the hose clamp is loosened, remove the two bolts that attach the water outlet

Fig. 59 The thermostat lifts from the housing

Fig. 58 The water outlet can now be removed to access the water pump

Fig. 60 Always replace the gasket on the water outlet to prevent leaks

Fig. 61 The thermostat is installed properly when the vent hole is aligned with the bolt hole as shown

11. Add coolant to the header tank as necessary.
12. Operate the engine and check for signs of leakage.

JH series engines

▶ See Figures 62 thru 67

On fresh water cooled JH series engines, the thermostat is located on the top of the fresh water pump, underneath the fresh water outlet connection.

✳✳ CAUTION

Never open, service or drain the fresh water cooling system when the engine is hot; serious burns can occur from the steam and hot coolant. Also, when draining engine coolant, keep in mind that cats and dogs are attracted to ethylene glycol antifreeze and could drink any that is left in an uncovered container. This will prove fatal in sufficient quantities. Always drain coolant into a sealable container.

1. Drain the coolant from the engine to a point where it is below the fresh water pump.
2. Loosen the clamps that attach the hose from the fresh water outlet to the header tank.
3. Remove the two bolts that secure the water outlet to the fresh water pump.
4. Remove the water outlet from the fresh water pump. If necessary, tap it lightly with a mallet to break the gasket.
5. Remove the thermostat(s) from the fresh water pump.

To install:

6. Clean the mating surfaces of the fresh water pump and water outlet.
7. Place the O-ring under the thermostat, (if equipped) and install the new thermostat in the fresh water pump, in the proper direction.

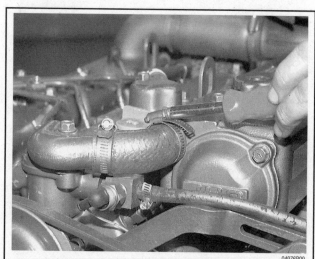

Fig. 63 Loosen the clamps that attach the water outlet hose to the header tank

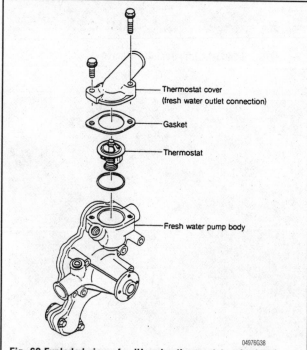

Fig. 62 Exploded view of a JH series thermostat and related components

- Thermostat cover (fresh water outlet connection)
- Gasket
- Thermostat
- Fresh water pump body

Fig. 64 Next, remove the two bolts that secure the fresh water outlet connection to the thermostat housing

Fig. 65 Tilt the hose and water outlet back, and pull the thermostat from the housing

Fig. 66 Excessive corrosion in the thermostat housing may indicate problems in the header tank

Fig. 67 Make sure the vent in the thermostat aligns with the bolt hole as shown

➥When installing the thermostat, make sure the vent hole aligns with the bolt hole as shown in the illustration. If the thermostat is not aligned properly, it will not seat in the housing properly, causing the water outlet to leak.

8. Using a new gasket, install the water outlet to the fresh water pump housing.
9. Install the two bolts and washers, and tighten them securely.
10. Tighten the clamps that attach the hose from the water outlet to the header tank.
11. Add coolant to the header tank as necessary.
12. Operate the engine and check for signs of leakage.

INSPECTION

The best way to inspect the thermostat is to remove it from the engine and visually inspect the mechanism. If the thermostat shows signs of contamination, corrosion, or wear marks, it is time to install a new thermostat.

For some people the cost involved in removing the thermostat is enough to warrant installing a new one each time it is serviced.

TESTING

▶ See Figure 68

Check the rating stamped on the flange or body of the thermostat. This will tell you its rated temperature (the temperature at which it should be fully open.

Test the thermostat by suspending it in container of water. Slowly bring the water to a boil and using a thermometer, not when the thermostat opens.

If the thermostat does not open near the rated temperature, it is faulty and should be replaced.

Fresh Water Pump

The fresh water pump is located on the front of the engine, and is operated by a belt driven from a pulley on the crankshaft. Unlike the raw water pump, the fresh water pump uses a metal impeller that does not require constant maintenance. The impeller of a fresh water pump does not touch the housing, but is closely fitted, and is quite efficient.

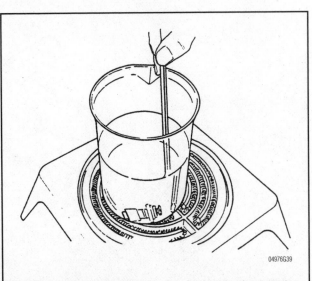

Fig. 68 Testing a thermostat using hot water and a thermometer

INSPECTION

▶ See Figure 69

Typically, the fresh water pump requires replacement when the seal fails, or the bearing(s) fail. If a bearing fails, the water pump will produce a grinding metallic sound. A bearing failure can be a cause of a overtightened belt, or quite simply, age. If the seal fails, the water pump will leak from the drain hole in the bottom of the pump. In most cases, replacement of the fresh water pump as a unit is most practical.

04976P46

Fig. 69 If water leaks from this hole on the bottom of the water pump, the internal seal has failed, and the pump should be replaced

REMOVAL & INSTALLATION

▶ See Figures 70 thru 78

✳✳ CAUTION

Never open, service or drain the fresh water cooling system when the engine is hot; serious burns can occur from the steam and hot coolant. Also, when draining engine coolant, keep in mind that cats and dogs are attracted to ethylene glycol antifreeze and could drink any that is left in an uncovered container. This will prove fatal in sufficient quantities. Always drain coolant into a sealable container.

1. Drain the fresh water cooling system. Make sure the draincocks are in the closed position after draining.
2. On JH series engines, remove the thermostat and water temperature sending units. Removing them with the pump attached to the engine will be easier.
3. Loosen the 4 bolts that secure the drive pulley to the water pump.
4. Loosen the water pump drive belt by loosening bolts on the alternator brackets.
5. Remove the drive belt from the water pump pulley.
6. Remove the 4 bolts that secure the water pump pulley.
7. Remove the hoses from the water pump. If necessary, remove any hose brackets to allow access to the pump bolts
8. Remove the bolts that attach the water pump to the engine.

➡When removing the water pump bolts, be sure to mark each bolt's location, since the bolts are of different lengths. If the bolts are not installed in their proper locations, the engine block can be damaged when the bolts are tightened.

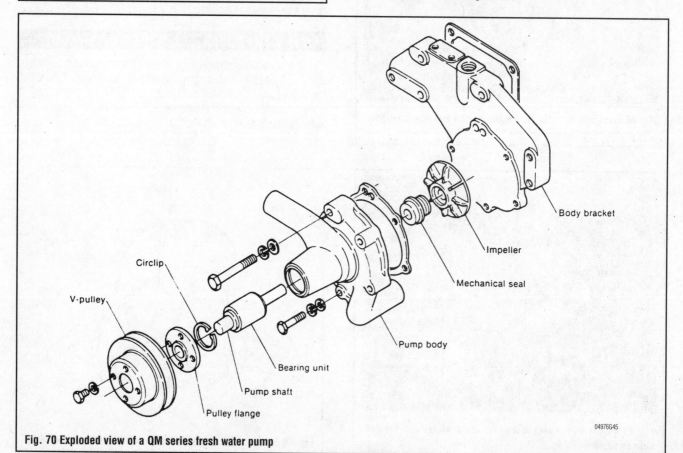

04976G45

Fig. 70 Exploded view of a QM series fresh water pump

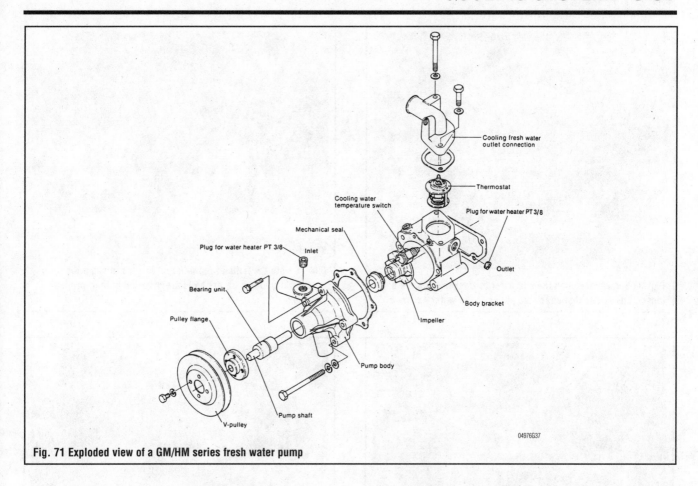

Fig. 71 Exploded view of a GM/HM series fresh water pump

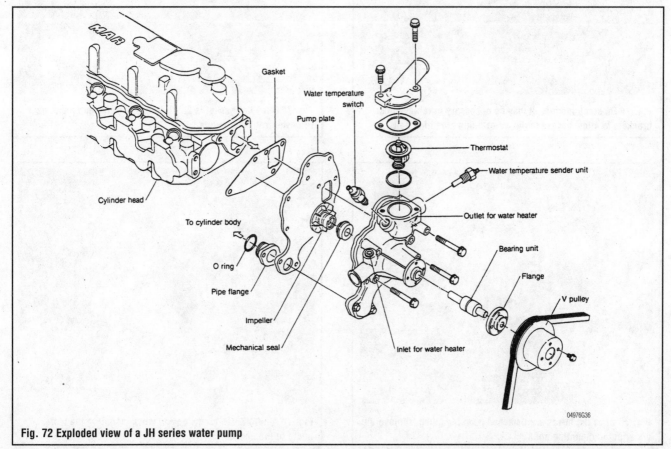

Fig. 72 Exploded view of a JH series water pump

Fig. 73 Before loosening the alternator to slacken the drive belt, loosen the 4 bolts that attach the pulley to the water pump

Fig. 76 Note the bolts are of different lengths; make a diagram if necessary to ensure the bolts are installed in their proper locations

Fig. 74 On some models, it may be necessary to remove hose brackets to allow access to the water pump mounting bolts

Fig. 77 Once the gasket is broke free, the water pump can be removed from the engine

Fig. 75 Once the hoses are detached from the pump, remove the water pump mounting bolts

Fig. 78 ALWAYS use a new gasket when installing the water pump to the engine

9. Remove the water pump from the engine.

10. On GM/HM engines, it is advisable to remove the thermostat housing and replace the gasket, since the water pump is already removed.

To install:

11. Clean the mating surfaces of the water pump and engine.

12. Using the appropriate gaskets, place the water pump on the engine and install the bolts. Make sure each bolt is in the proper location.

13. Tighten the water pump bolts securely.

14. Clean any corrosion from the inside of each hose to ensure a leak-free connection. Attach all hoses to the water pump. Replace any hose brackets as necessary.

15. Install the water pump pulley to the water pump. While holding the pulley by hand, tighten the bolts. (After the belt is installed and tightened, the bolts can be given an additional tightening)

16. Install the drive belt to the water pump pulley.

17. Using a belt tensioner, apply proper tension to the belt.

18. Refill the cooling system as necessary.

19. Operate the engine and check for leaks.

Header Tank

▶ See Figures 79, 80, 81 and 82

Many engines use a header tank that combines a water cooled exhaust manifold and the heat exchanger. The pressure cap on the header tank is identical to an automotive radiator cap.

INSPECTION

The header tank itself requires little maintenance other than a good visual inspection for cracks and leaks. The tank is usually made from aluminum which is vulnerable to corrosion. If white powdery deposits are noticed around the water hose fittings, remove the header tank and clean the corrosion to prevent further damage.

REMOVAL & INSTALLATION

QM & GM/HM series engines

▶ See Figures 83 thru 95

1. Turn the seacock to the **OFF** position.

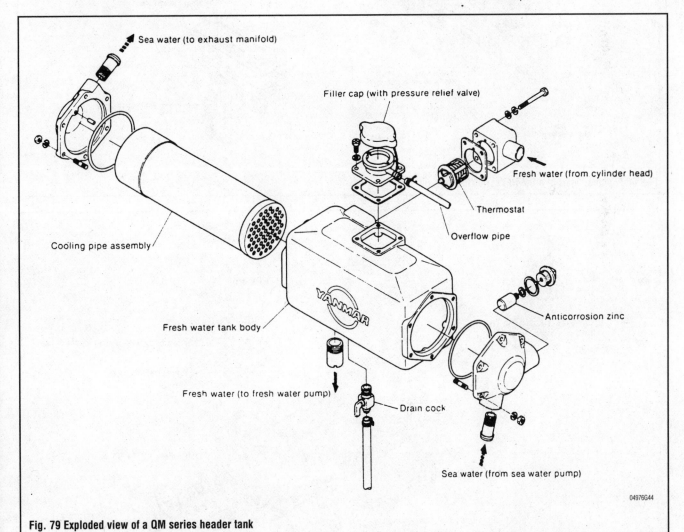

Fig. 79 Exploded view of a QM series header tank

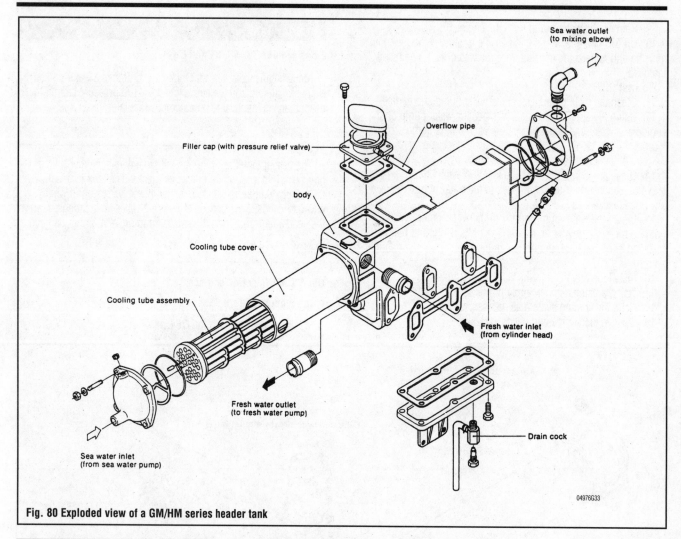

Sea water outlet
(to mixing elbow)

Filler cap (with pressure relief valve)

Overflow pipe

body

Cooling tube cover

Cooling tube assembly

Fresh water inlet
(from cylinder head)

Fresh water outlet
(to fresh water pump)

Drain cock

Sea water inlet
(from sea water pump)

04976G33

Fig. 80 Exploded view of a GM/HM series header tank

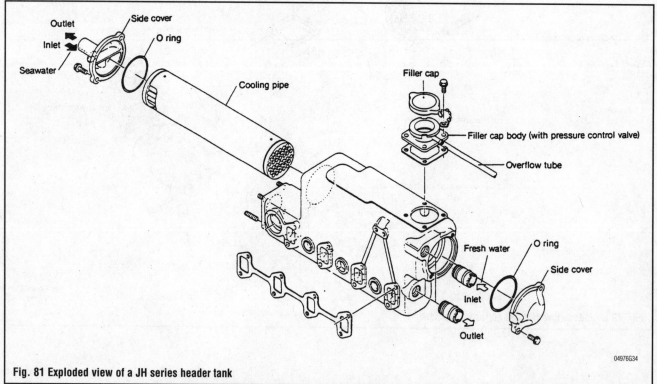

Outlet

Side cover

Inlet

O ring

Seawater

Cooling pipe

Filler cap

Filler cap body (with pressure control valve)

Overflow tube

Fresh water

O ring

Side cover

Inlet

Outlet

04976G34

Fig. 81 Exploded view of a JH series header tank

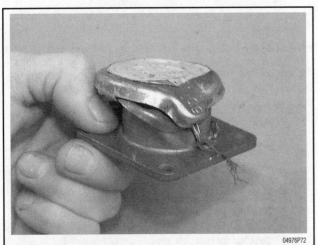

Fig. 82 This is what happens when the header tank is used as a step ladder out of the engine compartment. This pressure cap and housing must be replaced

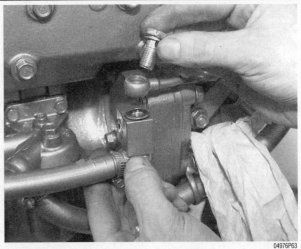

Fig. 84 On some engines, it may be necessary to remove additional components, such as fuel lines

2. Using the draincocks on the engine, drain the raw water from the engine by opening the draincocks. After the engine is drained, be sure to close the draincocks.

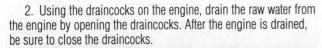

⁂ CAUTION

Never open, service or drain the fresh water cooling system when the engine is hot; serious burns can occur from the steam and hot coolant. Also, when draining engine coolant, keep in mind that cats and dogs are attracted to ethylene glycol antifreeze and could drink any that is left in an uncovered container. This will prove fatal in sufficient quantities. Always drain coolant into a sealable container.

3. Drain the coolant from the header tank.
4. Remove the hoses from the coolant tank.
5. Remove any accessories that are attached to the header tank, such as cable brackets, fuel filter brackets, fuel lines, etc.
6. Remove the four bolts that attach the exhaust elbow to the header tank.

Fig. 85 The fuel filter is attached to the header tank by two bolts

Fig. 83 Before loosening any bolts, remove all hoses that attach to the header tank

Fig. 86 Once the two bolts are removed, lower the fuel filter

Fig. 87 Next, remove the 4 bolts that attach the exhaust mixing elbow to the header tank

7. Remove the bolts and nuts that attach the header tank to the cylinder head.

➡ When removing the header tank bolts, be sure to mark each bolt's location, since some of the bolts are different lengths. If the bolts are not installed in their proper locations, the engine block and/or cylinder head can be damaged when the bolts are tightened.

8. Remove the header tank from the engine.

To install:

9. Clean the mating surfaces of the cylinder head and header tank.

10. Using new gaskets, install the header tank to the cylinder head. Make sure to install the bolts in their original locations. Make sure to install all the header tank and bracket bolts before tightening.

11. Once all the bolts are installed, they can be tightened.

12. Using a new gasket, attach the exhaust elbow to the header tank. Tighten the four bolts securely.

13. Install any accessories that were attached to the header tank, such as cable brackets, fuel filter brackets, fuel lines, etc. If new gaskets are required, make sure to replace them accordingly.

Fig. 88 Once the bolts are removed, remove the mixing elbow from the header tank

Fig. 90 Once all the components are detached from the header tank, remove the bolts and nuts on the side of the tank

Fig. 89 Although this exhaust elbow looks different, it is removed in the same manner

Fig. 91 Carefully remove the header tank from the cylinder head (note cylinder head studs)

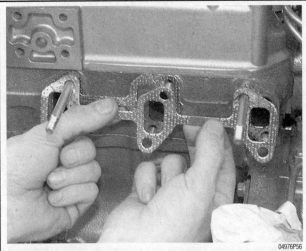

Fig. 92 After the header tank is from the cylinder head, remove the old gasket from the cylinder head

Fig. 93 Use a Scotch-Brite® pad to clean the mating surfaces of both the cylinder head and the header tank

Fig. 94 The same scraping and cleaning of the old gasket material should be done on the exhaust elbow mating surface

Fig. 95 As with all engine component assembly, always use a new gasket

14. Clean any corrosion from the inside of each hose to ensure a leak-free connection. Attach all hoses to the header tank.

15. Add coolant to the header tank.

16. If necessary, bleed the air in the fuel system.

17. Turn the seacock to the **ON** position, and operate the engine. Check for any signs of leakage.

JH series engines

♦ See Figure 96

1. Turn the seacock to the **OFF** position.

2. Using the draincocks on the engine, drain the raw water from the engine by opening the draincocks. After the engine is drained, be sure to close the draincocks.

3. Drain the coolant from the header tank.

※※ CAUTION

Never open, service or drain the fresh water cooling system when the engine is hot; serious burns can occur from the steam and hot coolant. Also, when draining engine coolant, keep in mind that cats and dogs are attracted to ethylene glycol antifreeze and could drink any that is left in an uncovered container. This will prove fatal in sufficient quantities. Always drain coolant into a sealable container.

4. On aftercooled turbo models, proceed as follows:

   a. drain the coolant from the aftercooler.

   b. Remove the air inlet and outlet hoses.

   c. Remove the coolant lines from the aftercooler.

   d. Unbolt the aftercooler from the header tank.

5. Remove the hoses from the coolant tank.

6. Remove any accessories that are attached to the header tank, such as cable brackets, fuel filter brackets, fuel lines, etc.

7. If necessary, remove the alternator to allow clearance to remove the header tank.

8. Remove the four bolts that attach the exhaust elbow to the header tank.

9. Remove the bolts that attach the header tank to the cylinder head.

➡When removing the header tank bolts, be sure to mark each bolt's location, since some of the bolts are different lengths. If the bolts

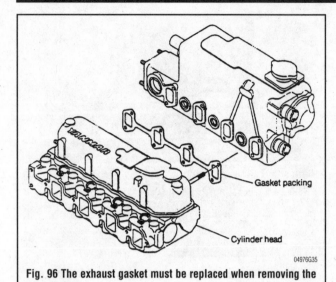

Fig. 96 The exhaust gasket must be replaced when removing the header tank

are not installed in their proper locations, the engine block and/or cylinder head can be damaged when the bolts are tightened.

10. Remove the header tank from the engine.

To install:

11. Clean the mating surfaces of the cylinder head and header tank.

12. Using new gaskets, install the header tank to the cylinder

head. Make sure to install the bolts in their original locations. Make sure to install all the header tank and bracket bolts before tightening.

13. Once all the bolts are installed, they can be tightened.

14. Using a new gasket, attach the exhaust elbow to the header tank. Tighten the four bolts securely.

15. On aftercooled turbo models, attach the aftercooler to the header tank, install the air inlet and outlet hoses, and coolant lines.

16. Install any accessories that were attached to the header tank, such as cable brackets, fuel filter brackets, fuel lines, etc. If new gaskets are required, make sure to replace them accordingly.

17. Clean any corrosion from the inside of each hose to ensure a leak-free connection. Attach all hoses to the header tank.

18. Install the fresh water and raw water hoses to the header tank.

19. Add coolant to the header tank.

20. Turn the seacock to the **ON** position, and operate the engine. Check for any signs of leakage.

Heat exchanger

▶ See Figure 97

Heat exchangers remove surplus heat from the engine. A heat exchanger consists of a cylinder with a number of small copper-nickel tubes running through it. The hot coolant passes through the exchanger while cold raw-water is pumped through the tubes. The cold water carries off the heat from coolant then discharges directly overboard or via a wet exhaust.

The heat exchanger is usually located inside the header tank.

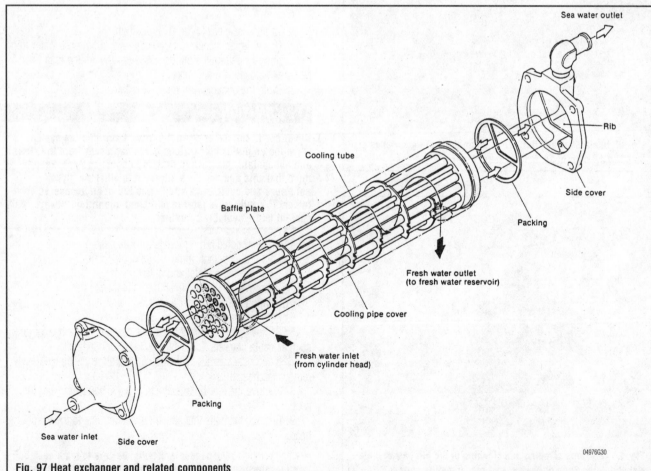

Fig. 97 Heat exchanger and related components

INSPECTION

See Figure 98

The easiest way to inspect the heat exchanger is to measure the temperature difference between the inlet and outlet pipes. The inlet pipe should be slightly above ambient water temperature and the outlet should feel warm.

If the outlet pipe feels hot, too little water is passing through the heat exchanger. Inspect the raw-water pump, strainer and exhaust injection elbow for restrictions.

If the outlet pipe feels cold, water flow is good but heat transfer is bad. Check the heat exchanger internal core for damage.

REMOVAL & INSTALLATION

See Figures 99 thru 105

1. Turn the seacock to the **OFF** position.
2. Using the draincocks on the engine, drain the raw water from the engine by opening the draincocks. After the engine is drained, be sure to close the draincocks.
3. Drain the coolant from the header tank.

CAUTION

Never open, service or drain the fresh water cooling system when the engine is hot; serious burns can occur from the steam and hot coolant. Also, when draining engine coolant, keep in mind that cats and dogs are attracted to ethylene glycol antifreeze and could drink any that is left in an uncovered container. This will prove fatal in sufficient quantities. Always drain coolant into a sealable container.

4. Remove any brackets or components that are in the way of the heat exchanger covers.
5. Remove the bolts that attach the heat exchanger covers to the header tank.

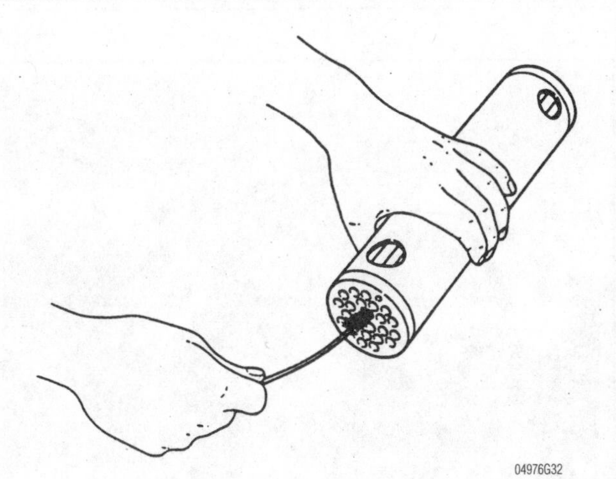
04976G32

Fig. 98 A bottle brush can be used to clean the heat exchanger tubes

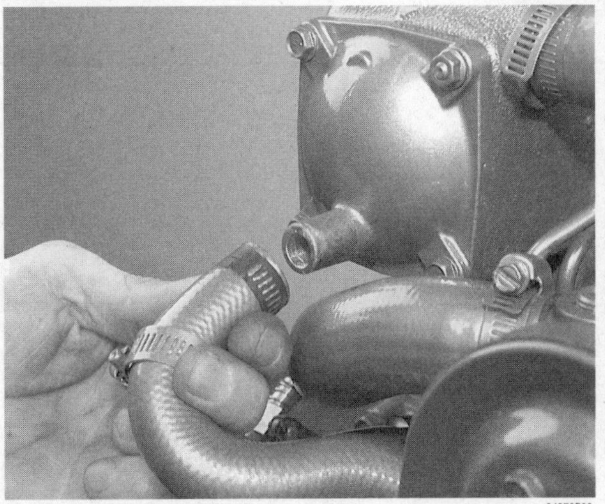
04976P33

Fig. 100 Remove the hoses from the header tank side covers

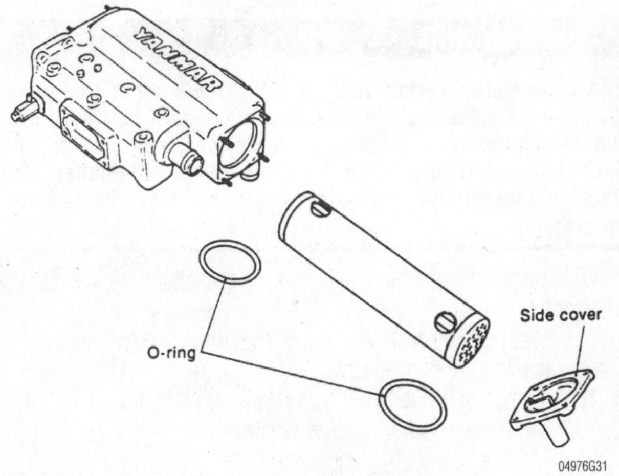

04976G31

Fig. 99 By removing the side covers from the header tank, the heat exchanger can be removed

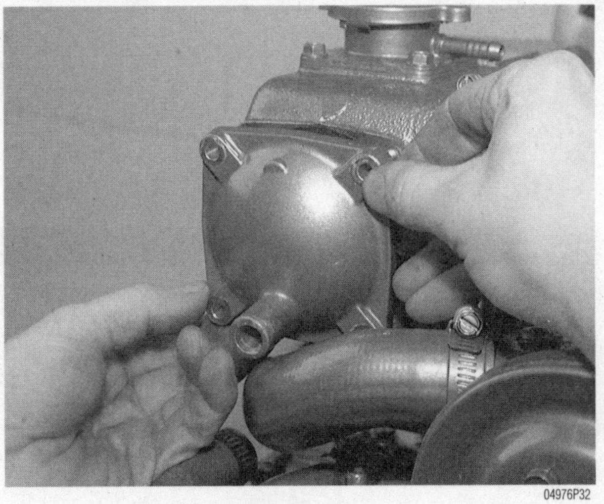
04976P32

Fig. 101 Remove the bolts that attach the side covers to the header tank, and remove the covers

Fig. 102 Carefully slide the heat exchanger out of the header tank

Fig. 104 Once the heat exchanger is in place, install the O-rings

Fig. 103 Before installing the heat exchanger, clean any debris from the O-ring mating surfaces on both sides of the tank

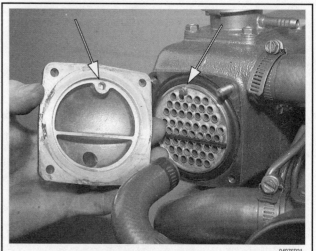

Fig. 105 Make sure the pin on the heat exchanger aligns with the hole in the side cover

6. Slide the heat exchanger out of the header tank.

To install:

7. Using steel wool or light sandpaper, carefully clean any buildup from the O-ring mating surfaces.

8. Place the heat exchanger into the header tank. Be sure to install a new O-ring on each end of the heat exchanger.

✳✳ WARNING

If the O-rings on BOTH sides of the heat exchanger are not replaced, the fresh water and raw water cooling circuits may mix if an O-ring leaks. This can cause the fresh water circuit to become contaminated with sea water, and the sea water circuit to discharge coolant overboard.

9. Once the heat exchanger is in place, install the O-ring on the covers.

✳✳ WARNING

Make absolutely certain that the heat exchanger is seated properly between the two covers on the header tank before tightening the bolts. If the heat exchanger is not positioned correctly, it will be crushed when the bolts are tightened, and leak coolant into the raw water circuit and be discharged overboard.

10. Tighten the bolts that attach the heat exchanger covers to the header tank.

11. Install any brackets or components that were removed to access the heat exchanger cover.

12. Fill the fresh water cooling system with coolant.

13. Turn the seacock to the **ON** position.

14. Start the engine, and check for leaks.

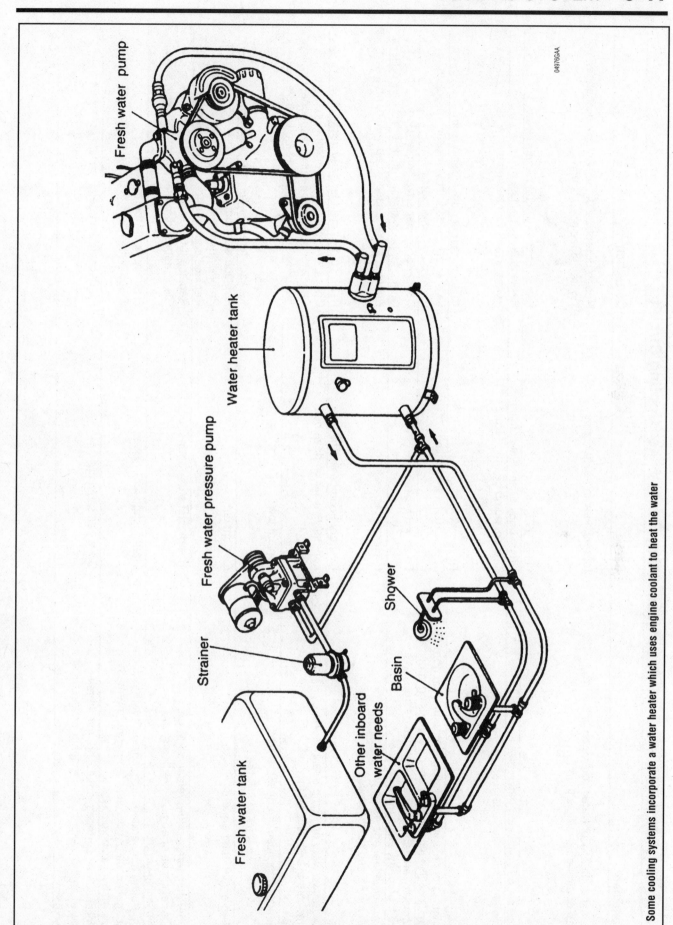

04976GAA

Fresh water pump

Water heater tank

Fresh water pressure pump

Strainer

Shower

Basin

Other inboard water needs

Fresh water tank

Some cooling systems incorporate a water heater which uses engine coolant to heat the water

Troubleshooting Fresh Water Cooling Circuits

Condition	Cause	Correction
Continuous flow of water from open pressure cap or through overflow tube	Internal leak in header tank	Repair or replace defective heat exchanger Tighten loose hose clamps
Coolant pump leaking externally	Oil in coolant Pump body or shaft seal failure	Repair blown cylinder head gasket Replace faulty coolant pump
Coolant pump shaft seal leaking	Coolant pump bearing failure	Replace faulty coolant pump
Engine not reaching operating temperture on normal circuit or engine will overheat on bypassing circuit	Drive belt too tight Thermostat not installed	Check for proper drive belt tension Investigate potential overheating problem Install proper thermostat
Engine overheats, electrical devices do not work properly, battery discharges	Drive belt broken	Check for proper drive belt tension Check for poor alignment of component pulleys
Excessive leakage as engine warms up and builds pressure	Pressure cap leaking	Replace faulty pressure cap seal or pressure cap
Excessive wear on pump bearings, large amounts of belt dust, thinning of belt	Cylinder head gasket blown Excessive wear on drive belt	Repair blown cylinder head gasket Check for proper drive belt tension Check for poor alignment of component pulleys Check for proper size and type of drive belt
Header tank needs frequent refilling	Coolant low	Natural expansion Repair external leaks Replace header tank, internal leaks
High engine temperature	Coolant dirty	Replace worn out coolant Properly maintain cooling system Replace coolant after engine overheat
Operating temperature remains low on normal circuit or engine overheats on bypassing circuit	Thermostat stuck open	Replace faulty thermostat Clean debris from thermostat and flush system
Operating temperature high, engine overheats on normal circuit or operating temperature remains low on bypassing circuit	Thermostat stuck closed	Replace faulty thermostat Clean debris from thermostat and flush system
Polished belt surfaces, squealing, overheating low alternator output, false reading on electrical devices	Drive belt slipping	Replace and properly tension drive belt

04976C03

7

ENGINE AND ENGINE OVERHAUL

ENGINE MECHANICAL

Diesel Engine History

The diesel engine bears the name of its inventor, Rudolph Diesel who on February 17, 1894, started and ran his "Rational Heat Engine" for less than a minute. This fist crude, single-cylinder device was based on a principle first demonstrated by something called a "pneumatic tinderbox" a curiosity of the time that resembled a glass bicycle pump and could ignite a piece of paper merely by compressing the air within its cylinder to the temperature where spontaneous combustion could occur. This compression-ignition principle is the basis for diesel engine operation.

Rudolph Diesel invented an engine that was four times as fuel efficient as the best steam engine of its day, using an idea that, although improved upon, remains basically unchanged in current production diesel engines. Only a small handful of new inventions are immediately recognized as revolutionary in their day and the diesel engine is among the historic few. Rudolph Diesel was hailed as a genius all over the world, particularly among the engineering community in America and was invited to be an honored guest at the 1915 San Francisco World's Fair.

Unfortunately, the military potential of the diesel engine was also recognized immediately by those in the Kaiser's intelligence community who feared that Diesel was about to put a powerful instrument of warfare into the hands of an eventual enemy. On September 29, 1913, while crossing the English Channel aboard the steamer Dresden, Rudolph Diesel mysteriously disappeared. A body was found floating eleven days later and papers taken indicated that it was Rudolph Diesel. Suicide was considered as a cause of death, but most believe Diesel was the victim of foul play, killed for the same reason he became famous, the awesome potential of his compression-ignition diesel engine.

Diesel Engine Advantages

The diesel engine offers a number of advantages over the older spark-ignition type of internal combustion engine. Beside being more thermally efficient and getting more power out of the combustion process, diesel engines are more reliable in operation, use less fuel, deliver more power per pound of engine, produce higher sustained torque at a lower rpm and utilize a fuel that is not only more economical, but also less of a fire hazard than gasoline. The fact that diesel fuel is more difficult to ignite, thus safer to handle is a big advantage when it comes to marine installations. Add to this the fact that diesel engines produce much less toxic exhaust emissions than gasoline engines, remarkably free of hydrocarbons and carbon monoxide and it's not hard to see why diesels have grown in popularity among manufacturer's when it comes to meeting emission standards.

Diesel engines have basically the same internal components as gasoline powered engines. The major differences are the fuel, the ignition of the fuel and the manner in which the fuel is directed into the combustion chambers. The diesel engines are of heavier construction to withstand the higher compression ratios and power impulses that the diesel develops. Diesel engines are manufactured from a one cylinder unit to as many as twenty-four cylinders.

The diesel engine lacks the conventional spark type ignition, but depends upon the heat of the compressed air within the cylinder to ignite a specially timed and injected, atomized spray of fuel into the cylinder as the piston nears its top dead center (TDC) position while on the compression stroke. As the air/fuel mixture is ignited and burns, expanding gases are formed with great pressure, which cause the piston to be forced downward in a power developing stroke.

The Diesel Four Stroke Cycle

♦ **See Figure 1**

In order for the four stroke cycle diesel engine to function properly, valves and injectors must act in direct relation to each other and to the four strokes of the engine. The intake and exhaust valves are camshaft operated, linked by tappets or cam followers, pushrods and rocker arms. The injectors are operated by either hydraulic or mechanical means, timed to the crankshaft and/or the camshaft rotation to provide the spray of fuel into the combustion chamber at the precise moment for efficient combustion.

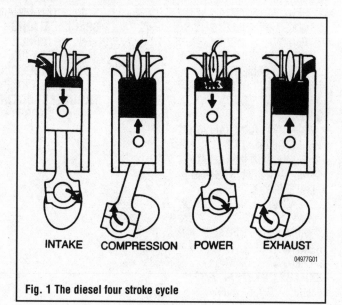

INTAKE   COMPRESSION   POWER   EXHAUST

04977G01

Fig. 1 The diesel four stroke cycle

INTAKE STROKE

During the intake stroke, the piston travels downward with the intake valve open and the exhaust valve closed. The downward travel of the piston allows and draws atmospheric air into the cylinder from the induction system. The intake charge consists of air only and contains no fuel mixture.

COMPRESSION STROKE

At the bottom of the intake stroke with the piston at Bottom Dead Center (BDC), the intake valve closes and the piston starts upwards on its compression stroke. The exhaust valve remains closed. At the end of the compression stroke, air in the combustion chamber has been forced by the piston to occupy a smaller space than it occupied at the beginning of the stroke. Thus, compression ratio is the direct proportion of the amount of space the air occupied in the combustion chamber before and after being compressed.

Diesel engine compression ratios range from approximately 14:1 to 22:1 in comparison to the gasoline engine compression of from 7.5:1 to 9.5:1. Compressing the air into a smaller space causes the temperature of the air to rise to a point high enough to ignite the injected fuel, which has a flash point lower than the temperature of the compressed air. The fuel is injected into the cylinder during the last part of the compression stroke and may continue over to the early part of the power stroke.

POWER STROKE

During the beginning of the power stroke, the piston is pushed downward by the burning and expanding gases. Both the intake and exhaust valves remain closed. As more fuel is added to the cylinder and burns, the gases become hotter and expand more rapidly, forcing the piston downward with much driving force and causing the crankshaft to rotate, in a power delivering action.

EXHAUST STROKE

As the piston reaches its Bottom Dead Center (BDC), the exhaust valve opens and the piston moves upward. The intake valve remains closed. The upward travel of the piston forces the burned gases from the combustion chamber through the open exhaust port and into the exhaust manifold. As the piston reaches the Top Dead Center (TDC) and starts its downward movement, the intake stroke is repeated and the cycling stroke continue in their proper sequence.

Firing Order

▶ See Figure 2

On all engines except for QM series, the number 1 cylinder is located closest to the flywheel. On QM series engines, the number 1 cylinder is located on the front of the engine, near the crankshaft drive pulley.
    The firing order for 4 cylinder engines is 1-3-4-2
    The firing order for 3 cylinder engines is 1-3-2
    The firing order for 2 cylinder engines is 1-2

Valve (Rocker) Cover

REMOVAL & INSTALLATION

▶ See Figures 3 thru 8

1. Remove the breather hose by loosening the clamp and sliding the hose from the fitting on the cover.
2. If necessary, remove any additional components attached to the valve cover.
3. Once all items are removed from the valve cover, remove the cap nut(s) on the top of the cover.
4. Remove the cover from the cylinder head. If necessary, tap the cover lightly with a mallet to break the gasket free from the cylinder head.
   To install:
5. Wipe the valve cover mating surface of the cylinder head with a clean rag.
6. Install a new gasket onto the valve cover. Make sure the gasket is placed in the groove around the entire perimeter of the cover.

Fig. 3 To remove the valve cover, simply remove the cap nuts on the top of the cover

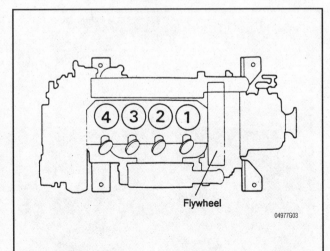

Fig. 2 On all engines except QM series, the number one cylinder is located closest to the flywheel

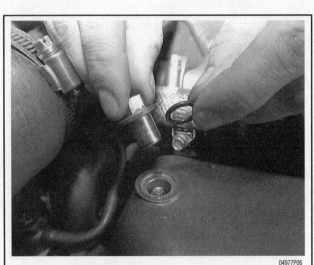

Fig. 4 Always check the rubber O-rings on the valve cover cap nuts and replace them if necessary to prevent leaks

Fig. 5 Once the cap nuts and vent hose are removed, the valve cover lifts from the cylinder head

Fig. 8 Make sure the gasket stays seated in the groove when lowering the cover (a common source of leaks when the cover is removed)

Fig. 6 The valve lash is easily adjusted once the valve cover is removed

7. Place the valve cover onto the cylinder head, in the proper direction.

8. Place a new O-ring onto each of the cap nuts, and thread them onto the valve cover studs.

9. Tighten the valve cover cap nut(s). Do not overtighten them.

10. Install the breather hose onto the fitting on the valve cover, and tighten the clamp as required.

11. If necessary, replace any components that were attached to the valve cover.

DECOMPRESSION LEVERS

▶ See Figures 9, 10 and 11

The decompression levers on the valve cover are used to keep the exhaust valves open while cranking the engine in emergency situations, such as extremely cold weather or a weak battery condition. On most models, these levers are maintenance-free, and do not require any attention. However, on some of the older QM series engines, the decompression levers can be adjusted.

Fig. 7 When installing the valve cover, check the condition of the gasket for hardness or cracking

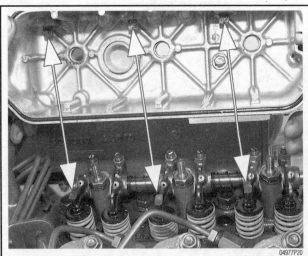
Fig. 9 The decompression levers on the valve cover keep the exhaust valves open for emergency starting

Fig. 10 In this position, the decompression levers are ENGAGED

Fig. 11 The decompression levers should be kept in this position during normal operation

Adjustment

▶ See Figure 12

1. Remove the decompression lever adjustment caps on the valve cover.
2. Starting with the number one cylinder, use the marks on the flywheel to set the piston at TDC on the compression stroke.
3. With the lever in the engaged position, loosen the locknut on the adjuster.
4. Screw the adjuster until it just touches the rocker arm.
5. After the screw touches the rocker arm, turn the adjuster ONE turn further (clockwise), then tighten the locknut while holding the adjuster stationary. The lift provided by the decompression lever should be 0.0314 in. (0.8 mm).
6. Confirm that the exhaust valves are not hitting the pistons when the decompression lever is engaged by carefully rotating the engine and feeling for resistance.
7. Install the decompression lever adjustment caps to the valve cover.

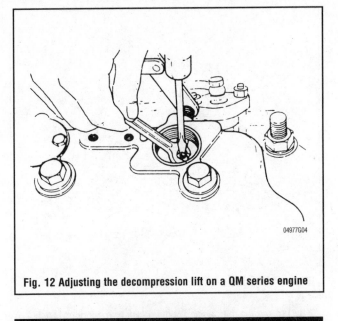

Fig. 12 Adjusting the decompression lift on a QM series engine

✳✳ WARNING

Do NOT use the decompression levers on a warm engine; damage to the valves and pistons may result.

Rocker Arm/Shafts

REMOVAL & INSTALLATION

QM Series Engines

▶ See Figures 13 and 14

1. Remove the valve cover from the cylinder head.
2. Mark each of the rocker arms and supports with paint or other means for proper assembly.
3. Loosen the lock nuts on each of the lash adjusters, and back the adjusters until the rockers are all free from tension of the valve springs.

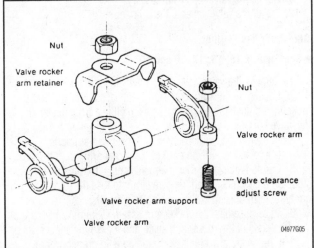

Fig. 13 Exploded view of a QM series rocker arm and shaft assembly

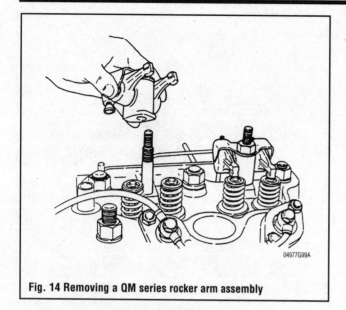

Fig. 14 Removing a QM series rocker arm assembly

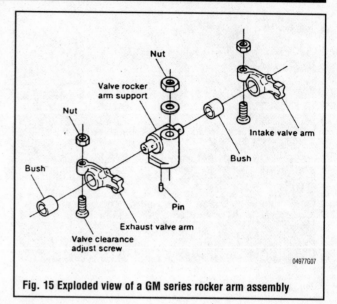

Fig. 15 Exploded view of a GM series rocker arm assembly

4. Loosen the nut on each of the rocker arm supports. Each support carries two rocker arms.

5. Unthread the nuts on each of the rocker arm support studs.

6. Carefully raise each of the rocker arm assemblies from the mounting studs on the cylinder head. Lay each assembly apart from each other so the parts are not mixed together.

To install:

7. Lower each of the rocker assemblies onto the mounting studs, in the correct location.

8. Apply a fresh coating of engine oil to the pushrod tips, and to the valve stem tips.

9. Hand thread each of the rocker arm support nuts onto the mounting studs. Make sure the pin on the bottom of each of the rocker arm supports is in place with the hole on the cylinder head. Make sure each of the pushrods is fully aligned in each of the lash adjusters.

➡ Make sure the lash adjusters are fully slackened, or the rockers will be tensioned when the rocker arm support nut is tightened.

10. Tighten the mounting nuts to 50.6–57.8 ft. lbs. (68.6–78.6 Nm).

11. Adjust the valve lash as necessary.

12. Install the valve cover to the cylinder head.

GM/HM Series Engines

▶ See Figures 15, 16, 17, 18 and 19

1. Remove the valve cover from the cylinder head.

2. Loosen the lock nuts on each of the lash adjusters, and loosen the adjusters until the rockers are all free from the tension of the valve springs.

3. Loosen the nuts on each of the rocker arm supports.

4. Unthread the nuts on each of the rocker arm support studs.

5. Carefully raise the rocker arm assembly from the mounting studs on the cylinder head. Be careful not to lose the locating pins on the bottom of each of the rocker arm supports.

6. To disassemble the rocker arm assembly, remove the circlips from the ends of the rocker arm shaft, and loosen the set screws on each of the rocker arm shafts. Keep the parts in proper order for reassembly.

To install:

7. Apply a fresh coating of engine oil to the pushrod tips, and to the valve stem tips.

8. Lower the rocker assembly onto the mounting studs, in the correct location. Make sure each of the pushrods are seated properly in the lash adjusters.

9. Hand thread each of the rocker arm support nuts onto the mounting studs. Make sure the pin on the bottom of each of the rocker arm supports is in place with the hole on the cylinder head.

➡ Make sure the lash adjusters are fully slackened, or the rockers will be tensioned when the rocker arm support nut is tightened.

10. Tighten the mounting nuts to 27 ft. lbs. (37 Nm).

11. Adjust the valve lash as necessary.

12. Install the valve cover to the cylinder head.

JH Series Engines

▶ See Figure 20

1. Remove the valve cover from the cylinder head.

2. Loosen the lock nuts on each of the lash adjusters, and back the adjusters until the rockers are all free from the tension of the valve springs.

3. Loosen the bolts on each of the rocker arm supports. Keep track of the location of each bolt, as they are different lengths.

4. Carefully raise the rocker arm assembly from the cylinder head.

5. To disassemble the rocker arm assembly, loosen the set screws on each of the rocker arm shafts. Keep the parts in proper order for reassembly.

To install:

6. Apply a fresh coating of engine oil to the pushrod tips, and to the valve stem tips.

7. Lower the rocker assembly onto the cylinder head. Make sure each of the pushrods are seated properly in the lash adjusters.

8. Hand thread each of the rocker arm support bolts into the cylinder head. Make sure the bolts are installed in their proper locations, since they are different lengths.

➡ Make sure the lash adjusters are fully slackened, or the rockers will be tensioned when the rocker arm support nut is tightened.

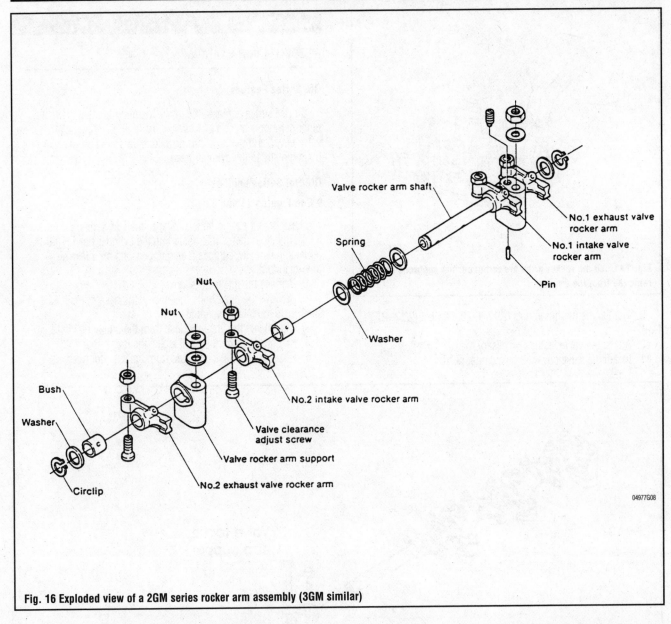

Valve rocker arm shaft

No.1 exhaust valve
rocker arm

No.1 intake valve
rocker arm

Pin

Spring

Washer

Nut

Nut

No.2 intake valve rocker arm

Valve clearance
adjust screw

Valve rocker arm support

Bush

Washer

Circlip

No.2 exhaust valve rocker arm

04977G08

Fig. 16 Exploded view of a 2GM series rocker arm assembly (3GM similar)

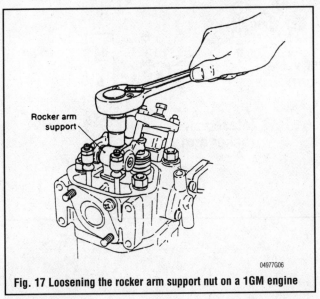

Rocker arm
support

04977G06

Fig. 17 Loosening the rocker arm support nut on a 1GM engine

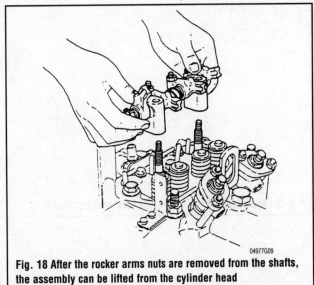

04977G09

**Fig. 18 After the rocker arms nuts are removed from the shafts,
the assembly can be lifted from the cylinder head**

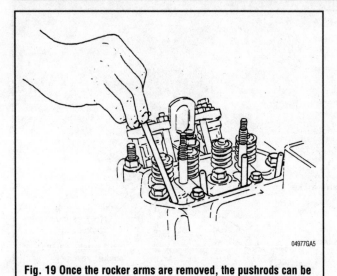

Fig. 19 Once the rocker arms are removed, the pushrods can be removed from the engine

9. Tighten the mounting nuts to 17.4–20.4 ft. lbs. (23.7–27.7 Nm).
10. Adjust the valve lash as necessary.
11. Install the valve cover to the cylinder head.

Intake Manifold

REMOVAL & INSTALLATION

QM Series Engines

On QM series engines, the intake manifold is integrated into other components. The 2QM intake manifold is integrated into the valve cover, and the 3QM has an extension on the air silencer that bolts directly to the cylinder head.

GM/HM Series Engines

♦ See Figures 21 and 22

The 1GM series engine is similar to the QM series, with the intake manifold that is integrated into the cylinder head. The remaining GM/HM engines are equipped with an intake air pipe/manifold.
1. Remove the intake air silencer.
2. Remove the three bolts that secure the base of the intake air silencer to the intake manifold.
3. Disconnect the breather hose from the intake manifold.
4. Remove the two bolts that attach the intake manifold.
5. Remove the intake manifold from the cylinder head.

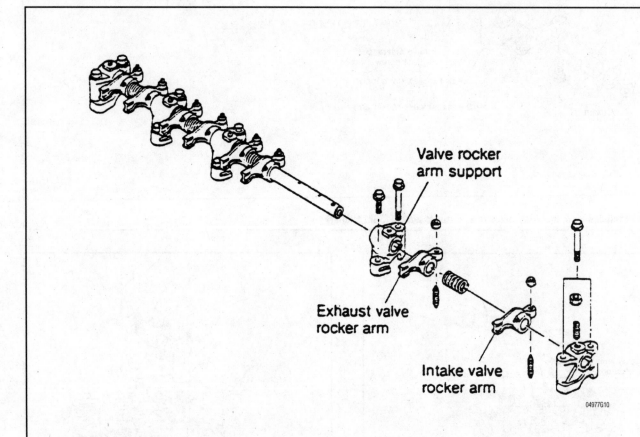

Fig. 20 Exploded view of a JH series rocker arm assembly

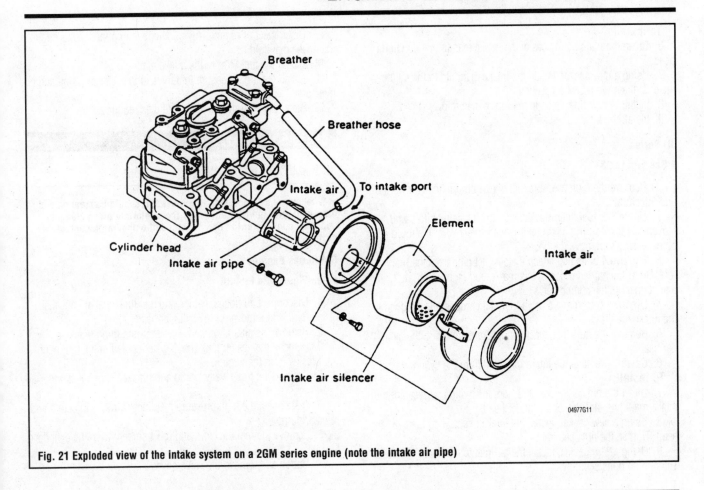

Fig. 21 Exploded view of the intake system on a 2GM series engine (note the intake air pipe)

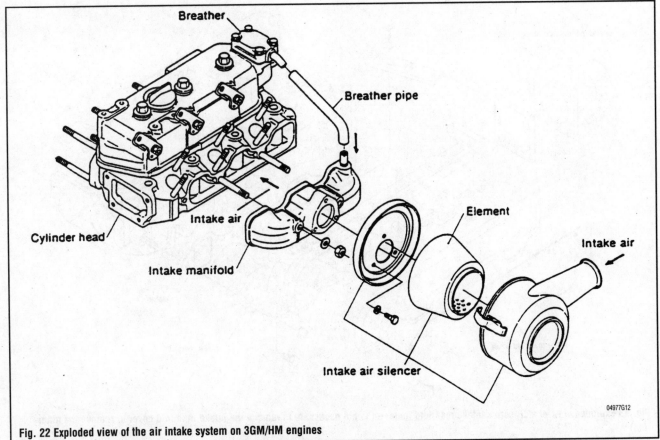

Fig. 22 Exploded view of the air intake system on 3GM/HM engines

To install:

6. Clean the mating surfaces of the intake manifold and cylinder head.

7. Using a new gasket, install the intake manifold to the cylinder head. Tighten the two bolts securely.

8. Install the breather hose to the intake manifold.

9. Install the intake air silencer.

JH Series

♦ **See Figure 23**

1. Remove the high pressure fuel lines from the injector pump and injectors.

2. Unbolt the fuel filter bracket from the intake manifold and move it aside, securing it temporarily with safety wire. It should not be necessary to remove the hoses.

3. Remove 4 bolts that attach the air inlet pipe from the manifold. On turbocharged models, this may require loosening the hoses that connect to the aftercooler.

4. Disconnect or remove any additional components attached to the intake manifold.

5. Remove the nuts that attach the intake manifold to the cylinder head.

6. Carefully remove the intake manifold from the cylinder head.

To install:

7. Clean the old gasket material from the mating surfaces of the intake manifold, air inlet, and cylinder head.

8. Using a new gasket, install the intake manifold to the cylinder head. Tighten the nuts securely.

9. Using a new gasket, install the air inlet to the intake manifold. Tighten the bolts securely.

10. Disconnect or remove any additional components attached to the intake manifold.

11. Attach the fuel filter to the intake manifold.

12. Install the high pressure fuel lines to the injector pump and injectors.

13. Bleed the air from the fuel system as required.

Cylinder Head

REMOVAL & INSTALLATION

➡ **Whenever a cylinder head is removed, it should be cleaned and inspected prior to installation. For more information on cleaning and inspection, refer to "Engine Reconditioning" at the end of this section.**

QM Series Engines

♦ **See Figures 24 thru 30**

1. Disconnect all hoses, cables and fuel lines that are attached or interfere with removal of the cylinder head.

2. Remove the alternator, and alternator mounting brackets.

3. Remove the header tank (fresh water cooled only) and mounting bracket from the engine.

4. Remove the fresh water pump and bracket from the cylinder head.

5. Disconnect the high pressure fuel lines from the injectors and injector pump.

6. Remove the fuel injectors and prechambers from the cylinder head.

7. Unbolt and remove the valve cover.

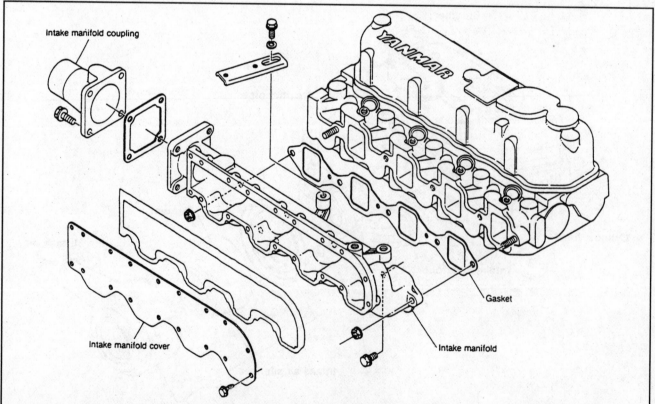

Fig. 23 Exploded view of a JH series intake manifold (note—it is not necessary to remove the intake manifold cover to remove the manifold)

04977G13

8. Remove the air silencer and backing plate from the cylinder head.

9. Remove the exhaust manifold.

10. Remove the oil lines that attach to the cylinder head.

11. Remove the rocker arms from the cylinder head.

12. Remove the pushrods from the engine, keeping them in order for reassembly.

13. Loosen the cylinder head nuts, in the reverse order of the tightening sequence. Loosen the nuts in stages, to avoid warpage of the cylinder head.

✳✳ WARNING

Failure to loosen the cylinder head nuts in the proper order may cause the cylinder head to become warped. On QM series engines, loosen the cylinder head nuts in the reverse order in which they are tightened.

14. Once all the cylinder head nuts are removed, twist the cylinder head horizontally to break the gasket free from the cylinder head.

15. Lift the cylinder head from the engine block.

To install:

16. Carefully clean the old gasket material from the cylinder head and block mating surfaces.

17. Check that all of the cylinder head studs are tight. If a loose stud is found, it can be tightened by locking two nuts against each other. Tighten the studs to 65–69 ft. lbs. (88–94 Nm).

18. Coat both sides of the head gasket with Three Bond® 50, and install the gasket in the proper direction. (Look for a "up" marking on the gasket, which means the word "up" can be read when the gasket is placed on the engine block)

➡**Make absolutely the gasket is of the proper thickness. Using the wrong gasket may cause the valves to hit the pistons, or may affect the compression ratio, decreasing engine performance.**

19. Carefully lower the cylinder head onto the head gasket.

20. Using fresh engine oil, lubricate the threads of the cylinder head mounting studs.

21. Hand thread each of the cylinder head nuts on the studs, checking for smoothness. If a nut binds, it should be replaced, since it can affect the torque setting.

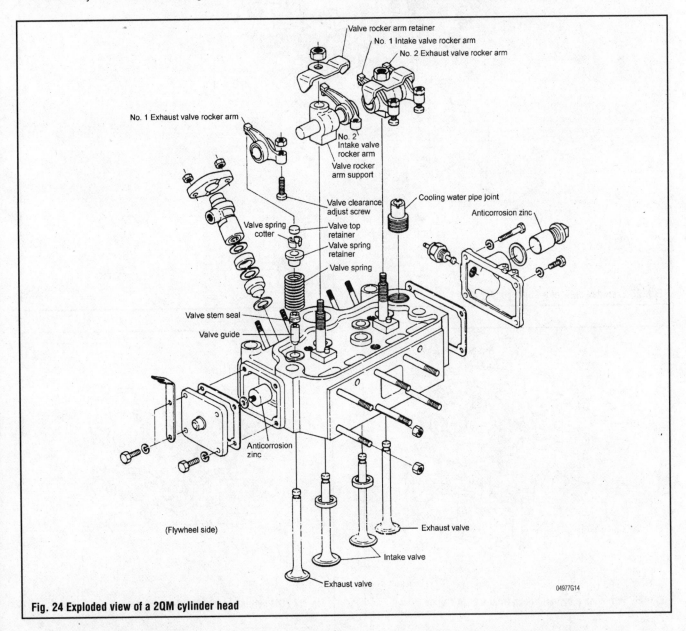

Fig. 24 Exploded view of a 2QM cylinder head

04977G14

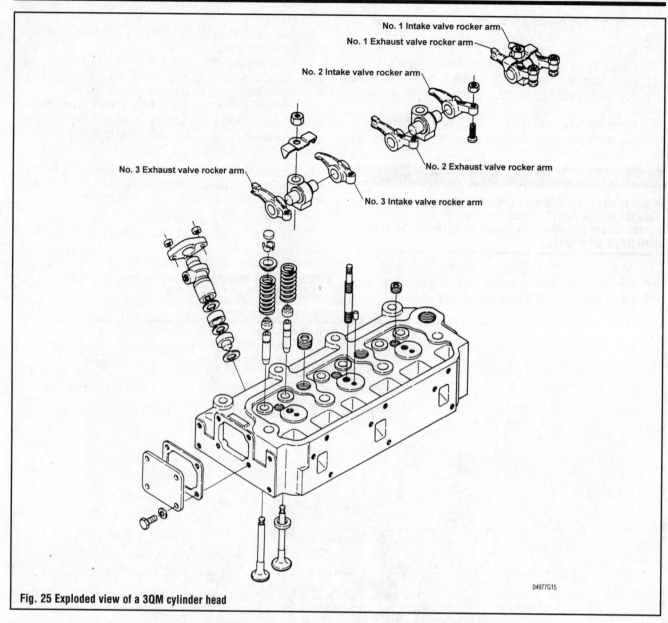

No. 1 Intake valve rocker arm
No. 1 Exhaust valve rocker arm
No. 2 Intake valve rocker arm
No. 2 Exhaust valve rocker arm
No. 3 Exhaust valve rocker arm
No. 3 Intake valve rocker arm

04977G15

Fig. 25 Exploded view of a 3QM cylinder head

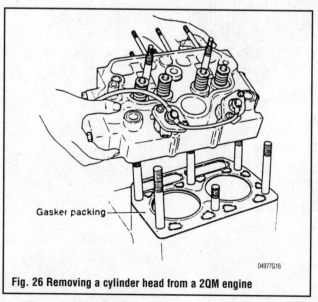

Gasker packing

04977G16

Fig. 26 Removing a cylinder head from a 2QM engine

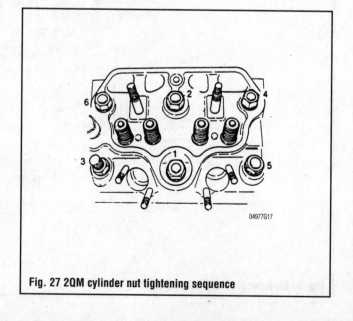

04977G17

Fig. 27 2QM cylinder nut tightening sequence

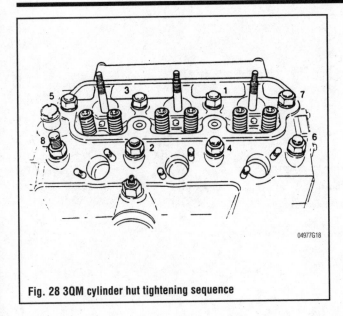

Fig. 28 3QM cylinder hut tightening sequence

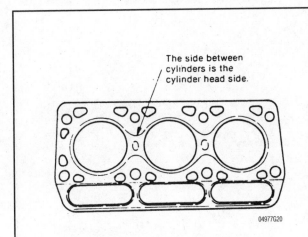

The side between cylinders is the cylinder head side.

Fig. 29 Make sure the head gasket is placed in the proper direction before installing the cylinder head

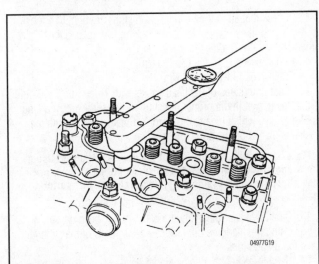

Fig. 30 A torque wrench MUST be used to tighten the cylinder head nuts

22. Using the illustrations as a guide, tighten the cylinder head nuts in the proper sequence as follows:

   a. Tighten the cylinder head nuts to 43 ft. lbs. (58 Nm) in the proper sequence.

   b. Again, in the proper sequence, tighten the cylinder head nuts to 87 ft. lbs. (118 Nm)

   c. Finally, tighten the cylinder head nuts to 116–130 ft. lbs. (158–177 Nm) in the proper sequence. Once the nuts are tightened to the final torque, check them once again to verify the torque.

23. Apply a fresh coating of engine oil to the pushrod tips.

24. Place the pushrods into the engine, in the proper order and direction.

25. Apply a fresh coating of engine oil to the valve stem tips.

26. Install the rocker arms to the cylinder head. Lubricate the rocker arms as necessary.

27. Install the oil lines that attach to the cylinder head. Use new copper gaskets.

28. Install the exhaust manifold.

29. Install the valve cover and air silencer assembly.

30. Install the prechambers and fuel injectors to the cylinder head.

31. Install the high pressure fuel lines to the injectors and injector pump.

32. Install the fresh water pump and bracket to the cylinder head.

33. Install the header tank (fresh water cooled only) and mounting bracket to the engine.

34. Connect all hoses, cables and fuel lines that were attached to the cylinder head.

35. Install the alternator, and alternator mounting brackets. Tension the alternator belt as necessary.

36. Bleed the air from the fuel system as required.

37. Fill the header tank with the proper ratio of coolant and water.

1GM Series Engine

▶ See Figures 31, 32, 33, 34 and 35

1. Remove the intake silencer, and silencer mounting plate.

2. Remove the high pressure and return lines from the fuel injector.

3. Remove the alternator, and alternator mounting brackets.

4. Unbolt and remove the valve cover.

5. Remove the exhaust manifold from the cylinder head.

6. Remove the oil lines that attach to the cylinder head.

7. Remove the rocker arms from the cylinder head.

8. Remove the pushrods from the engine, keeping track of their order and direction.

9. Loosen the cylinder head nuts, in the reverse order of the tightening sequence. Loosen the nuts in stages, to avoid warpage of the cylinder head.

✳✳ WARNING

Failure to loosen the cylinder head nuts in the proper order may cause the cylinder head to become warped. Loosen the cylinder head nuts in the reverse order in which they are tightened.

10. Once all the cylinder head nuts are removed, twist the cylinder head horizontally to break the gasket free from the cylinder head.

11. Lift the cylinder head from the engine block.

To install:

12. Carefully clean the old gasket material from the cylinder head and block mating surfaces.

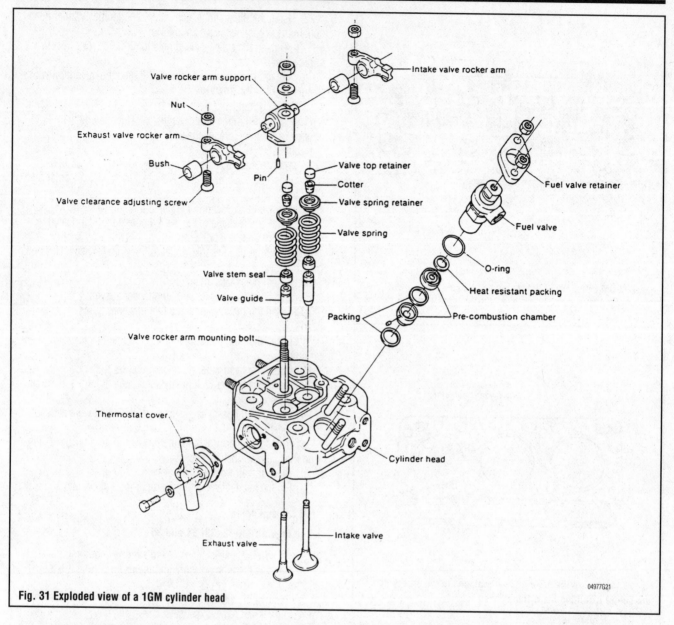

Fig. 31 Exploded view of a 1GM cylinder head

04977G21

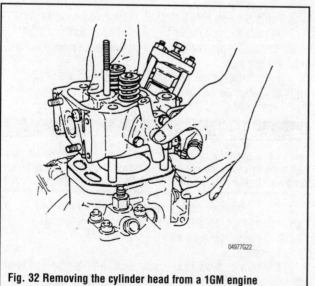

04977G22

Fig. 32 Removing the cylinder head from a 1GM engine

13. Check that all of the cylinder head studs are tight. If a loose stud is found, it can be tightened by locking two nuts against each other. Tighten the studs as follows:
    a. 1GM: 18–22 ft. lbs. (24–30 Nm)
    b. 1GM10: 43 ft. lbs. (58 Nm)

14. Coat both sides of the head gasket with Three Bond® 50, and install the gasket in the proper direction. (Look for a "up" marking on the gasket, which means the word "up" can be read when the gasket is placed on the engine block)

➡**Make absolutely the gasket is of the proper thickness. Using the wrong gasket may cause the valves to hit the pistons, or may affect the compression ratio, decreasing engine performance.**

15. Carefully lower the cylinder head onto the head gasket.

16. Using fresh engine oil, lubricate the threads of the cylinder head mounting studs.

17. Hand thread each of the cylinder head nuts on the studs, checking for smoothness. If a nut binds, it should be replaced, since it can affect the torque setting.

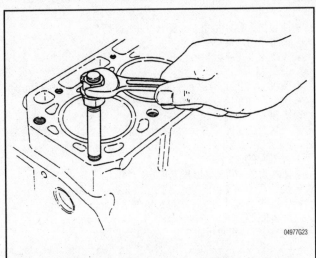

Fig. 33 By locking two nuts together on a stud, the stud can be tightened to the proper torque

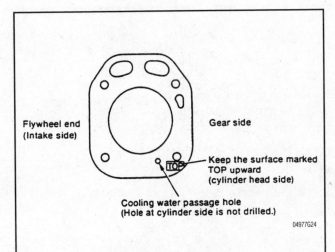

Flywheel end
(Intake side)

Gear side

Keep the surface marked
TOP upward
(cylinder head side)

Cooling water passage hole
(Hole at cylinder side is not drilled.)

04977G24

Fig. 34 Make sure that the head gasket is installed in the proper direction before installing the cylinder head

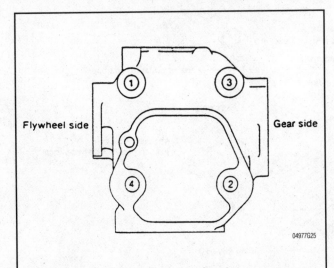

Flywheel side

Gear side

04977G25

Fig. 35 1GM Cylinder head nut/bolt tightening sequence

18. Using the illustrations as a guide, tighten the cylinder head nuts in the proper sequence in stages, working to the final torque of 54.2 ft lbs. (73 Nm)

19. Apply a fresh coating of engine oil to the pushrod tips, and to the tips of the valve stems.

20. Place the pushrods into the engine, in the proper order and direction.

21. Install the rocker arms to the cylinder head. Lubricate the rocker arms as necessary.

22. Install the oil lines that attach to the cylinder head. Use new copper gaskets.

23. Install the exhaust manifold.

24. Install the high pressure fuel line to the injector and injector pump.

25. Install the alternator, and alternator mounting brackets. Tension the alternator belt as necessary.

26. Install the valve cover and air silencer assembly.

27. Bleed the air from the fuel system as required.

2GM and 3GM/HM Series Engines

▶ **See Figures 36 thru 45**

1. Disconnect all hoses, cables and fuel lines that are attached or interfere with removal of the cylinder head.

2. Remove the alternator, and alternator mounting brackets.

3. Remove the header tank (fresh water cooled only) from the cylinder head.

4. Remove the exhaust manifold from the cylinder head. (raw water cooled only)

5. Remove the fresh water pump and thermostat housing from the cylinder head.

6. Disconnect the high pressure fuel lines from the injectors and injector pump.

7. Remove the fuel injectors and prechambers from the cylinder head.

8. Remove the oil lines that attach to the cylinder head.

9. Unbolt and remove the valve cover.

10. Remove the rocker arms from the cylinder head.

11. Remove the pushrods from the engine, keeping them in order for reassembly.

12. Loosen the cylinder head nuts, in the reverse order of the tightening sequence. Loosen the nuts in stages, to avoid warpage of the cylinder head.

✴✴ WARNING

Failure to loosen the cylinder head nuts in the proper order may cause the cylinder head to become warped.

13. Once all the cylinder head nuts are removed, twist the cylinder head horizontally to break the gasket free from the cylinder head.

14. Lift the cylinder head from the engine block.

To install:

15. Carefully clean the old gasket material from the cylinder head and block mating surfaces.

16. Check that all of the cylinder head studs are tight. If a loose stud is found, it can be tightened by locking two nuts against each other. Tighten the studs as follows:

   a. 2GM, 3GM, 3HM: 29–33 ft.lbs (39–45 Nm)

   b. 2GM20, 3GM30: 58 ft. lbs. (79 Nm)

   c. 3GM35: 72 ft. lbs. (98 Nm)

17. Coat both sides of the head gasket with Three Bond® 50, and install the gasket in the proper direction. (Look for a "up" marking

on the gasket, which means the word "up" can be read when the gasket is placed on the engine block.)

➡**Make absolutely the gasket is of the proper thickness. Using the wrong gasket may cause the valves to hit the pistons, or may affect the compression ratio, decreasing engine performance.**

18. Carefully lower the cylinder head onto the head gasket.

19. Using fresh engine oil, lubricate the threads of the cylinder head mounting studs.

20. Hand thread each of the cylinder head nuts on the studs, checking for smoothness. If a nut binds, it should be replaced, since it can affect the torque setting.

21. Using the illustrations as a guide, tighten the nuts and bolts in three steps. First, tighten them to one third of the total torque. Next, go over the sequence again, torquing the nuts and bolts to two thirds of the total torque. Finally, tighten the nuts and bolts to their full torque settings.

    a. 2GM:
- Nuts 1, 2, 3, 4, 5, 6—tighten to 72.3 ft. lbs. (98.3 Nm)
- Bolts 7,8—tighten to 18.1 ft. lbs. (26.6 Nm)

    b. 2GM20:
- Nuts 1, 3, 5—tighten to 86.8 ft. lbs. (118 Nm)
- Bolts 7, 8—tighten to 21.7 ft. lbs. (29.5 Nm)
- Bolts 2, 4, 6—tighten to 86.8 ft. lbs. (118 Nm)

    c. 3GM:
- Nuts 1, 2, 3, 4, 5, 6, 7, 8—tighten to 72.3 ft. lbs. (98.3 Nm)
- Bolts 9, 10, 11—tighten to 18.1 ft. lbs. (26.6 Nm)

    d. 3HM:
- Nuts 1, 2, 3, 4, 5, 6, 7, 8—tighten to 94 ft. lbs. (127.8 Nm)
- Bolts 9, 10, 11—tighten to 21.7 ft. lbs. (29.5 Nm)

    e. 3GM30:
- Nuts 5, 7—tighten to 86.8 ft. lbs. (118 Nm)
- Bolts 9, 10, 11—tighten to 21.7 ft. lbs. (29.5 Nm)
- Bolts 1, 2, 3, 4, 6, 8—tighten to 86.8 ft. lbs. (118 Nm)

    f. 3HM35:
- Nuts 5, 7—tighten to 94 ft. lbs. (127.8 Nm)
- Bolts 9, 10, 11—tighten to 21.7 ft. lbs. (29.5 Nm)
- Bolts 1, 2, 3, 4, 6, 8—tighten to 94 ft. lbs. (127.8 Nm)

22. Apply a fresh coating of engine oil to the pushrod tips, and to the valve stem tips.

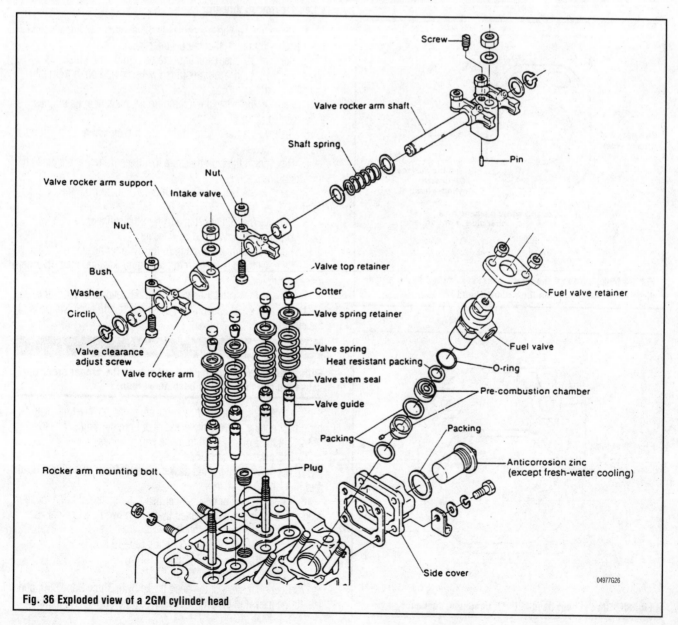

Fig. 36 Exploded view of a 2GM cylinder head

04977G26

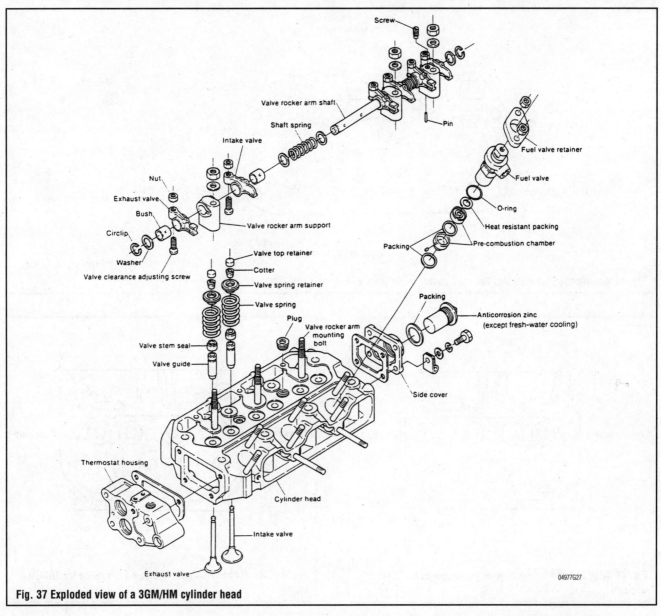

Fig. 37 Exploded view of a 3GM/HM cylinder head

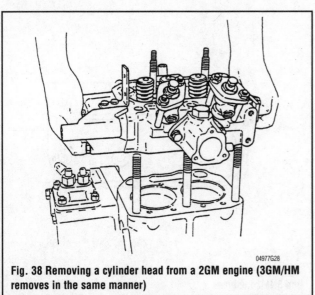

Fig. 38 Removing a cylinder head from a 2GM engine (3GM/HM removes in the same manner)

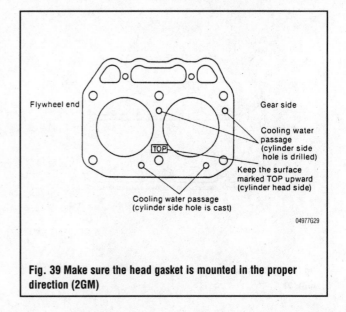

Fig. 39 Make sure the head gasket is mounted in the proper direction (2GM)

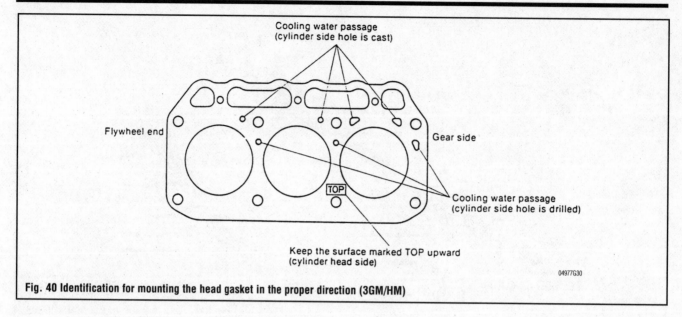

Fig. 40 Identification for mounting the head gasket in the proper direction (3GM/HM)

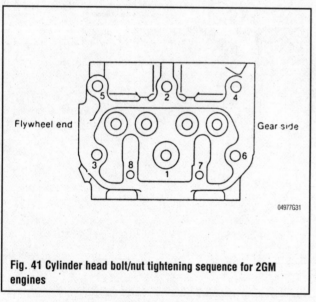

Fig. 41 Cylinder head bolt/nut tightening sequence for 2GM engines

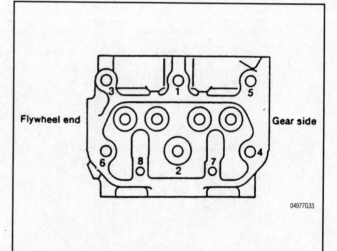

Fig. 43 Cylinder head bolt/nut tightening sequence for 2GM20 engines

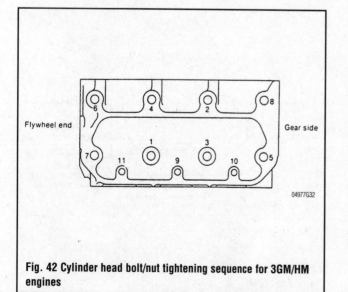

Fig. 42 Cylinder head bolt/nut tightening sequence for 3GM/HM engines

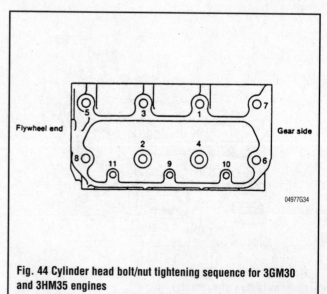

Fig. 44 Cylinder head bolt/nut tightening sequence for 3GM30 and 3HM35 engines

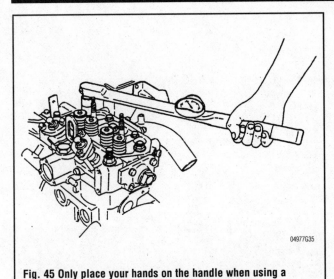

04977G35

Fig. 45 Only place your hands on the handle when using a torque wrench; the torque readings can be affected

23. Install the pushrods into the engine.
24. Install the rocker arms to the cylinder head. Lubricate the rocker arms as necessary.
25. Adjust the valve lash as required.
26. Install the valve cover.
27. Install the oil lines that attach to the cylinder head.
28. Install the fuel injectors and prechambers to the cylinder head.
29. Connect the high pressure fuel lines to the injectors and injector pump.
30. Install the fresh water pump and thermostat housing to the cylinder head.
31. Install the header tank (fresh water cooled only) to the cylinder head.
32. Install the exhaust manifold to the cylinder head. (raw water cooled only)
33. Install the alternator, and alternator mounting brackets.
34. Connect all hoses, cables and fuel lines that were disconnected to allow for removal of the cylinder head.
35. Bleed the air from the fuel system as required.
36. Fill the header tank with the proper ratio of coolant and water.

JH Series Engines

♦ **See Figures 46, 47 and 48**

1. Disconnect all hoses, cables and fuel lines that are attached or interfere with removal of the cylinder head.
2. Remove the alternator, and alternator mounting brackets.
3. Remove the aftercooler (turbocharged models) from the engine.
4. Remove the turbocharger, oil lines, and associated brackets from the engine.
5. Remove the header tank and mounting bracket from the engine.
6. Disconnect the high pressure fuel lines from the injectors and injector pump.
7. Remove the fuel injectors from the cylinder head.
8. Remove the intake manifold from the cylinder head.
9. Remove the fresh water pump and bracket.
10. Remove the oil lines that attach to the cylinder head.
11. Unbolt and remove the valve cover.

12. Remove the rocker arms from the cylinder head.
13. Remove the pushrods from the engine, keeping them in order for reassembly.
14. Loosen the cylinder head bolts and nuts, in the reverse order of the tightening sequence. Loosen the nuts in stages, to avoid warpage of the cylinder head.

⁜ WARNING

Failure to loosen the cylinder head nuts in the proper order may cause the cylinder head to become warped.

15. Once all the cylinder head fasteners are removed, carefully lift the cylinder head from the engine block. Be sure not to lose the positioning pins.
   To install:
16. Carefully clean the old gasket material from the cylinder head and block mating surfaces.

⁜ WARNING

Make sure to clean all debris and fluids from the cylinder bolt holes. If debris or fluid is inside a bolt hole, the bolt will not be able to be fully tightened, and the cylinder head gasket may leak.

17. Install the gasket onto the engine block, in the proper direction. (Look for a "up" marking on the gasket, which means the word "up" can be read when the gasket is placed on the engine block)

➡**Make absolutely the gasket is of the proper thickness. Using the wrong gasket may cause the valves to hit the pistons, or may affect the compression ratio, decreasing engine performance.**

18. Install the positioning pins.
19. Carefully lower the cylinder head onto the head gasket.
20. Using fresh engine oil, lubricate the threads of the cylinder head bolts.
21. Hand thread each of the cylinder head bolts into the engine block, checking for smoothness. If a bolt binds, it should be replaced, since it can affect the torque setting.
22. Using the illustrations as a guide, tighten the cylinder head nuts in the proper sequence as follows:
   a. Tighten the cylinder head bolts to 25–32 ft. lbs. (34–43.5 Nm) in the proper sequence.
   b. Again, in the proper sequence, tighten the cylinder head bolts to 54–61.5 ft. lbs. (73.4–83.6 Nm). Once the nuts are tightened to the final torque, check them once again to verify the torque.
23. Apply a fresh coating of engine oil to the pushrod tips, and to the valve stem tips.
24. Install the pushrods, in the proper direction and order.
25. Install the rocker arms to the cylinder head. Lubricate the rocker arms as necessary.
26. Adjust the valve lash as necessary.
27. Install the valve cover.
28. Install the oil lines that attach to the cylinder head.
29. Install the fresh water pump and bracket.
30. Install the intake manifold to the cylinder head.
31. Install the fuel injectors into the cylinder head.
32. Connect the high pressure fuel lines to the injectors and injector pump.
33. Install the header tank and mounting bracket.
34. Install the turbocharger, oil lines, and associated brackets to the engine. (turbo models only)

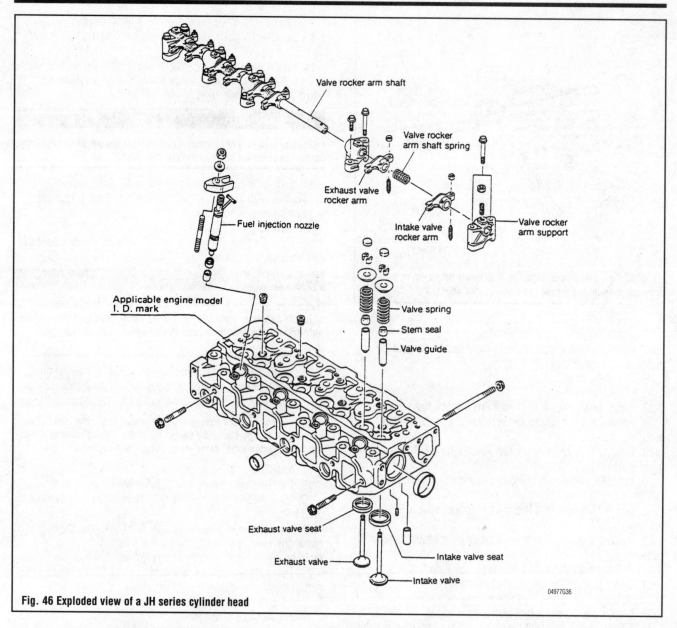

Valve rocker arm shaft

Valve rocker arm shaft spring

Exhaust valve rocker arm

Fuel injection nozzle

Intake valve rocker arm

Valve rocker arm support

Applicable engine model I. D. mark

Valve spring

Stem seal

Valve guide

Exhaust valve seat

Exhaust valve

Intake valve seat

Intake valve

04977G36

Fig. 46 Exploded view of a JH series cylinder head

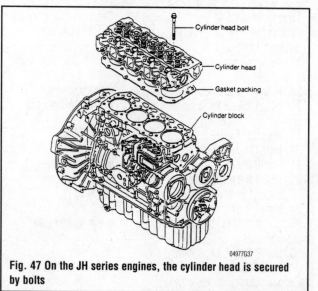

Cylinder head bolt

Cylinder head

Gasket packing

Cylinder block

04977G37

Fig. 47 On the JH series engines, the cylinder head is secured by bolts

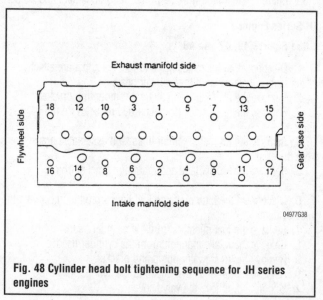

Exhaust manifold side

Flywheel side

Gear case side

18 12 10 3 1 5 7 13 15

16 14 8 6 2 4 9 11 17

Intake manifold side

04977G38

Fig. 48 Cylinder head bolt tightening sequence for JH series engines

35. Install the aftercooler (turbo models only) to the engine.
36. Install the alternator, and alternator mounting brackets.
37. Connect all hoses, cables and fuel lines that were removed for removal of the cylinder head.

Oil Pan

REMOVAL & INSTALLATION

QM & GM/HM Series engines

▶ See Figures 49 and 50

1. On QM series engines, unbolt the dipstick tube from the engine and oil pan.
2. On 1GM series engines, remove the oil intake pipe by removing the plug.
3. Remove the bolts that attach the oil pan to the engine block.

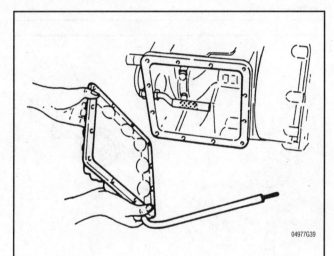

Fig. 49 Removing the oil pan from a 2QM engine (3QM removes in the same manner)

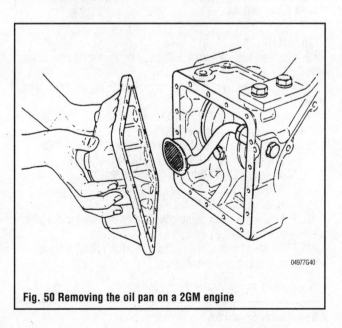

Fig. 50 Removing the oil pan on a 2GM engine

➡Keep track of the location of each bolt as it is removed, since some of the bolts are of different lengths. Installing an oversize bolt can crack the engine block if the bolt is tightened.

4. Using a mallet, tap on the oil pan until the gasket breaks free.
To install:
5. Clean the gasket material from the mating surfaces of the oil pan and engine block.
6. Using a new gasket, install the oil pan to the engine. Make sure to install the bolts in their proper locations. Torque the bolts as follows:
    a. QM series engines: 17–21 ft. lbs. (23–29 Nm)
    b. GM series engines: 6.5 ft. lbs. (8.8 Nm)
7. On 1GM series engines, install the oil intake pipe and plug.
8. On QM series engines, attach the dipstick tube to the engine and oil pan.

JH Series engines

▶ See Figure 51

1. Remove the brackets that attach the oil pan the flywheel housing.
2. Unbolt the dipstick tube from the engine and oil pan.
3. On turbocharged models, remove the oil return line from the oil pan.
4. Remove the bolts that attach the oil pan to the engine block.

➡Keep track of the location of each bolt as it is removed, since some of the bolts are of different lengths. Installing an oversize bolt can crack the engine block if the bolt is tightened.

5. Using a mallet, tap on the oil pan until the gasket breaks free.
To install:
6. Clean the gasket material from the mating surfaces of the oil pan and engine block.
7. Using a new gasket, install the oil pan to the engine. Make sure to install the bolts in their proper locations.
8. Install the turbocharger oil return line to the oil pan.
9. Install the brackets that secure the oil pan to the flywheel housing.

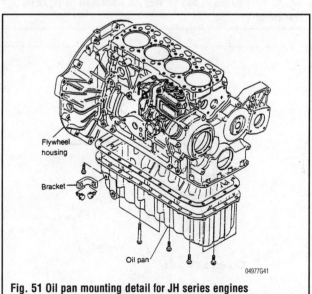

Fig. 51 Oil pan mounting detail for JH series engines

Crankshaft Pulley And Seal

REMOVAL & INSTALLATION

▶ **See Figures 52, 53, 54 and 55**

1. Loosen the alternator mounting brackets, and remove the drive belt from the crankshaft pulley.

2. Loosen the raw water pump mounting bracket, and remove the drive belt from the crankshaft pulley.

3. Remove the crankshaft pulley bolt/nut. By threading two bolts into the front of the pulley, a pry bar can be used to hold the pulley stationary while the bolt/nut is loosened. Make sure the bolts are completely threaded into the pulley, or the threads may be damaged. Otherwise, an air operated impact wrench is the best tool to use for removing the pulley bolt/nut.

➡**3HM engines use a crankshaft bolt with counterclockwise threads.**

4. Once the bolt/nut is removed, install a suitable puller tool and remove the crankshaft pulley. Be careful not to lose the key when the pulley is removed.

Fig. 52 A single nut or bolt secures the crankshaft pulley to the crankshaft

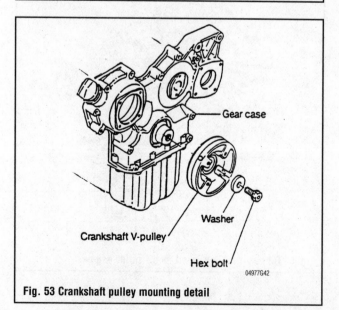

Fig. 53 Crankshaft pulley mounting detail

Gear case

Washer

Crankshaft V-pulley

Hex bolt

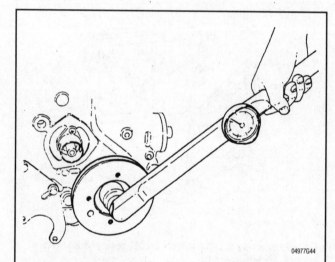

Fig. 54 In most cases, a gear puller must be used to remove the crankshaft pulley

Fig. 55 Always use a torque wrench when tightening the crankshaft pulley bolt/nut

To install:

5. Lubricate the crankshaft pulley seal on the timing gear cover with oil.

6. Install the key, and slide the crankshaft pulley onto the crankshaft.

7. Apply Three Bond® 3B8-005 (or equivalent) on the bolt/nut and install the crankshaft pulley bolt/nut.

8. Tighten the crankshaft bolt/nut as follows:
   a. 2QM series: 101–108.5 ft. lbs. (137.4–147.6 Nm)

9. 3QM series: 60–71.6 ft. lbs. (82–97.4 Nm)
   a. GM/HM series: 72.3 ft. lbs. (113 Nm)
   b. JH series: 83–90.4 ft. lbs. (123 Nm)

10. Place the raw water pump drive belt on the crankshaft pulley, and properly tension the belt.

11. Place the alternator belt on the crankshaft pulley, and properly tension the belt with the alternator.

CRANKSHAFT PULLEY SEAL REPLACEMENT

▶ **See Figures 56 and 57**

1. Remove the crankshaft pulley from the crankshaft.

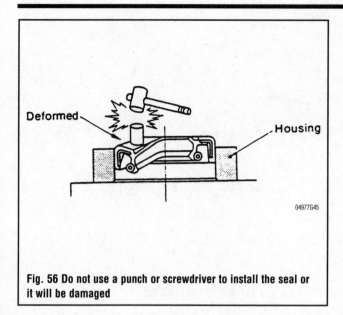

Fig. 56 Do not use a punch or screwdriver to install the seal or it will be damaged

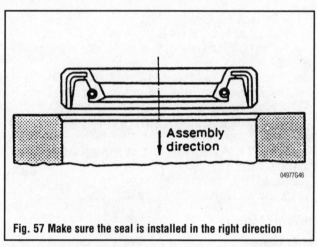

Fig. 57 Make sure the seal is installed in the right direction

2. Using a seal puller, or other suitable tool, carefully pry the old seal from the timing gear cover.

To install:

3. Using a seal installation tool or a large socket, carefully drive the new seal into the timing case. The seal should be flush with the surface of the timing gear cover.

✳✳ WARNING

Do not use a screwdriver or a punch to install the crankshaft pulley seal.

4. Apply a light coating of grease to the inner lip of the crankshaft pulley seal.

5. Install the crankshaft pulley.

Timing Gear Cover

REMOVAL & INSTALLATION

QM Series Engines

▶ **See Figure 58**

1. Remove the alternator.
2. Remove the crankshaft pulley.

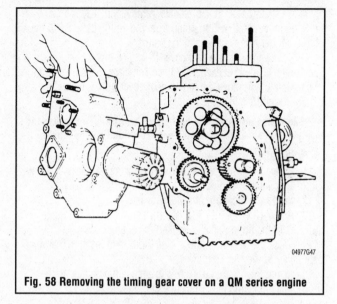

Fig. 58 Removing the timing gear cover on a QM series engine

3. Remove the raw water pump from the timing gear cover.
4. Remove the fuel lift pump from the timing gear cover.
5. If equipped, remove the manual starter mechanism.
6. Remove the covers on the side of the timing gear cover and engine block (above the injector pump) and disconnect the governor link and spring.
7. Remove the timing gear cover-to-engine block bolts. Keep track of the location of each bolt, since some are of different lengths.
8. Carefully remove the timing gear cover from the engine block. Do not lose the locating pins.

➡**When removing the timing gear cover, be careful not to lose the locating pins between the timing gear cover and engine block.**

To install:

9. Clean the gasket material from the mating surfaces of the timing gear cover and engine block.
10. Install the locating pins, if removed.
11. Apply Three Bond® 3B8-005 (or equivalent) to the mating surfaces of the timing gear cover and the engine block.
12. Place the gasket on the engine block, and install the timing gear cover to the engine block.
13. Install the timing gear cover-to-engine block bolts in their proper locations and tighten them to 17.4–21 ft. lbs. (24–28.6 Nm)
14. Connect the governor link and spring. Install the covers on the side of the timing gear cover and engine block (above the injector pump).
15. If equipped, install the manual starter mechanism.
16. Install the fuel lift pump to the timing gear cover.
17. Install the raw water pump. Make sure to properly tension the belt.
18. Install the crankshaft pulley.
19. Install the alternator. Tension the belt properly.

GM/HM Series Engines

▶ **See Figures 59, 60 and 61**

1. Remove the alternator.
2. Remove the crankshaft pulley.
3. Remove the raw water pump from the timing gear cover.
4. Disconnect the oil line on the side of the timing gear cover.
5. Disconnect the engine stop lever and regulator cables.
6. Remove the straight pin from the manual starting handle.

7. On fresh water cooled models, remove the header tank.

8. Remove the injection pump from the timing case. Keep track of the number and thickness of the shims.

9. Remove the timing gear cover-to-engine block bolts. Keep track of the location of each bolt, since some are of different lengths.

10. Carefully remove the timing gear cover from the engine block. Do not lose the locating pins.

➡When removing the timing gear cover, be careful not to lose the locating pins between the timing gear cover and engine block.

To install:

11. Clean the gasket material from the mating surfaces of the timing gear cover and engine block.

12. Install the locating pins, if removed.

13. Apply Three Bond® 3B8-005 (or equivalent) to the mating surfaces of the timing gear cover and the engine block.

14. Place the gasket on the engine block, and install the timing gear cover to the engine block.

15. Install the timing gear cover-to-engine block bolts in their proper locations and tighten them as follows:

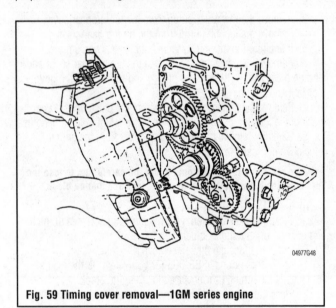

Fig. 59 Timing cover removal—1GM series engine

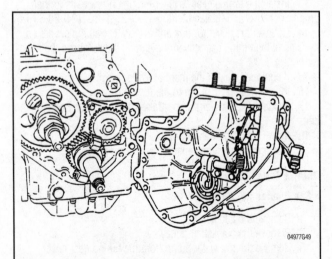

Fig. 60 Removing the timing cover on a 2 and 3GM/HM series engines

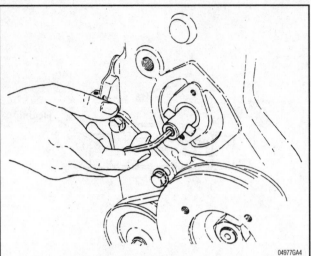

Fig. 61 The manual starting shaft pin is removed by loosening the set screw in the center of the shaft

   a. M6 bolts: 6.5 ft. lbs. (8.8 Nm)
   b. M8 bolts: 18 ft. lbs. (24 Nm)

16. Install the injection pump. Make sure to use the proper number of shims.

17. On fresh water cooled models, install the header tank.

18. Install the straight pin to the manual starting handle.

19. Connect the engine stop lever and regulator cables.

20. Connect the oil line on the side of the timing gear cover. Use new copper gaskets.

21. Install the raw water pump onto the timing gear cover. Tension the belt properly.

22. Install the crankshaft pulley.

23. Install the alternator. Tension the belt properly.

JH Series Engines

▶ See Figure 62

1. Remove the alternator from the timing gear cover.

2. Remove the crankshaft pulley.

3. Remove the raw water pump from the rear of the timing gear cover.

4. Remove the oil pan from the bottom of the engine.

5. Remove the timing gear cover-to-engine block bolts.

6. Carefully remove the timing gear cover from the engine. Do not pry on the timing gear cover; use a mallet and lightly tap on the cover until the gasket breaks loose.

➡When removing the timing gear cover, be careful not to lose the locating pins between the timing gear cover and engine block.

To install:

7. Clean the gasket material from the mating surfaces of the timing gear cover and engine block.

8. Install the locating pins, if removed.

9. Apply Three Bond® 3B8-005 (or equivalent) to the mating surfaces of the timing gear cover and the engine block.

10. Place the gasket on the engine block, and install the timing gear cover to the engine block.

11. Install the timing gear cover-to-engine block bolts in their proper locations and tighten them to 17.4–21 ft. lbs. (24–28.6 Nm).

12. Using a new gasket, install the oil pan to the bottom of the engine.

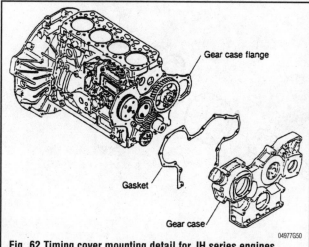

Fig. 62 Timing cover mounting detail for JH series engines (Cylinder head does not have to be removed to remove timing cover)

13. Install the raw water pump to the rear of the timing case.
14. Install the crankshaft pulley.
15. Install the alternator, and tension the belt.

Oil Pump

▶ See Figure 63

REMOVAL & INSTALLATION

QM Series Engines

▶ See Figures 64, 65 and 66

1. Remove the timing gear cover from the engine block.
2. Remove the three bolts that secure the oil pump body to the engine.
3. Remove the oil pump. Make sure not to lose the locating pin.
To install:
4. Pack the pump with petroleum jelly to prevent cavitation of the oil upon engine startup.

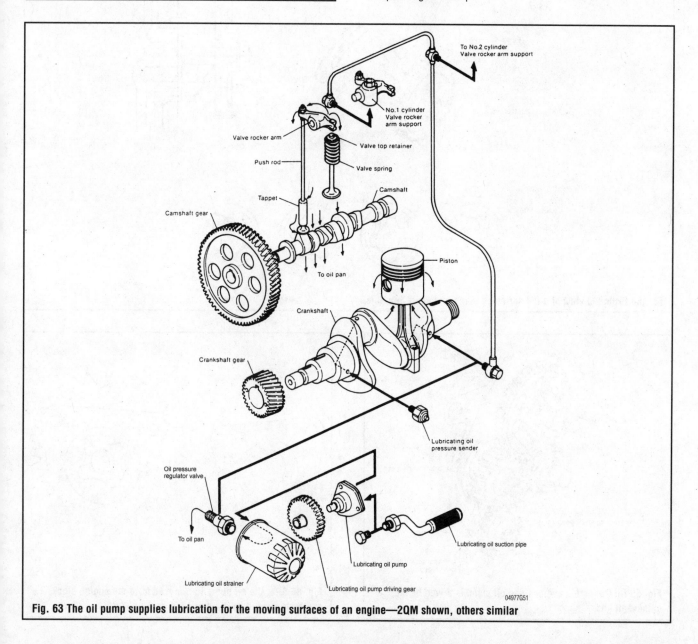

Fig. 63 The oil pump supplies lubrication for the moving surfaces of an engine—2QM shown, others similar

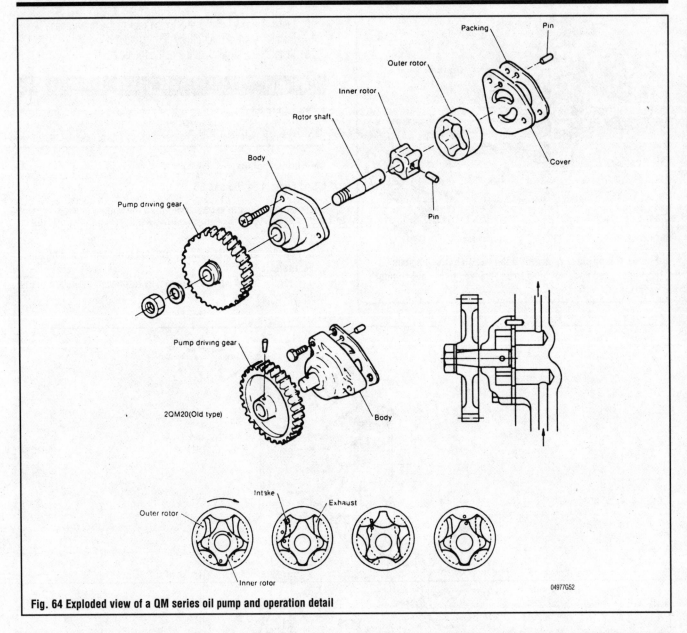

Fig. 64 Exploded view of a QM series oil pump and operation detail

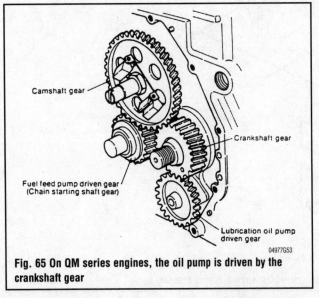

Fig. 65 On QM series engines, the oil pump is driven by the crankshaft gear

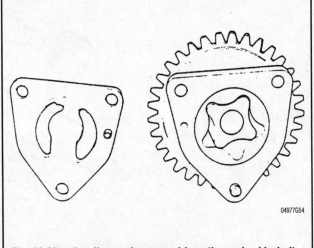

Fig. 66 After the oil pump is removed from the engine block, it can be inspected by removing the backing plate

5. Using a new gasket, install the oil pump to the engine block. Make sure the locating pin is in position. Tighten the bolts securely.

6. Install the timing gear cover to the engine block.

GM Series Engines

▶ **See Figures 67, 68, 69 and 70**

1. Remove the timing gear cover from the engine.

2. On 1GM engines, remove the governor sleeve and thrust bearing.

3. On 1 GM engines, remove the nut that secures the governor weight support, and remove the support.

4. Loosen the oil pump to engine block bolts.

5. Remove the oil pump from the engine block.

To install:

6. Clean the old gasket material from the mating surfaces.

7. Pack the pump with petroleum jelly to prevent cavitation of the oil upon engine startup.

8. Using a new gasket, place the oil pump onto the engine block. Make sure alignment pins are properly located.

9. Tighten the oil pump-to-engine bolts to 6.5 ft. lbs. (8.8 Nm)

10. On 1 GM engines, install the governor weight support, and torque the nut to 58–72.3 ft. lbs. (78.8–98 Nm)

11. On 1GM engines, install the governor sleeve and thrust bearing.

12. Install the timing gear cover.

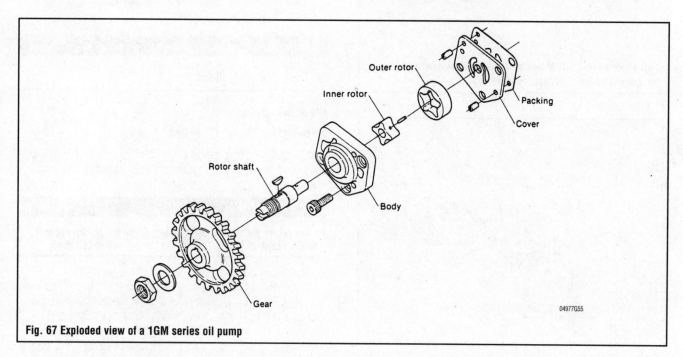

Fig. 67 Exploded view of a 1GM series oil pump

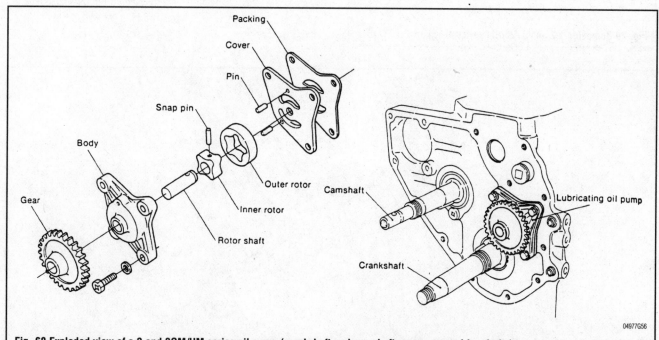

Fig. 68 Exploded view of a 2 and 3GM/HM series oil pump (crankshaft and camshaft gears removed for clarity)

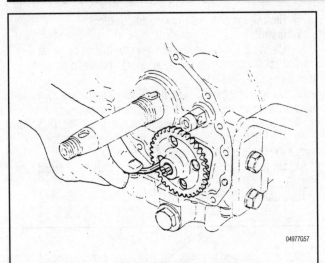

Fig. 69 Removing a 1GM series oil pump. Note how the holes in the gear are used to access the fasteners

Fig. 70 Removing a 2 and 3 GM/HM series oil pump

JH Series Engines

▶ **See Figures 71 and 72**

1. Remove the oil pan from the engine.
2. Remove the timing gear cover from the front of the engine.
3. Remove the 4 bolts that secure the oil pump to the engine block.

To install:

4. Clean the gasket material from the oil pump mating surface on the engine block.
5. Pack the pump with petroleum jelly to prevent cavitation of the oil upon engine startup.
6. Using a new gasket, install the oil pump to the engine block. Tighten the bolts securely.
7. Install the timing gear cover to the engine.
8. Install the oil pan to the engine.

Timing Gears

REMOVAL & INSTALLATION

QM Series Engines

▶ **See Figures 73, 74, 75, 76 and 77**

1. Using the timing marks on the flywheel, align the engine to Top Dead Center (TDC) on the compression stroke of the number one cylinder.

❊❊ WARNING

Do not rotate the engine too quickly; combustion can occur if there is residual fuel in the injector pump or the cylinders.

2. Remove the timing gear cover from the engine.
3. Verify that the timing marks on the camshaft gear and crankshaft gear are aligned. Using paint or other means of identification, highlight the alignment marks on each of the gears. Take special note of the relationship of each of the gears before removal.
4. Once all of the gears are matchmarked for alignment upon installation, remove the oil shielding washer from the crankshaft gear, and slide the crankshaft gear from the crankshaft. Do not lose the gear key.

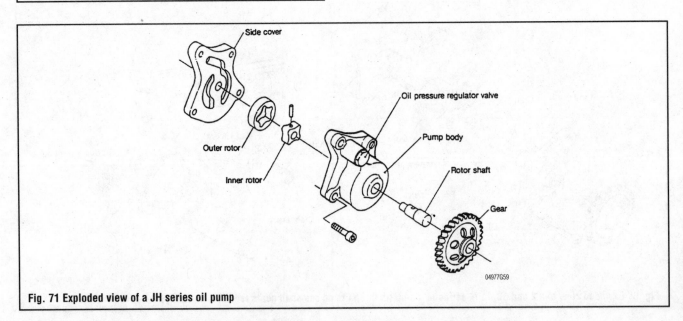

Fig. 71 Exploded view of a JH series oil pump

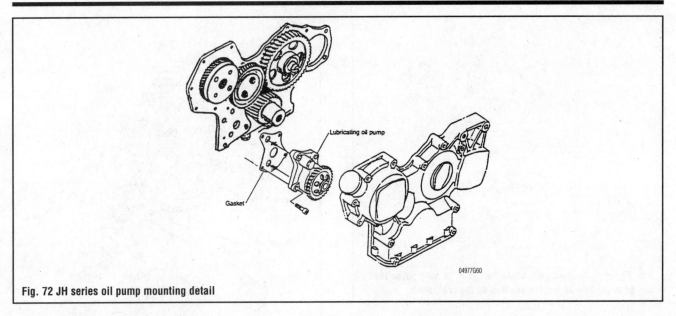

Fig. 72 JH series oil pump mounting detail

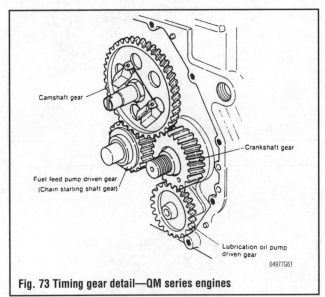

Fig. 73 Timing gear detail—QM series engines

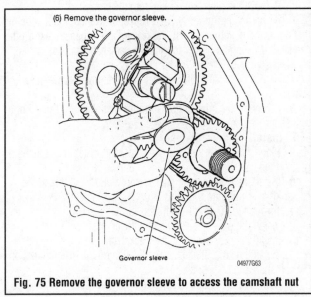

Fig. 75 Remove the governor sleeve to access the camshaft nut

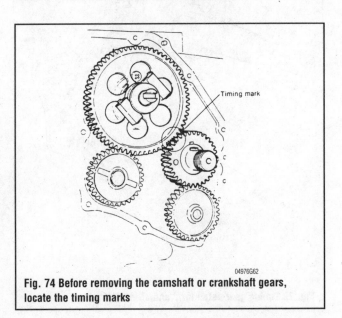

Fig. 74 Before removing the camshaft or crankshaft gears, locate the timing marks

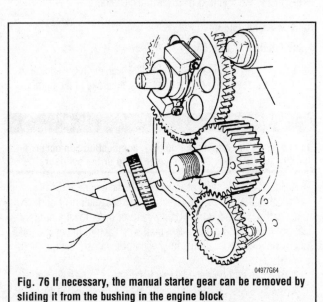

Fig. 76 If necessary, the manual starter gear can be removed by sliding it from the bushing in the engine block

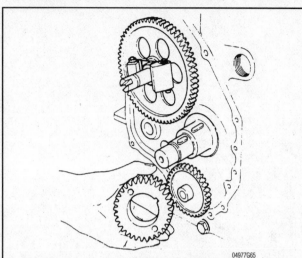

Fig. 77 When removing the crankshaft gear, be sure not to lose the gear key that holds it in position on the crankshaft

5. Remove the governor sleeve from the crankshaft.

6. Using an appropriately sized socket or wrench, loosen the nut that holds the governor weight bracket and the camshaft gear on the camshaft. Use a gear holder to keep the camshaft stationary when removing the nut.

7. Remove the camshaft gear nut, and the governor weight assembly.

8. Remove the camshaft gear from the camshaft. Be careful not to lose the gear key.

To install:

9. Place the gear key on the camshaft, and install the camshaft gear on the camshaft.

10. Install the governor weight assembly, and hand thread the camshaft gear nut onto the camshaft.

11. Torque the nut to 43–58 ft. lbs. (58.5–78.9 Nm).

12. Install the governor sleeve onto the crankshaft.

13. Install the gear key onto the crankshaft.

14. Install the crankshaft gear, while aligning the matchmarks on the camshaft and crankshaft gears.

15. Apply a coating of fresh engine oil to the gears.

16. Install the timing gear cover to the engine.

GM/HM Series Engines

▶ See Figures 78 thru 86

1. Using the timing marks on the flywheel, align the engine to Top Dead Center (TDC) on the compression stroke of the number one cylinder.

※※ WARNING

Do not rotate the engine too quickly; combustion can occur if there is residual fuel in the injector pump or the cylinders.

2. Remove the timing gear cover from the engine.

3. Verify that the timing marks on the camshaft gear and crankshaft gear are aligned. Using paint or other means of identification, highlight the alignment marks on each of the gears. Take special note of the relationship of each of the gears before removal.

4. Using a tool hold the flywheel stationary, loosen the nut that holds the fuel injector cam and cam gear on the camshaft.

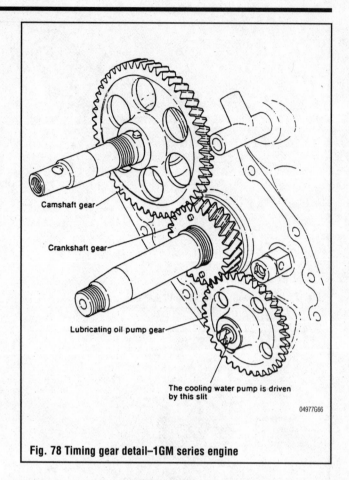

Fig. 78 Timing gear detail–1GM series engine

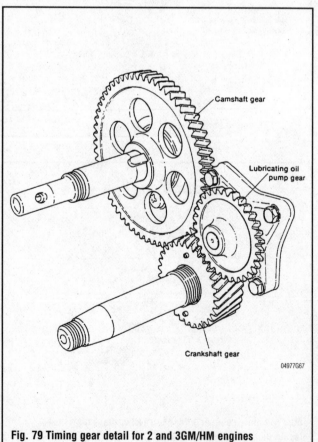

Fig. 79 Timing gear detail for 2 and 3GM/HM engines

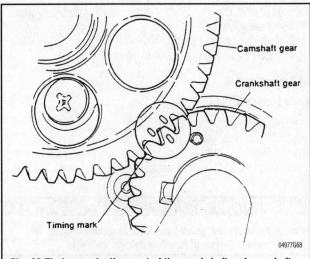

Fig. 80 Timing mark alignment of the crankshaft and camshaft gears

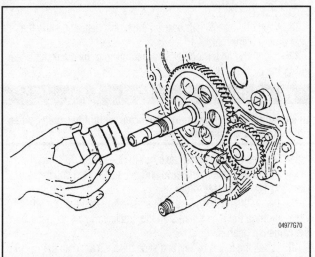

Fig. 82 Once the camshaft nut is removed, the fuel cam can be removed from the camshaft . . .

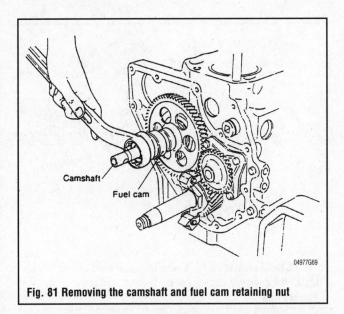

Fig. 81 Removing the camshaft and fuel cam retaining nut

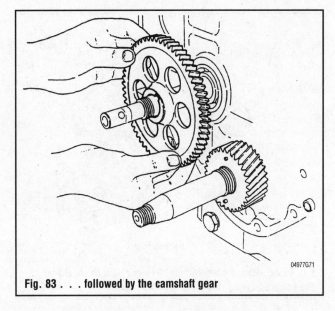

Fig. 83 . . . followed by the camshaft gear

5. Mark the direction of the fuel injector cam (or verify the "O" marking on the outside of the cam), then slide it from the camshaft. Be careful not to lose the fuel cam key.

6. Remove the governor sleeve and needle bearing collar from the crankshaft.

7. Using an appropriately sized socket or wrench, loosen the nut that holds the governor weight bracket and the crankshaft gear on the crankshaft. Use a gear holder on the flywheel if necessary to keep the crankshaft stationary when removing the nut.

✳✳ WARNING

Use caution not to damage the governor weight assembly when loosening the crankshaft gear nut.

8. Remove the crankshaft gear nut, and the governor weight assembly.

9. Remove the camshaft gear from the camshaft. Be careful not to lose the camshaft gear key.

10. Remove the crankshaft gear from the crankshaft. Be careful not to lose the gear key.

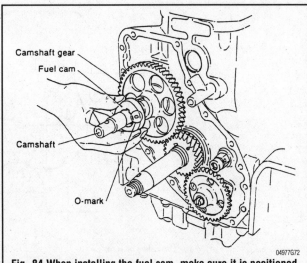

Fig. 84 When installing the fuel cam, make sure it is positioned properly

To install:

11. Install the crankshaft gear on the crankshaft, and slide the gear on the crankshaft.

12. Install the governor weight assembly and the crankshaft gear nut to the crankshaft.

✳✳ WARNING

Use caution not to damage the governor weight assembly when tightening the crankshaft gear nut.

13. Tighten the crankshaft gear nut to 58–72.3 ft. lbs.

14. Install the governor sleeve and needle bearing collar to the crankshaft in the proper order.

15. Install the gear key and the camshaft gear to the camshaft while aligning the matchmarks on the crankshaft gear to those on the camshaft gear.

16. Install the fuel cam key, and slide the fuel cam onto the camshaft in the proper direction (with the "O" marking facing outward).

17. Install the camshaft gear and fuel cam retaining nut, and tighten it to 50.6–57.9 ft. lbs.

18. Verify that the timing marks on the camshaft gear and crankshaft gear are aligned.

19. Apply a coating of fresh engine oil to the gears.

20. Install the timing gear cover to the engine.

JH Series Engines

▶ See Figures 87 and 88

Removal of the camshaft gear requires that the camshaft be removed from the engine, and the gear heated in oil to 356–392 degrees Fahrenheit (180–200 degrees Celsius) and pressed from the camshaft. Refer to the appropriate heading regarding camshaft removal.

✳✳ WARNING

Attempting to pry the gears from their respective shafts with pullers or other tools will result in engine damage.

As with the camshaft gear, removal of the crankshaft gear requires that the crankshaft be removed from the engine, and the

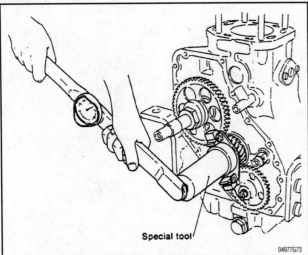

Fig. 85 As with the camshaft nut, A torque wrench should be used to properly tighten the crankshaft nut

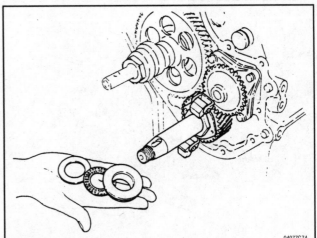

Fig. 86 Before installing the timing gear cover, make sure to install the governor sleeve, needle bearing, and collar in the proper order

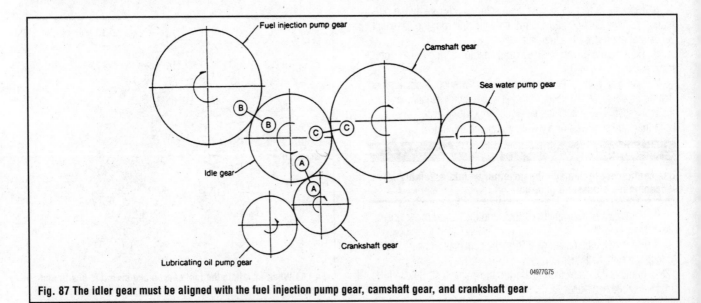

Fig. 87 The idler gear must be aligned with the fuel injection pump gear, camshaft gear, and crankshaft gear

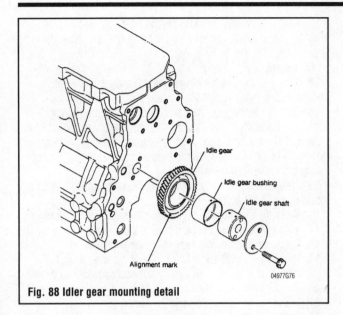

Fig. 88 Idler gear mounting detail

gear heated in oil to 356–392 degrees Fahrenheit (180–200 degrees Celsius) and pressed from the camshaft.

The idler gear (between the camshaft gear and injector pump gear) can be removed by removing the two bolts that retain the idler gear shaft to the engine once the timing gear cover is removed. Before removal it is recommended that the timing marks (A, B, and C) be aligned before removal of the idler gear. It should be noted that it may take several revolutions of the engine to make the timing gears fully align. Make sure to turn the engine slowly, and loosen the fuel lines to the injectors to prevent the engine from starting. Refer to the camshaft removal procedure for more information.

Camshaft, Bearings And Lifters

REMOVAL & INSTALLATION

QM Series Engines

▶ See Figures 89, 90, 91 and 92

1. Remove the gear case cover from the engine.
2. Remove the manual starter gear, if equipped.
3. Remove the governor sleeve from the camshaft.
4. Remove the rocker arm assembly and pushrods from the cylinder head.
5. Remove the fuel lift pump from the side of the engine block.
6. Rotate the engine upside down on the engine stand.

❈❈ WARNING

The engine must be rotated upside down to allow sufficient clearance between the lifters and the camshaft lobes. Do not attempt to remove the camshaft with the engine right side up, or the camshaft and lifters will be damaged.

7. Remove the bottom cover (oil pan) from the engine.
8. Verify the matchmarks on the camshaft and crankshaft gears. Use paint to highlight the marks.
9. Turn the camshaft until the camshaft bearing set screw can be seen through one of the holes in the camshaft gear.
10. Remove the screw that retains the ball bearing on the camshaft.

➡**To remove the camshaft on QM series engines, it is not necessary to remove the camshaft gear.**

11. Carefully remove the camshaft from the engine block.

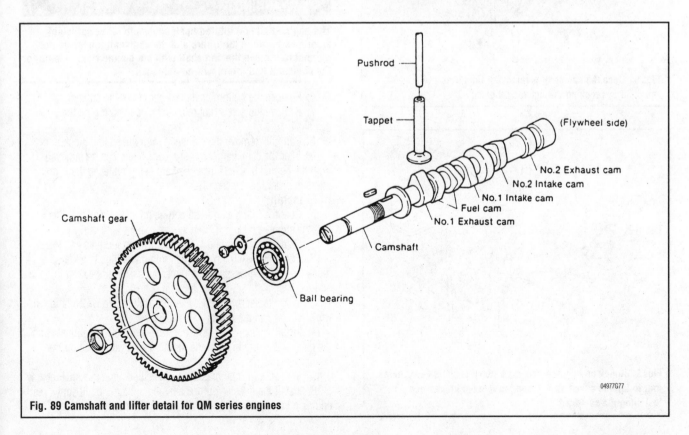

Fig. 89 Camshaft and lifter detail for QM series engines

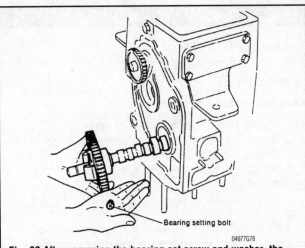

Fig. 90 After removing the bearing set screw and washer, the camshaft can be removed from the engine (note the engine is upside down)

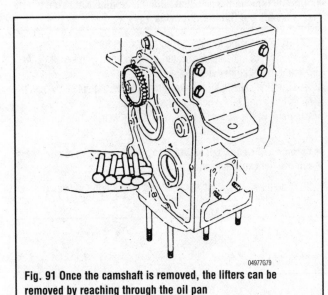

Fig. 91 Once the camshaft is removed, the lifters can be removed by reaching through the oil pan

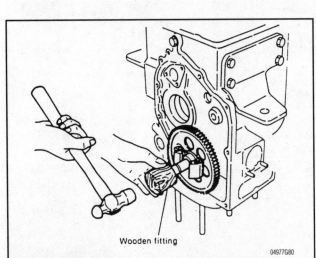

Fig. 92 It may be necessary to use a piece of wood and a hammer to tap the camshaft ball bearing into the block when installing the camshaft

12. After the camshaft is removed, the lifters can be removed from the engine block. Unless they are being replaced, keep the lifters in order for reassembly.

To install:

13. Coat the lifters with fresh engine oil, and install the lifters into the engine block.

14. Coat the bearing journals and camshaft lobes with fresh engine oil. Carefully install the camshaft into the engine block. Use a block of wood if necessary to tap the camshaft ball bearing into the engine block. When installing the gear, make sure to align the timing marks between the two gears.

15. Install the screw that retains the ball bearing which holds in camshaft. tighten it securely.

16. Verify the matchmarks on the camshaft and crankshaft gears.

17. Install the bottom cover (oil pan) to the engine.

18. Rotate the engine right side up on the engine stand.

19. Install the fuel lift pump to the side of the engine block.

20. Install the rocker arm assembly and pushrods to the cylinder head be sure to pre-lubricate them accordingly.

21. Install the governor sleeve to the camshaft.

22. Install the manual starter gear, if equipped.

23. Install the gear case cover to the engine.

GM/HM Series Engines

▶ **See Figures 93, 94, 95 and 96**

1. Remove the gear case cover from the engine.

2. Remove the governor sleeve from the camshaft.

3. Remove the camshaft gear and fuel injector pump cam from the camshaft.

4. Remove the rocker arm assembly and pushrods from the cylinder head.

5. Rotate the engine upside down on the engine stand.

✸✸ WARNING

The engine must be rotated upside down to allow sufficient clearance between the lifters and the camshaft lobes. Do not attempt to remove the camshaft with the engine right side up, or the camshaft and lifters will be damaged.

6. Remove the bottom cover (oil pan) from the engine.

7. Remove the screw that retains the ball bearing on the camshaft.

8. Carefully remove the camshaft from the engine block.

9. After the camshaft is removed, the lifters can be removed from the engine block. Unless they are being replaced, keep the lifters in order for reassembly.

To install:

10. Coat the lifters with fresh engine oil, and install the lifters into the engine block.

11. Using fresh engine oil, coat the bearing journals and lobes.

12. Carefully install the camshaft into the engine block. Use a block of wood if necessary to tap the camshaft ball bearing into the engine block.

13. Install the screw that retains the ball bearing which holds in camshaft. tighten it securely.

14. Install the camshaft gear onto the camshaft. When installing the gear, make sure to align the timing marks with those on the crankshaft gear.

15. Verify the matchmarks on the camshaft and crankshaft gears.

16. Install the injector pump cam, and tighten the nut apply fresh engine oil as necessary to provide lubrication upon starting.

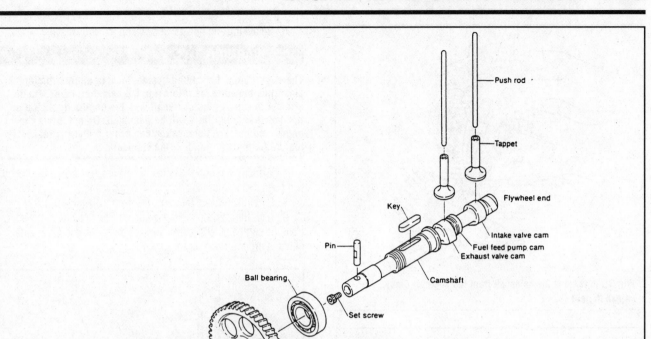

Fig. 93 1GM series camshaft detail

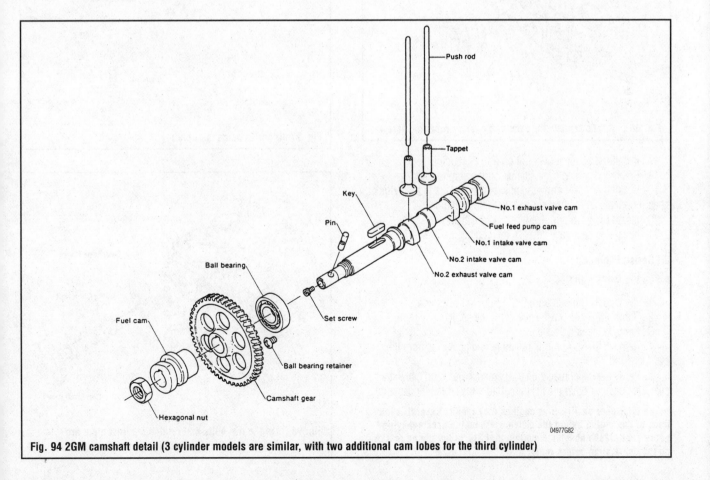

Fig. 94 2GM camshaft detail (3 cylinder models are similar, with two additional cam lobes for the third cylinder)

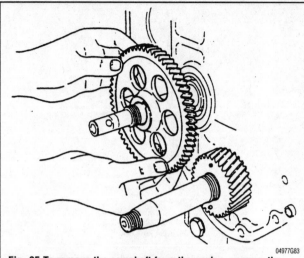

Fig. 95 To remove the camshaft from the engine, remove the camshaft gear . . .

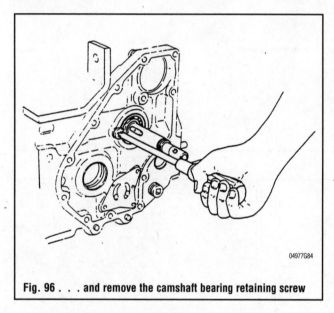

Fig. 96 . . . and remove the camshaft bearing retaining screw

17. Install the bottom cover (oil pan) to the engine.
18. Rotate the engine right side up on the engine stand.
19. Install the rocker arm assembly and pushrods to the cylinder head. Lubricate all moving surfaces with fresh engine oil.
20. Install the governor sleeve to the camshaft.
21. Install the gear case cover to the engine.

JH Series Engines

▶ See Figures 97 and 98

1. Remove the oil pan from the bottom of the engine.
2. Remove the gear case cover from the engine.
3. Remove the valve cover.
4. Remove the rocker arm assembly and pushrods from the cylinder head.
5. Verify the matchmarks on the camshaft gear and crankshaft gear. Use paint or other means to highlight the marks for assembly.

➡The idler gear on JH series engines may require several revolutions of the engine before the matchmarks on the accompanying gears align. Make absolutely certain that ALL of the timing marks on the gears align before removal.

6. Rotate the engine upside down on the engine stand.

❈❈ WARNING

The engine must be rotated upside down to allow sufficient clearance between the lifters and the camshaft lobes. Do not attempt to remove the camshaft with the engine right side up, or the camshaft and lifters will be damaged. Do not rotate the engine too quickly; combustion can occur if there is residual fuel in the injector pump or the cylinders.

7. Working through the holes in the camshaft gear, remove the two bolts that secure the camshaft thrust plate to the engine block.
8. Carefully remove the camshaft from the engine block.
9. After the camshaft is removed, the lifters can be removed from the engine block. Unless they are being replaced, keep the lifters in order for reassembly.

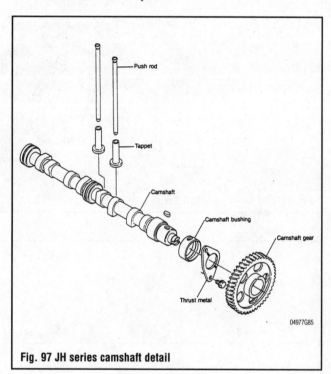

Fig. 97 JH series camshaft detail

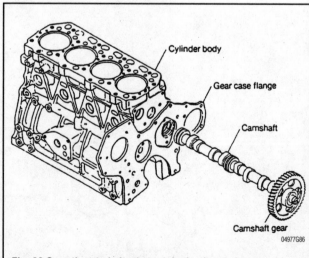

Fig. 98 Once the two bolts that retain the thrust plate are removed, the camshaft can be removed from the engine

To install:

10. Coat the lifters with fresh oil, and install the lifters into the engine block.

11. Lubricate the camshaft journals and lobes with fresh oil, and carefully install the camshaft into the engine block.

12. Install the two bolts to the camshaft thrust plate.

13. Install the camshaft gear onto the camshaft. When installing the gear, make sure to align the timing marks with those on the crankshaft gear.

14. Verify the matchmarks on the camshaft and crankshaft gears.

15. Rotate the engine right side up on the engine stand.

16. Install the rocker arm assembly and pushrods to the cylinder head.

17. Install the governor sleeve to the camshaft.

18. Install the gear case cover to the engine.

19. Install the oil pan to the bottom of the engine.

Flywheel

REMOVAL & INSTALLATION

▶ **See Figures 99 thru 105**

1. Remove the transmission from the engine.

2. Remove the damper disc from the flywheel or clutch assembly from the flywheel. Note the direction in which the disc is mounted; it must be installed in the same manner.

3. On 2QM models, remove the large nut that secures the flywheel. Use a special flywheel holder, and a large socket with a breaker bar.

4. On all other models, remove the bolts that secure the flywheel to the crankshaft. Use a special flywheel holder tool to hold the tool stationary while loosening the bolts.

5. Carefully remove the flywheel from the end of the crankshaft. A locating pin is used between the flywheel and crankshaft. Be careful not to lose the pin when removing the flywheel.

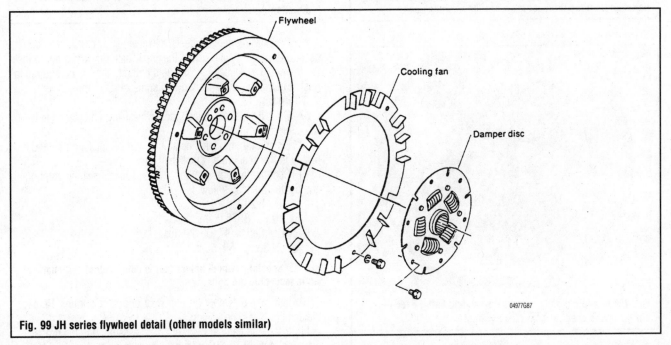

Fig. 99 JH series flywheel detail (other models similar)

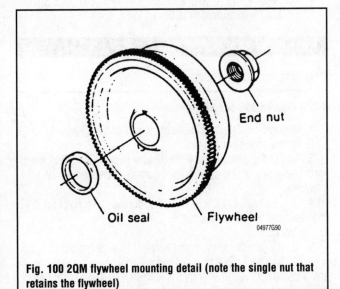

Fig. 100 2QM flywheel mounting detail (note the single nut that retains the flywheel)

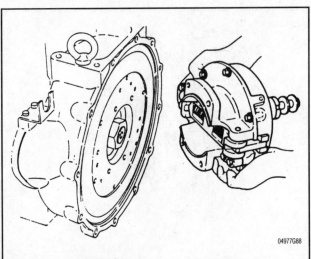

Fig. 101 Some older models have a clutch assembly that attaches directly to the flywheel

Fig. 102 To remove the flywheel, the damper disc must be unbolted from the flywheel to access the mounting bolts

Fig. 103 In addition to the flywheel mounting bolts, a locating pin (arrow) is used to secure the flywheel

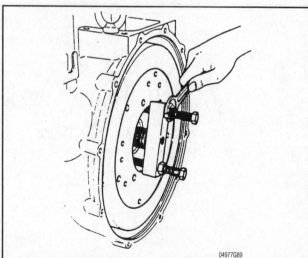

Fig. 104 The flywheel on 2QM engines requires a special tool to remove it from the crankshaft

Fig. 105 Make sure to install the damper disc in the proper direction

6. On 2QM models, a special flywheel removal tool must be used to draw the flywheel from the crankshaft. Be careful not to lose the flywheel key. Usually, installing a bolt into the key is required to push it out of the crankshaft, but in some cases, it may fall out.

To install:

7. On 2QM models, install the flywheel key, and place the flywheel on the crankshaft.

8. On all other models, install the flywheel locating pin, and place the flywheel on the crankshaft.

9. Coat the flywheel nut/bolts with a suitable sealing compound, and tighten them as follows:

    a. 2QM: 290 ft. lbs.

    b. 3QM: 68–76 ft. lbs.

    c. GM Series: 47–50.6 ft. lbs.

    d. JH Series: 50–58 ft. lbs.

➡Use a special flywheel holder tool to hold the tool stationary while loosening the bolts.

10. Inspect the damper disc for wear or broken springs. Replace the damper disc if necessary.

11. Install the damper disc (or clutch assembly) in the proper direction to the flywheel. Torque the bolts to 18 ft. lbs. (24.5 Nm).

12. Install the transmission to the engine.

Rear Main Seal

REMOVAL & INSTALLATION

▸ **See Figures 106 and 107**

1. Remove the transmission from the rear of the engine.

2. Remove the flywheel.

3. On 2QM engines, remove the flywheel key by threading a bolt through the key until it pushes itself from the recession on the crankshaft.

4. Using a seal removal tool or other suitable tool, carefully pry the seal from the rear of the engine.

To install:

5. Lightly grease the inner and outer surfaces of the seal.

6. Using a seal driver or other suitable tool, drive the seal into the housing on the rear of the engine. Make sure the seal lips on the crankshaft surface are not folded over when installing the seal.

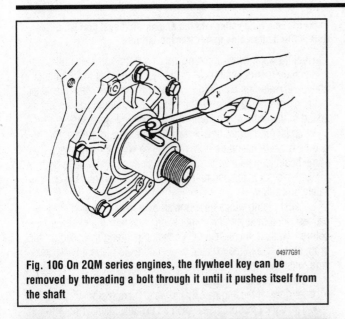

Fig. 106 On 2QM series engines, the flywheel key can be removed by threading a bolt through it until it pushes itself from the shaft

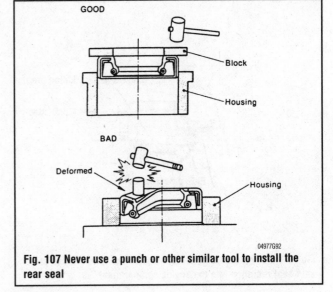

Fig. 107 Never use a punch or other similar tool to install the rear seal

✳✳ WARNING

Do not use a screw driver or a punch to install the seal! it will be damaged, and the seal will leak. The seal must be kept level when driven into the housing.

7. Once the seal has been installed, install the flywheel.
8. Install the transmission to the engine.

EXHAUST SYSTEM

Exhaust Pipes

▶ See Figures 108, 109, 110, 111 and 112

The exhaust systems used on Yanmar engines are of the wet type. Exhaust gas and cooling water are gathered in the water mixing elbow of the exhaust port for discharge together. Accordingly, a heat-resisting rubber hose can be used for the exhaust pipe. This permits a simple exhaust piping layout.

It is necessary to arrange the piping to allow for inspection of the whole system. Also, a suitable arrangement is necessary to prevent sea water from flowing back into the engine. An antisiphon valve

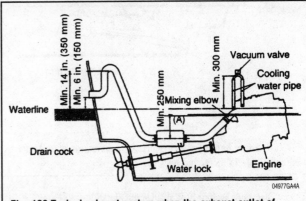

Fig. 109 Typical exhaust system when the exhaust outlet of engine is below the waterline and a vacuum valve is used

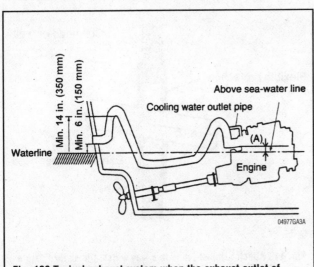

Fig. 108 Typical exhaust system when the exhaust outlet of engine is above the waterline

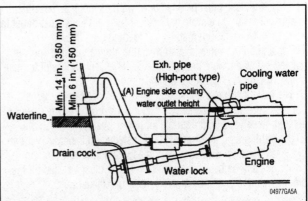

Fig. 110 Typical exhaust system when a high-port exhaust water mixing elbow is used

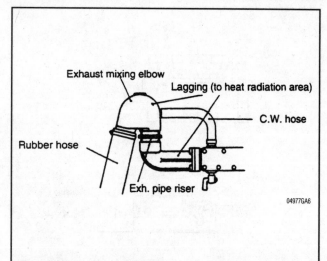

Fig. 111 The water mixing elbow must be well insulated to prevent heat radiation into the engine compartment

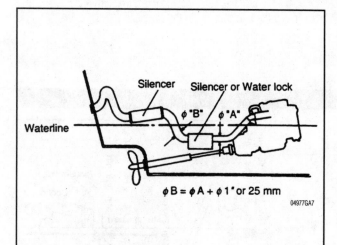

$$\phi B = \phi A + \phi 1" \text{ or } 25 \text{ mm}$$

Fig. 112 Sharp bends in the exhaust, silencers and water locks all contribute to increased back pressure

must used in the system to prevent water remaining in the hose from flowing back to the engine side when stopping the engine or immediately after starting.

The antisiphon valve must be fixed at the lowest possible position, and the hose must be tilted downward as much as possible. It is also necessary to elevate the exhaust hose at the exhaust outlet to more than 14 in. (350mm)above the loading draft line.

The antisiphon valve is used when the engine's exhaust port is located below the waterline. This device is especially useful for sailboats which do not use their engine frequently.

The installation of a antisiphon valve does not cause a large output loss due to the increase of exhaust back pressure. The use of a antisiphon valve is inappropriate for power boats which use the engine's full output. If a water lock is installed on their exhaust piping, back pressure will increase and engine output will drop. In power boats, a high-port water mixing elbow must be used for the piping layout to prevent the reverse flow of exhaust water.

➡Be sure to install a drain plug or cock at the bottom of the antisiphon valve. This is to prevent possible engine damage due to the

entry of remaining water into the engine's exhaust port when the hull is tilted stern side up for transportation.

Cooling water exhaust piping varies according to the hull shape and engine room position. When the water outlet of the engine is above the waterline no additions need to be made. When the water outlet of the engine is below the waterline a vacuum valve must be added to the water mixing elbow of the cooling water pipe.

Wrap lagging around the water mixing elbow to prevent the engine room temperature from rising. The engine room space in some boats is generally narrow, so be sure to provide lagging to prevent excessive temperature rise in the engine room and also as a safety precaution.

Exhaust piping which develops an excessive back pressure causes incomplete engine combustion, the emission of abnormally colored exhaust, increased fuel consumption, engine speed (output) drop, and a rise in the exhaust temperature. To prevent back pressure, observe the following points with regard to the exhaust piping:

• Do not use a hose with a diameter smaller than that of the exhaust hose of the engine exhaust water mixing elbow.

• The back pressure rises when an exhaust silencer or antisiphon valve is used. Make the hose diameter at the outlet of the exhaust silencer or antisiphon valve 1 in. (25 mm) larger than that of the hose on the inlet side in such cases.

• Avoid sharp bends in the exhaust piping hose.

• Excessive rise in back pressure causes surging in turbocharged engines, and this may result in breakage of the turbocharger.

Accordingly, the antisiphon valve should not be used for turbocharged engines.

➡When it is feared that the exhaust piping is causing excessive back pressure, it is necessary to de-rate the engine output and select a more suitable propeller.

EXHAUST PIPE CAUTIONS

◆ See Figure 113

1. Use an all weather, heat-resistant and oil proof rubber hose.
When the hose passes through the cabin or other sections where people go in and out, use a hose with high reliability.

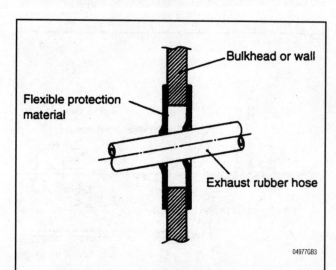

Fig. 113 Attach exhaust pipes in a way which allows for flexible movement during engine vibrations

2. When bending the hose, bend it in a large curve so that the hose diameter is not altered.

3. Keep the hose away from substances which may damage or distort it.

4. When the hose goes through a bulkhead or other walls, protect the hose from distortion or friction damage.

5. Attach the hose in a way which allows for the flexible movement of the hose during engine vibrations.

MEASURING EXHAUST GAS BACKPRESSURE

◆ See Figures 114 and 115

Before measuring the exhaust back pressure, it is necessary to prepare a spacer for the manometer. When the manometer is not available, install a vinyl hose and measure the water level difference with a scale.

1. On naturally aspirated engines, install the spacer for the manometer between the exhaust water mixing elbow and exhaust manifold.

2. On turbocharged engines, install the spacer for the manometer between the exhaust water mixing elbow and turbocharger.

3. Connect the vinyl hose of the spacer to the vinyl hose of the manometer at one side, and pour water into the hose.

4. Start the engine, raise the load gradually and measure the back pressure. The water level difference of the manometer hose shows the back pressure.

➡ The fitting at the end of the hose of the spacer which takes out the back pressure must be heat-resistant to prevent melting.

MEASURING EXHAUST GAS TEMPERATURE

◆ See Figures 116 and 117

One of the best ways to measure engine output is to use exhaust gas temperature. Exhaust gas will rise as engine output increases and fall as it decreases. The temperature will also peak at a predetermined level where engine output is most efficient.

A thermometer, thermocouple and tachometer must be procured in order to determine the exhaust temperature. Prior to measure-

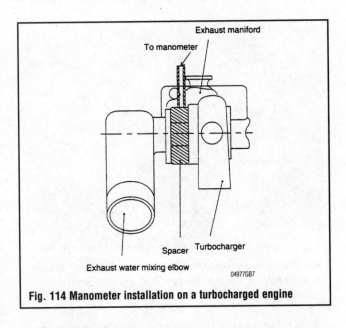

Fig. 114 Manometer installation on a turbocharged engine

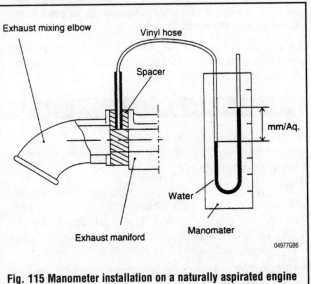

Fig. 115 Manometer installation on a naturally aspirated engine

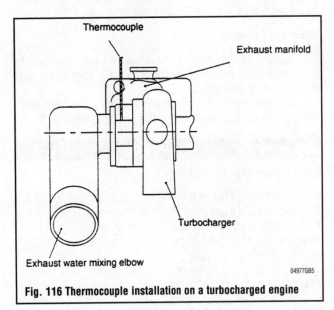

Fig. 116 Thermocouple installation on a turbocharged engine

Fig. 117 Thermocouple installation on a naturally aspirated engine

ment, be sure to measure the ambient temperature (open air, engine room and sea-water) and the temperatures of the engine's lube oil and fresh cooling water.

For exhaust temperature values, ask you local marine professional what is appropriate for you local climate. Temperature values are based on the atmospheric conditions of your local and can vary wildly between different areas.

➡When measuring temperatures, keep the top end of the thermocouple in The center of the exhaust port, or the temperature measurements will fluctuate.

Naturally Aspirated Engines

1. Make a slit in the gasket packing installed between the exhaust water mixing elbow and exhaust manifold. Insert the thermocouple into the slit and fasten the tip in the center of the exhaust port.

➡Use a thermocouple with a cable diameter of below 0.09 in. (2 mm). If the diameter is larger , more packing is required.

2. Start the engine. Measure the engine speed and exhaust temperature at quarter, half, three-quarter and full throttle under load.

Turbocharged Engine

1. Remove the blind plug located between the exhaust manifold and turbocharger inlet, insert the thermocouple, and fasten the tip in the center of the exhaust port.

2. Start the engine. Measure the engine speed and exhaust temperature quarter, half, three-quarter and full throttle under load.

Scuppers

▶ **See Figures 118, 119 and 120**

The scuppers should be at least 6 in. (150mm) higher than the load waterline of the boat. Use sea-water resistant material for the metal fixtures of the scupper. Ensure complete sealing with a silicone sealing agent at the scupper location in the hull, and on the threaded parts of the scupper.

Since the bow rises during cruising, and the stern falls. The exhaust port on the stern should be fitted to be 6 in. (150mm) above the waterline during cruising.

Fig. 118 Scuppers should be manufactured from a sea water resistant material such as stainless or plastic

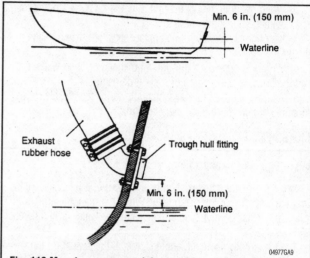

Fig. 119 Mount scuppers a minimum of 6 in. (150mm) higher than the loaded waterline of the boat

Fig. 120 Installing a butterfly (flapper) cover on the end of the scupper helps to prevent sea-water from entering into the exhaust port when the engine is stopped

➡When the exhaust port falls below the waterline, the back pressure rises, engine output is lowered and this causes abnormally colored exhaust and engine trouble.

Installation of a butterfly (flapper) cover on the end of the scupper helps to prevent sea-water from entering into the exhaust port when the engine is stopped.

Exhaust Manifold

REMOVAL & INSTALLATION

▶ **See Figures 121 and 122**

This procedure only applies to raw water cooled engines (except for QM series engines, which use a separate header tank). On fresh water cooled engines, the exhaust manifold is integrated into the header tank, for cooling of the exhaust gas. Refer to the procedures in the cooling section of this manual for more information.

➡1GM and 2GM engines do not have exhaust manifolds; the exhaust elbow attaches directly to the cylinder head. Refer to the

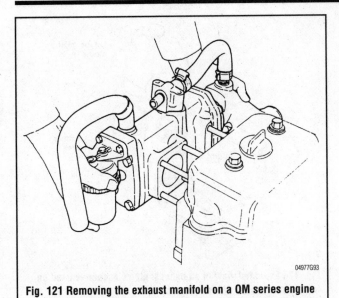

Fig. 121 Removing the exhaust manifold on a QM series engine

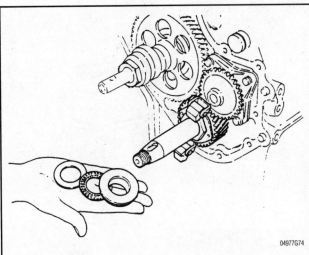

Fig. 122 Remove the nuts that hold the exhaust manifold, and slide it from the studs on the cylinder head

Exhaust Elbow removal and installation procedure for more information.

1. Turn the seacock to the **OFF** position.
2. Using the draincock, drain the water from the exhaust manifold.
3. Disconnect the exhaust elbow from the manifold.
4. Remove the water hoses that attach to the exhaust manifold.
5. Remove any components that are attached to or interfere with sliding the exhaust manifold from the mounting studs on the cylinder head.
6. Remove the nuts that secure the exhaust manifold to the cylinder head.
7. Carefully slide the exhaust manifold from the mounting studs on the cylinder head.

To install:
8. Clean the gasket material from the mating surfaces of the cylinder head and exhaust manifold.
9. Place a new gasket on the cylinder head.

10. Slide the exhaust manifold onto the studs the mounting studs on the cylinder head.
11. Install the nuts that secure the exhaust manifold to the cylinder head and tighten them securely and evenly.
12. Attach any components that were attached to or interfered with sliding the exhaust manifold from the mounting studs on the cylinder head.
13. Install the water hoses that attach to the exhaust manifold.
14. Attach the exhaust elbow to the manifold.

INSPECTION

▶ **See Figure 123**

Inspect exhaust manifolds for cracks, especially on wet exhaust systems where cooling water exits through exhaust. If a crack forms in these systems, it is possible to suck water into the engine causing sever engine damage. Salt or corrosion noticed during inspection indicates a leak in the manifold.

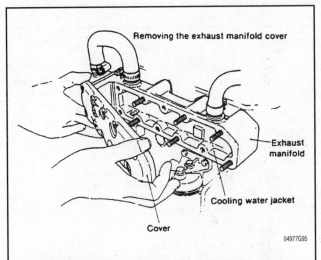

Fig. 123 The side cover on the manifold can be removed to inspect the cooling water jacket for corrosion

Exhaust Elbow

▶ **See Figures 124 thru 129**

The exhaust elbow (also known as a mixing elbow) on Yanmar engines can be divided into two types: Angled, and U-type. The installation of the engine in the vessel ultimately dictates which type of elbow is used. Although the two types are different in design, both are removed in the same manner.

✳✳ WARNING

Do NOT change the type of exhaust elbow that is used on your engine. Water may "backfill" through the exhaust system and cause severe engine damage.

1. Turn the seacock to the **OFF** position, if it is not already closed.
2. Drain the cooling water from the exhaust manifold (or header tank).
3. Disconnect the cooling water hose(s) from the fitting on the exhaust elbow.

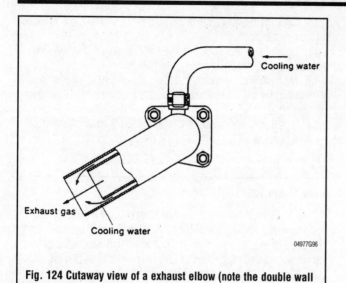

Fig. 124 Cutaway view of a exhaust elbow (note the double wall construction)

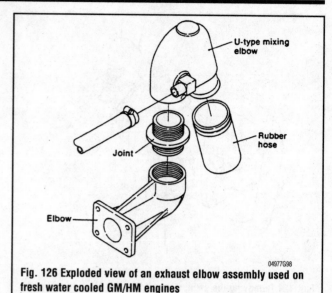

Fig. 126 Exploded view of an exhaust elbow assembly used on fresh water cooled GM/HM engines

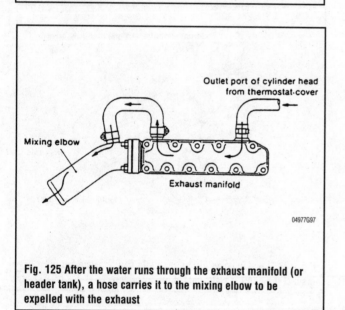

Fig. 125 After the water runs through the exhaust manifold (or header tank), a hose carries it to the mixing elbow to be expelled with the exhaust

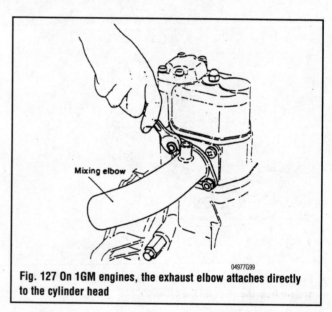

Fig. 127 On 1GM engines, the exhaust elbow attaches directly to the cylinder head

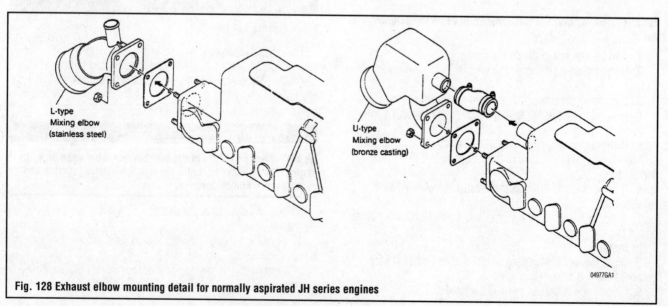

Fig. 128 Exhaust elbow mounting detail for normally aspirated JH series engines

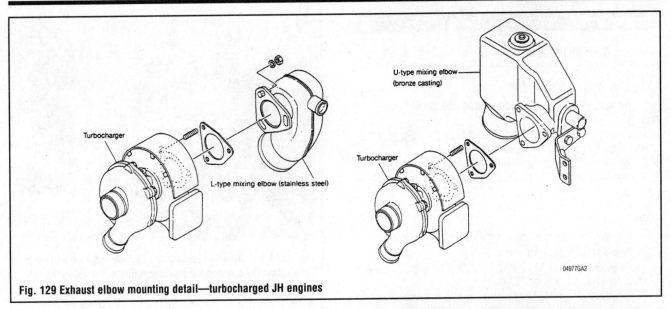

Fig. 129 Exhaust elbow mounting detail—turbocharged JH engines

4. Disconnect the rubber exhaust hose from the elbow.

5. Remove the nuts and/or bolts that secure the exhaust elbow to the manifold (or turbocharger).

6. On turbocharged models, remove the exhaust elbow bracket.

➡ **On 1GM and 2GM raw water cooled engines, the exhaust elbow is attached directly to the cylinder head.**

7. Clean the gasket material from the mating surfaces of the exhaust manifold and mixing elbow (or cylinder head, or turbocharger, depending on the engine).

8. Using a new gasket, install the exhaust elbow. Tighten the bolts/nuts securely and evenly.

9. On turbocharged models, attach the exhaust elbow bracket.

10. Connect the rubber exhaust hose to the elbow.

11. Connect the cooling water hose(s) from the fitting on the exhaust elbow.

INSPECTION

▶ **See Figures 130 and 131**

Exhaust elbows are easily restricted with carbon and corrosion deposits especially if the engine is used for trolling or allowed to idle for long periods of time while charging batteries.

Just as in the inspection of the exhaust manifold, salt or corrosion indicates a leak caused by more severe corrosion elsewhere in the elbow. Soot deposits by the mounting flange indicate a loose joint or gasket failure. Also, inspect all bolts and fasteners for proper tightness. Heat and vibration tend to loosen them.

➡ **Engines fitted with threaded joints on the exhaust elbow can rarely be disassembled due to the extent of corrosion on the threads. If the elbow has been inspected, and needs replacement, it is recommended that the elbow must be replaced as a unit. The use of pipe wrenches may allow enough force to be applied to break or distort the threads that secure the elbow to the exhaust manifold or header tank.**

Fig. 130 Here's an example of a new exhaust elbow. Notice the divider that separates the exhaust gas and cooling water

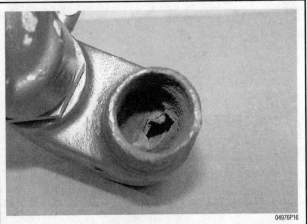

Fig. 131 This exhaust elbow is severely corroded. Corrosion of this extent can cause engine problems

Turbocharger

REMOVAL & INSTALLATION

▶ **See Figures 132, 133 and 134**

1. Remove the air cleaner/silencer from the turbocharger inlet.

2. If equipped, open the draincock and drain the water from the turbocharger. Disconnect the hoses.

3. Remove the exhaust elbow from the turbocharger.

4. Remove the oil return line that attaches the turbocharger to the oil pan.

5. Disconnect the oil inlet line from the top of the turbocharger.

6. Remove the hose that attaches the air turbine outlet to the intake manifold (or aftercooler)

7. Remove any remaining brackets, hoses, cables etc. that interfere with removal of the turbocharger.

8. Remove the four bolts that secure the turbocharger assembly to the header tank (which is the exhaust manifold).

9. Remove the turbocharger from the header tank.

To install:

10. If necessary, clean the gasket material from the mating surfaces of the turbocharger and header tank.

11. Using a new gasket, install the turbocharger onto the header tank. Do not tighten the nuts yet.

12. Using a new gasket, install the oil return line to the bottom of the turbocharger. Tighten the bolts securely, but not excessively, or the threads may be damaged.

➡ **The entire turbocharger assembly will require replacement if the return line threads are damaged.**

13. Tighten the nuts that secure the turbocharger to the header tank.

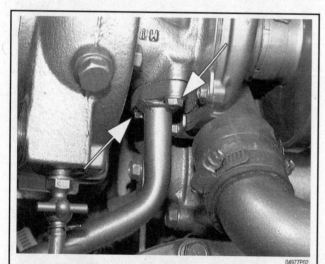

Fig. 132 Remove the two bolts that attach the oil return line to the turbocharger

Fig. 133 If the turbocharger is water cooled, make sure to drain the coolant (by opening the draincock) before removing the turbocharger

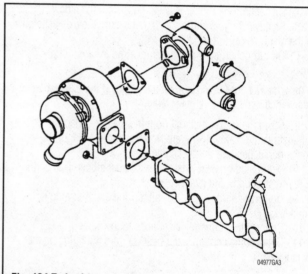

Fig. 134 Turbocharger mounting detail—JH engines

14. Using a small funnel or a baster, fill the oil inlet hole on the top of the turbocharger with fresh engine oil. Wait until the oil has drained through the return line, and repeat the process.

15. Connect the oil inlet line the top of the turbocharger. Use new copper gaskets.

16. Install the hose that attaches the air turbine outlet to the intake manifold (or aftercooler)

17. Install the exhaust elbow to the turbocharger.

18. If equipped, attach the cooling water hoses to the turbocharger.

19. Connect the air cleaner/silencer to the turbocharger inlet.

20. Install any additional components that were removed to allow for removal of the turbocharger.

INSPECTION

▶ See Figure 135

➡It is important to remember that equipment should not be stored in the engine compartment. Many times equipment will move around during rough weather, damaging vital engine components. This is the case with the turbocharger oil line. Many oil lines have been ripped from their mounting or kinked when equipment is pressed against them.

To inspect the inlet side of the turbocharger, simply remove the air silencer/cleaner to view the turbine.

To inspect the exhaust side of the turbocharger, remove the exhaust elbow from the turbocharger. If the turbine has excessive carbon accumulation or other damage on the impeller blades or housing, it should be serviced by a qualified turbocharger repair facility. Although the turbocharger is a relatively simple component, it is very precise, and requires special tools for disassembly.

04973P13

Fig. 135 By removing the air silencer, the inlet side of the turbocharger can be inspected for damage to the impeller blades and housing

SELSTK02

The exhaust outlet of this sailboat uses a flapper to prevent seawater entry

Troubleshooting Turbochargers

Condition	Cause	Correction
Engine performance problems	Restricted air filter	Clean air filter, maintain filter more frequently
	Turbocharger dirty	Decarbon turbocharger
	Intercooler dirty	Clean intercooler
	Air leak from intake pipe downstream of turbo	Find and repair leak
	Turbocharger binding	Decarbon turbocharger
	Restricted exhaust gas flow	Find and repair cause of restriction
	Exhaust gas leak before turbocharger	Find and repair leak
	Turbocharger seized	Investigate source of problem and then replace faulty turbocharger
Excessive black smoke	Restricted air filter	Clean air filter, maintain filter more frequently
	Turbocharger dirty	Decarbon turbocharger
	Intercooler dirty	Clean intercooler
	Air leak from intake pipe downstream of turbo	Find and repair leak
	Turbocharger binding	Decarbon turbocharger
	Restricted exhaust gas flow	Find and repair cause of restriction
	Exhaust gas leak before turbocharger	Find and repair leak

04977C15

ENGINE RECONDITIONING

Determining Engine Condition

Anything that generates heat and/or friction will eventually burn or wear out (for example, a light bulb generates heat, therefore its life span is limited). With this in mind, a running engine generates tremendous amounts of both; friction is encountered by the moving and rotating parts inside the engine and heat is created by friction and combustion of the fuel. However, the engine has systems designed to help reduce the effects of heat and friction and provide added longevity. The oiling system reduces the amount of friction encountered by the moving parts inside the engine, while the cooling system reduces heat created by friction and combustion. If either system is not maintained, a break-down will be inevitable. Therefore, you can see how regular maintenance can affect the service life of your engine. If you do not drain, flush and refill your cooling system at the proper intervals, deposits will begin to accumulate, thereby reducing the amount of heat it can extract from the coolant. The same applies to your oil and filter; if it is not changed often enough it becomes laden with contaminates and is unable to properly lubricate the engine. This increases friction and wear.

There are a number of methods for evaluating the condition of your engine. A compression test can reveal the condition of your pistons, piston rings, cylinder bores, head gasket, valves and valve seats. An oil pressure test can warn you of possible engine bearing, or oil pump failures. Excessive oil consumption, evidence of oil in the engine air intake area and/or bluish smoke from the exhaust may indicate worn piston rings, worn valve guides and/or valve seals.

COMPRESSION TEST

▶ **See Figure 136**

A noticeable lack of engine power, excessive oil consumption and/or poor fuel mileage measured over an extended period are all indicators of internal engine wear. Worn piston rings, scored or worn cylinder bores, blown head gaskets, sticking or burnt valves, and worn valve seats are all possible culprits. A check of each cylinder's compression will help locate the problem.

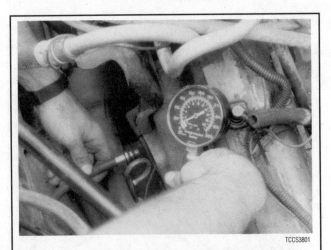

TCCS3801

Fig. 136 A special diesel compression gauge should be used to test your engine's compression. Diesel gauges can withstand the extreme cylinder pressure developed by a diesel engine

The procedure for checking compression is located in the "Maintenance and Tune-up" section of this manual.

OIL PRESSURE TEST

Check for proper oil pressure at the sending unit passage with an externally mounted mechanical oil pressure gauge (as opposed to the dash-mounted gauge). A tachometer may also be needed, as some specifications may require running the engine at a specific rpm.

1. With the engine cold, locate and remove the oil pressure sending unit.
2. Following the manufacturer's instructions, connect a mechanical oil pressure gauge and, if necessary, a tachometer to the engine.
3. Start the engine and allow it to idle.
4. Check the oil pressure reading when cold and record the number. You may need to run the engine at a specified rpm, so check the specifications.
5. Run the engine until normal operating temperature is reached.
6. Check the oil pressure reading again with the engine hot and record the number. Turn the engine **OFF**.
7. If the cold pressure is well above the specification, and the hot reading was lower than the specification, you may have the wrong viscosity oil in the engine. Change the oil, making sure to use the proper grade and quantity, then repeat the test.

Low oil pressure readings could be attributed to internal component wear, pump related problems, a low oil level, or oil viscosity that is too low. High oil pressure readings could be caused by an overfilled crankcase, too high of an oil viscosity or a faulty pressure relief valve.

Buy Or Rebuild?

Now that you have determined that your engine is worn out, you must make some decisions. The question of whether or not an engine is worth rebuilding is largely a subjective matter and one of personal worth. Is the engine a popular one, or is it an obsolete model? Are parts available? Would it be less expensive to buy a new engine, have your engine rebuilt by a pro or rebuild it yourself? If you have considered all these matters and more, and have still decided to rebuild the engine, then it is time to decide how you will rebuild it.

➡ **The editors at Seloc feel that most engine machining should be performed by a professional machine shop. Don't think of it as wasting money, rather, as an assurance that the job has been done right the first time. There are many expensive and specialized tools required to perform such tasks as boring and honing an engine block or having a valve job done on a cylinder head. Even inspecting the parts requires expensive micrometers and gauges to properly measure wear and clearances. Also, a machine shop can deliver to you clean, and ready to assemble parts, saving you time and aggravation. Your maximum savings will come from performing the removal, disassembly, assembly and installation of the engine and purchasing or renting only the tools required to perform the above tasks. Depending on the particular circumstances, you may save 40 to 60 percent of the cost doing these yourself.**

A complete rebuild or overhaul of an engine involves replacing all of the moving parts (pistons, rods, crankshaft, camshaft, etc.) with new ones and machining the non-moving wearing surfaces of the block and heads. Unfortunately, this may not be cost effective.

For instance, your crankshaft may have been damaged or worn, but it can be machined undersize for a minimal fee.

So, as you can see, you can replace everything inside the engine, but, it is wiser to replace only those parts which are really needed, and, if possible, repair the more expensive ones. Later in this section, we will break the engine down into its two main components: the cylinder head and the engine block. We will discuss each component, and the recommended parts to replace during a rebuild on each.

Engine Overhaul Tips

Most engine overhaul procedures are fairly standard. In addition to specific parts replacement procedures and specifications for your individual engine, this section is also a guide to acceptable rebuilding procedures. Examples of standard rebuilding practice are given and should be used along with specific details concerning your particular engine.

Competent and accurate machine shop services will ensure maximum performance, reliability and engine life. In most instances it is more profitable for the do-it-yourself mechanic to remove, clean and inspect the component, buy the necessary parts and deliver these to a shop for actual machine work.

Much of the assembly work (crankshaft, bearings, piston rods, and other components) is well within the scope of the do-it-yourself mechanic's tools and abilities. You will have to decide for yourself the depth of involvement you desire in an engine repair or rebuild.

TOOLS

The tools required for an engine overhaul or parts replacement will depend on the depth of your involvement. With a few exceptions, they will be the tools found in a mechanic's tool box. More in-depth work will require some or all of the following:
• A dial indicator (reading in thousandths) mounted on a universal base
• Micrometers and telescope gauges
• Jaw and screw-type pullers
• Scraper
• Valve spring compressor
• Ring groove cleaner
• Piston ring expander and compressor
• Ridge reamer
• Cylinder hone or glaze breaker
• Plastigage®
• Engine stand

The use of most of these tools is illustrated in this section. Many can be rented for a one-time use from a local parts jobber or tool supply house.

Occasionally, the use of special tools is called for. See the information on Special Tools and the Safety Notice in the front of this manual before substituting another tool.

CLEANING

▶ See Figures 137, 138 and 139

Before the engine and its components are inspected, they must be thoroughly cleaned. You will need to remove any engine varnish, oil sludge and/or carbon deposits from all of the components to insure an accurate inspection. A crack in the engine block or cylinder head can easily become overlooked if hidden by a layer of sludge or carbon.

Most of the cleaning process can be carried out with common hand tools and readily available solvents or solutions. Carbon deposits can be chipped away using a hammer and a hard wooden chisel. Old gasket material and varnish or sludge can usually be removed using a scraper and/or cleaning solvent. Extremely stubborn deposits may require the use of a power drill with a wire brush. If using a wire brush, use extreme care around any critical machined surfaces (such as the gasket surfaces, bearing saddles, cylinder bores, etc.).

Always follow any safety recommendations given by the manufacturer of the tool and/or solvent. You should always wear eye protection during any cleaning process involving scraping, chipping or spraying of solvents.

An alternative to the mess and hassle of cleaning the parts yourself is to drop them off at a local machine shop. They will, more than likely, have the necessary equipment to properly clean all of the parts for a nominal fee.

✷✷ CAUTION

Always wear eye protection during any cleaning process involving scraping, chipping or spraying of solvents.

TCCS3211

Fig. 137 Use a ring expander tool to remove the piston rings

TCCS3208

Fig. 138 Clean the piston ring grooves using a ring groove cleaner tool, or . . .

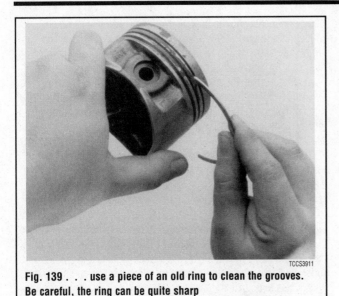

Fig. 139 . . . use a piece of an old ring to clean the grooves. Be careful, the ring can be quite sharp

Remove any oil galley plugs, freeze plugs and/or pressed-in bearings and carefully wash and degrease all of the engine components including the fasteners and bolts. Small parts such as the valves, springs, etc., should be placed in a metal basket and allowed to soak. Use pipe cleaner type brushes, and clean all passageways in the components. Use a ring expander and remove the rings from the pistons. Clean the piston ring grooves with a special tool or a piece of broken ring. Scrape the carbon off of the top of the piston. You should never use a wire brush on the pistons. After preparing all of the piston assemblies in this manner, wash and degrease them again.

✸✸ WARNING

Use extreme care when cleaning around the cylinder head valve seats. A mistake or slip may cost you a new seat.

When cleaning the cylinder head, remove carbon from the combustion chamber with the valves installed. This will avoid damaging the valve seats.

REPAIRING DAMAGED THREADS

▶ **See Figures 140, 141, 142, 143 and 144**

Several methods of repairing damaged threads are available. Heli-Coil® (shown here), Keenserts® and Microdot® are among the most widely used. All involve basically the same principle—drilling out stripped threads, tapping the hole and installing a prewound insert—making welding, plugging and oversize fasteners unnecessary.

Two types of thread repair inserts are usually supplied: a standard type for most inch coarse, inch fine, metric course and metric fine thread sizes and a spark lug type to fit most spark plug port sizes. Consult the individual tool manufacturer's catalog to determine exact applications. Typical thread repair kits will contain a selection of prewound threaded inserts, a tap (corresponding to the outside diameter threads of the insert) and an installation tool. Spark plug inserts usually differ because they require a tap equipped with pilot threads and a combined reamer/tap section. Most manufacturers also supply blister-packed thread repair inserts separately in addition to a master kit containing a variety of taps and inserts plus installation tools.

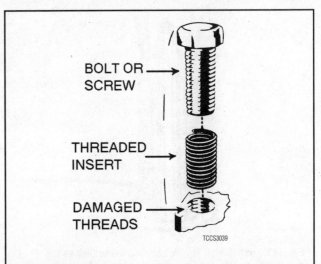

Fig. 140 Damaged bolt hole threads can be replaced with thread repair inserts

Fig. 141 Standard thread repair insert (left), and spark plug thread insert

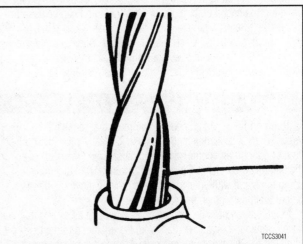

Fig. 142 Drill out the damaged threads with the specified size bit. Be sure to drill completely through the hole or to the bottom of a blind hole

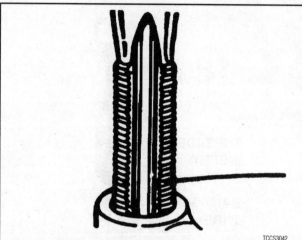

Fig. 143 Using the kit, tap the hole in order to receive the thread insert. Keep the tap well oiled and back it out frequently to avoid clogging the threads

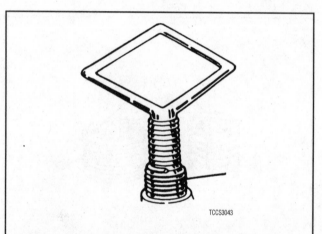

Fig. 144 Screw the insert onto the installer tool until the tang engages the slot. Thread the insert into the hole until it is ¼–½ turn below the top surface, then remove the tool and break off the tang using a punch

Before attempting to repair a threaded hole, remove any snapped, broken or damaged bolts or studs. Penetrating oil can be used to free frozen threads. The offending item can usually be removed with locking pliers or using a screw/stud extractor. After the hole is clear, the thread can be repaired, as shown in the series of accompanying illustrations and in the kit manufacturer's instructions.

Engine Preparation

To properly rebuild an engine, you must first remove it from the vessel, then disassemble and diagnose it. Ideally you should place your engine on an engine stand. This affords you the best access to the engine components. Follow the manufacturer's directions for using the stand with your particular engine. Remove the flywheel before installing the engine to the stand.

Now that you have the engine on a stand, and assuming that you have drained the oil and coolant from the engine, it's time to strip it of all but the necessary components. Before you start disassembling the engine, you may want to take a moment to draw some pictures, or fabricate some labels or containers to mark the locations of vari-

ous components and the bolts and/or studs which fasten them. Modern day engines use a lot of little brackets and clips which hold wiring harnesses and such, and these holders are often mounted on studs and/or bolts that can be easily mixed up. The manufacturer spent a lot of time and money designing your engine, and they wouldn't have wasted any of it by haphazardly placing brackets, clips or fasteners. If it's present when you disassemble it, put it back when you assemble, you will regret not remembering that little bracket which holds a wire harness out of the path of a rotating part.

You should begin by unbolting any accessories still attached to the engine. Then, unfasten any manifolds (intake or exhaust) which were not removed during the engine removal procedure. Finally, remove any covers remaining on the engine such as the rocker arm, front or timing cover and oil pan. Some front covers may require the vibration damper and/or crank pulley to be removed beforehand. The idea is to reduce the engine to the bare necessities (cylinder head, valve train, engine block, crankshaft, pistons and connecting rods), plus any other 'in block' components such as oil pumps, balance shafts and auxiliary shafts.

Finally, remove the cylinder head from the engine block and carefully place on a bench. Disassembly instructions for each component follow later in this section.

Cylinder Head

There are two basic types of cylinder heads used on today's engines: the Overhead Valve (OHV) and the Overhead Camshaft (OHC). The latter can also be broken down into two subgroups: the Single Overhead Camshaft (SOHC) and the Dual Overhead Camshaft (DOHC). Generally, if there is only a single camshaft on a head, it is just referred to as an OHC head. Also, an engine with an OHV cylinder head is also known as a pushrod engine.

Diesel cylinder heads are made of cast iron and have steel valves and seats. Most use two valves per cylinder, while the more hi-tech engines will utilize a multi-valve configuration. When the valve contacts the seat, it does so on precision machined surfaces, which seals the combustion chamber. All cylinder heads have a valve guide for each valve. The guide centers the valve to the seat and allows it to move up and down within it. The clearance between the valve and guide can be critical. Too much clearance and the engine may consume oil, lose vacuum and/or damage the seat. Too little, and the valve can stick in the guide causing the engine to run poorly if at all, and possibly causing severe damage. The last component all cylinder heads have are valve springs. The spring holds the valve against its seat. It also returns the valve to this position when the valve has been opened by the valve train or camshaft. The spring is fastened to the valve by a retainer and valve locks (sometimes called keepers).

An ideal method of rebuilding the cylinder head would involve replacing all of the valves, guides, seats, springs, etc. with new ones. However, depending on how the engine was maintained, often this is not necessary. A major cause of valve, guide and seat wear is an improperly tuned engine. An engine that is running too rich, will often wash the lubricating oil out of the guide with gasoline, causing it to wear rapidly. Conversely, an engine which is running too lean will place higher combustion temperatures on the valves and seats allowing them to wear or even burn. Springs fall victim to engine rpm and heat. Generally, the valves, guides, springs and seats in a cylinder head can be machined and re-used, saving you money. However, if a valve is burnt, it may be wise to replace all of the valves, since they were all operating in the same environment.

The same goes for any other component on the cylinder head. Think of it as an insurance policy against future problems related to that component.

Unfortunately, the only way to find out which components need replacing, is to disassemble and carefully check each piece. After the cylinder head is disassembled, thoroughly clean all of the components.

DISASSEMBLY

▶ **See Figures 145, 146 and 147**

Before disassembling the cylinder head, you may want to fabricate some containers to hold the various parts, as some of them can be quite small (such as keepers) and easily lost. Also keeping yourself and the components organized will aid in assembly and reduce confusion. Where possible, try to maintain a components original location; this is especially important if there is not going to be any machine work performed on the components.

1. If you haven't already removed the rocker arms and/or shafts, do so now.
2. Position the head so that the springs are easily accessed.
3. Use a valve spring compressor tool, and relieve spring tension from the retainer.

➡ **Due to engine varnish, the retainer may stick to the valve locks. A gentle tap with a hammer may help to break it loose.**

4. Remove the valve locks from the valve tip and/or retainer. A small magnet may help in removing the locks.
5. Lift the valve spring, tool and all, off of the valve stem.
6. If equipped, remove the valve seal. If the seal is difficult to remove with the valve in place, try removing the valve first, then the seal. Follow the steps below for valve removal.
7. Position the head to allow access for withdrawing the valve.

➡ **Cylinder heads that have seen a lot of miles and/or abuse may have mushroomed the valve lock grove and/or tip, causing difficulty in removal of the valve. If this has happened, use a metal file to carefully remove the high spots around the lock grooves and/or tip. Only file it enough to allow removal.**

8. Remove the valve from the cylinder head.
9. If equipped, remove the valve spring shim. A small magnetic tool or screwdriver will aid in removal.

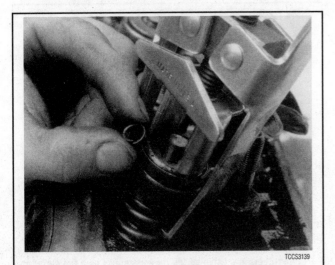

Fig. 145 Be careful not to lose the small valve locks (keepers)

Fig. 146 Remove the valve seal from the valve stem—O-ring type seal shown

Fig. 147 Removing an umbrella/positive type seal

10. Repeat Steps 3 though 9 until all of the valves have been removed.

INSPECTION

Now that all of the cylinder head components are clean, it's time to inspect them for wear and/or damage. To accurately inspect them, you will need some specialized tools:

- A 0–1 in. micrometer for the valves
- A dial indicator or inside diameter gauge for the valve guides
- A spring pressure test gauge

If you do not have access to the proper tools, you may want to bring the components to a shop that does.

Valves

▶ **See Figures 148 and 149**

The first thing to inspect are the valve heads. Look closely at the head, margin and face for any cracks, excessive wear or burning. The margin is the best place to look for burning. It should have a

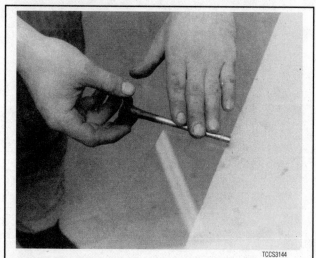

Fig. 148 Valve stems may be rolled on a flat surface to check for bends

squared edge with an even width all around the diameter. When a valve burns, the margin will look melted and the edges rounded. Also inspect the valve head for any signs of tulipping. This will show as a lifting of the edges or dishing in the center of the head and will usually not occur to all of the valves. All of the heads should look the same, any that seem dished more than others are probably bad. Next, inspect the valve lock grooves and valve tips. Check for any burrs around the lock grooves, especially if you had to file them to remove the valve. Valve tips should appear flat, although slight rounding with high mileage engines is normal. Slightly worn valve tips will need to be machined flat. Last, measure the valve stem diameter with the micrometer. Measure the area that rides within the guide, especially towards the tip where most of the wear occurs. Take several measurements along its length and compare them to each other. Wear should be even along the length with little to no taper. If no minimum diameter is given in the specifications, then the stem should not read more than 0.001 in. (0.025mm) below the specification. Any valves that fail these inspections should be replaced.

Springs, Retainers and Valve Locks

▶ See Figures 150 and 151

The first thing to check is the most obvious, broken springs. Next check the free length and squareness of each spring. If applicable, insure to distinguish between intake and exhaust springs. Use a ruler and/or carpenter's square to measure the length. A carpenter's square should be used to check the springs for squareness. If a spring pressure test gauge is available, check each springs rating and compare to the specifications chart. Check the readings against the specifications given. Any springs that fail these inspections should be replaced.

The spring retainers rarely need replacing, however they should still be checked as a precaution. Inspect the spring mating surface and the valve lock retention area for any signs of excessive wear. Also check for any signs of cracking. Replace any retainers that are questionable.

Valve locks should be inspected for excessive wear on the outside contact area as well as on the inner notched surface. Any locks which appear worn or broken and its respective valve should be replaced.

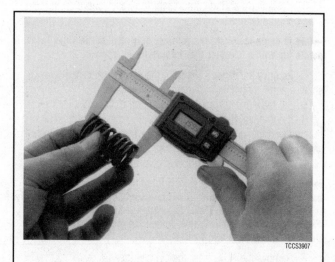

Fig. 150 Use a caliper to check the valve spring free-length

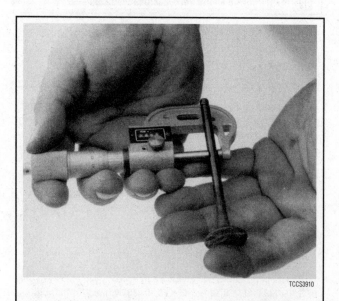

Fig. 149 Use a micrometer to check the valve stem diameter

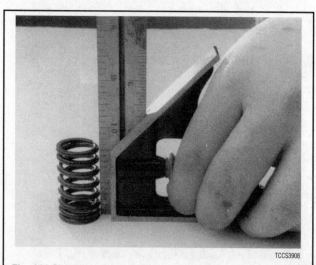

Fig. 151 Check the valve spring for squareness on a flat surface; a carpenter's square can be used

Cylinder Head

There are several things to check on the cylinder head: valve guides, seats, cylinder head surface flatness, cracks and physical damage.

VALVE GUIDES

Now that you know the valves are good, you can use them to check the guides, although a new valve, if available, is preferred. Before you measure anything, look at the guides carefully and inspect them for any cracks, chips or breakage. Also if the guide is a removable style (as in most aluminum heads), check them for any looseness or evidence of movement. All of the guides should appear to be at the same height from the spring seat. If any seem lower (or higher) from another, the guide has moved. Mount a dial indicator onto the spring side of the cylinder head. Lightly oil the valve stem and insert it into the cylinder head. Position the dial indicator against the valve stem near the tip and zero the gauge. Grasp the valve stem and wiggle towards and away from the dial indicator and observe the readings. Mount the dial indicator 90 degrees from the initial point and zero the gauge and again take a reading. Compare the two readings for a out of round condition. Check the readings against the specifications given. An Inside Diameter (I.D.) gauge designed for valve guides will give you an accurate valve guide bore measurement. If the I.D. gauge is used, compare the readings with the specifications given. Any guides that fail these inspections should be replaced or machined.

VALVE SEATS

A visual inspection of the valve seats should show a slightly worn and pitted surface where the valve face contacts the seat. Inspect the seat carefully for severe pitting or cracks. Also, a seat that is badly worn will be recessed into the cylinder head. A severely worn or recessed seat may need to be replaced. All cracked seats must be replaced. A seat concentricity gauge, if available, should be used to check the seat run-out. If run-out exceeds specifications the seat must be machined (if no specification is given use 0.002 in. or 0.051mm).

CYLINDER HEAD SURFACE FLATNESS

▶ See Figures 152 and 153

After you have cleaned the gasket surface of the cylinder head of any old gasket material, check the head for flatness.

Place a straightedge across the gasket surface. Using feeler gauges, determine the clearance at the center of the straightedge and across the cylinder head at several points. Check along the centerline and diagonally on the head surface. If the warpage exceeds 0.003 in. (0.076mm) within a 6.0 in. (15.2cm) span, or 0.006 in. (0.152mm) over the total length of the head, the cylinder head must be resurfaced. After resurfacing the heads of a V-type engine, the intake manifold flange surface should be checked, and if necessary, milled proportionally to allow for the change in its mounting position.

CRACKS AND PHYSICAL DAMAGE

▶ See Figure 154

Generally, cracks are limited to the combustion chamber, however, it is not uncommon for the head to crack in a spark plug hole, port, outside of the head or in the valve spring/rocker arm area. The first area to inspect is always the hottest: the exhaust seat/port area.

A visual inspection should be performed, but just because you

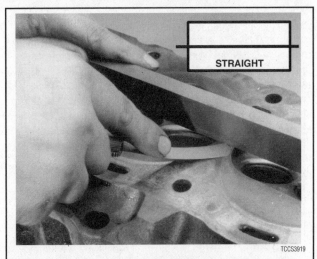

Fig. 152 Check the head for flatness across the center of the head surface using a straightedge and feeler gauge

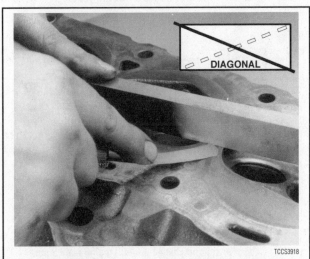

Fig. 153 Checks should also be made along both diagonals of the head surface

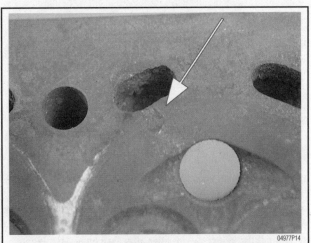

Fig. 154 The groove between the combustion chamber and the cooling passage of this cylinder head was caused from a faulty head gasket

don't see a crack does not mean it is not there. Some more reliable methods for inspecting for cracks include Magnaflux®, a magnetic process or Zyglo®, a dye penetrant. Magnaflux® is used only on ferrous metal (cast iron) heads. Zyglo® uses a spray on fluorescent mixture along with a black light to reveal the cracks. It is strongly recommended to have your cylinder head checked professionally for cracks, especially if the engine was known to have overheated and/or leaked or consumed coolant. Contact a local shop for availability and pricing of these services.

Physical damage is usually very evident. For example, a broken mounting ear from dropping the head or a bent or broken stud and/or bolt. All of these defects should be fixed or, if unrepairable, the head should be replaced.

REFINISHING & REPAIRING

Many of the procedures given for refinishing and repairing the cylinder head components must be performed by a machine shop. Certain steps, if the inspected part is not worn, can be performed yourself inexpensively. However, you spent a lot of time and effort so far, why risk trying to save a couple bucks if you might have to do it all over again?

Valves

Any valves that were not replaced should be refaced and the tips ground flat. Unless you have access to a valve grinding machine, this should be done by a machine shop. If the valves are in extremely good condition, as well as the valve seats and guides, they may be lapped in without performing machine work.

It is a recommended practice to lap the valves even after machine work has been performed and/or new valves have been purchased. This insures a positive seal between the valve and seat.

LAPPING THE VALVES

➡Before lapping the valves to the seats, read the rest of the cylinder head section to insure that any related parts are in acceptable enough condition to continue.

➡Before any valve seat machining and/or lapping can be performed, the guides must be within factory recommended specifications.

1. Invert the cylinder head.
2. Lightly lubricate the valve stems and insert them into the cylinder head in their numbered order.
3. Raise the valve from the seat and apply a small amount of fine lapping compound to the seat.
4. Moisten the suction head of a hand-lapping tool and attach it to the head of the valve.
5. Rotate the tool between the palms of both hands, changing the position of the valve on the valve seat and lifting the tool often to prevent grooving.
6. Lap the valve until a smooth, polished circle is evident on the valve and seat.
7. Remove the tool and the valve. Wipe away all traces of the grinding compound and store the valve to maintain its lapped location.

✳✳ WARNING

Do not get the valves out of order after they have been lapped. They must be put back with the same valve seat with which they were lapped.

Springs, Retainers and Valve Locks

There is no repair or refinishing possible with the springs, retainers and valve locks. If they are found to be worn or defective, they must be replaced with new (or known good) parts.

Cylinder Head

Most refinishing procedures dealing with the cylinder head must be performed by a machine shop. Read the sections below and review your inspection data to determine whether or not machining is necessary.

VALVE GUIDE

➡If any machining or replacements are made to the valve guides, the seats must be machined.

Unless the valve guides need machining or replacing, the only service to perform is to thoroughly clean them of any dirt or oil residue.

There are only two types of valve guides used on engines: the replaceable-type (all aluminum heads) and the cast-in integral-type (most cast iron heads). There are four recommended methods for repairing worn guides.

- Knurling
- Inserts
- Reaming oversize
- Replacing

Knurling is a process in which metal is displaced and raised, thereby reducing clearance, giving a true center, and providing oil control. It is the least expensive way of repairing the valve guides. However, it is not necessarily the best, and in some cases, a knurled valve guide will not stand up for more than a short time. It requires a special knurlizer and precision reaming tools to obtain proper clearances. It would not be cost effective to purchase these tools, unless you plan on rebuilding several of the same cylinder head.

Installing a guide insert involves machining the guide to accept a bronze insert. One style is the coil-type which is installed into a threaded guide. Another is the thin-walled insert where the guide is reamed oversize to accept a split-sleeve insert. After the insert is installed, a special tool is then run through the guide to expand the insert, locking it to the guide. The insert is then reamed to the standard size for proper valve clearance.

Reaming for oversize valves restores normal clearances and provides a true valve seat. Most cast-in type guides can be reamed to accept an valve with an oversize stem. The cost factor for this can become quite high as you will need to purchase the reamer and new, oversize stem valves for all guides which were reamed. Oversizes are generally 0.003 to 0.030 in. (0.076 to 0.762mm), with 0.015 in. (0.381mm) being the most common.

To replace cast-in type valve guides, they must be drilled out, then reamed to accept replacement guides. This must be done on a fixture which will allow centering and leveling off of the original valve seat or guide, otherwise a serious guide-to-seat misalignment may occur making it impossible to properly machine the seat.

Replaceable-type guides are pressed into the cylinder head. A hammer and a stepped drift or punch may be used to install and remove the guides. Before removing the guides, measure the protrusion on the spring side of the head and record it for installation. Use the stepped drift to hammer out the old guide from the combustion chamber side of the head. When installing, determine whether or not the guide also seals a water jacket in the head, and if it does, use the recommended sealing agent. If there is no water jacket, grease the valve guide and its bore. Use the stepped drift, and hammer the new guide into the cylinder head from the spring side of the cylinder head. A stack of washers the same thickness as the measured protrusion may help the installation process.

VALVE SEATS

➡ **Before any valve seat machining can be performed, the guides must be within factory recommended specifications.**

➡ **If any machining or replacements were made to the valve guides, the seats must be machined.**

If the seats are in good condition, the valves can be lapped to the seats, and the cylinder head assembled. See the valves section for instructions on lapping.

If the valve seats are worn, cracked or damaged, they must be serviced by a machine shop. The valve seat must be perfectly centered to the valve guide, which requires very accurate machining.

CYLINDER HEAD SURFACE

If the cylinder head is warped, it must be machined flat. If the warpage is extremely severe, the head may need to be replaced. In some instances, it may be possible to straighten a warped head enough to allow machining. In either case, contact a professional machine shop for service.

CRACKS AND PHYSICAL DAMAGE

Certain cracks can be repaired in both cast iron and aluminum heads. For cast iron, a tapered threaded insert is installed along the length of the crack. Aluminum can also use the tapered inserts, however welding is the preferred method. Some physical damage can be repaired through brazing or welding. Contact a machine shop to get expert advice for your particular dilemma.

ASSEMBLY

The first step for any assembly job is to have a clean area in which to work. Next, thoroughly clean all of the parts and components that are to be assembled. Finally, place all of the components onto a suitable work space and, if necessary, arrange the parts to their respective positions.

1. Lightly lubricate the valve stems and insert all of the valves into the cylinder head. If possible, maintain their original locations.
2. If equipped, install any valve spring shims which were removed.
3. If equipped, install the new valve seals, keeping the following in mind:
   • If the valve seal presses over the guide, lightly lubricate the outer guide surfaces.
   • If the seal is an O-ring type, it is installed just after compressing the spring but before the valve locks.
4. Place the valve spring and retainer over the stem.
5. Position the spring compressor tool and compress the spring.
6. Assemble the valve locks to the stem.
7. Relieve the spring pressure slowly and insure that neither valve lock becomes dislodged by the retainer.
8. Remove the spring compressor tool.
9. Repeat Steps 2 through 8 until all of the springs have been installed.

Engine Block

GENERAL INFORMATION

A thorough overhaul or rebuild of an engine block would include replacing the pistons, rings, bearings, timing belt/chain assembly and oil pump. For OHV engines also include a new camshaft and lifters. The block would then have the cylinders bored and honed oversize (or if using removable cylinder sleeves, new sleeves installed) and the crankshaft would be cut undersize to provide new wearing surfaces and perfect clearances. However, your particular engine may not have everything worn out. What if only the piston rings have worn out and the clearances on everything else are still within factory specifications? Well, you could just replace the rings and put it back together, but this would be a very rare example. Chances are, if one component in your engine is worn, other components are sure to follow, and soon. At the very least, you should always replace the rings, bearings and oil pump. This is what is commonly called a "freshen up".

Cylinder Ridge Removal

Because the top piston ring does not travel to the very top of the cylinder, a ridge is built up between the end of the travel and the top of the cylinder bore.

Pushing the piston and connecting rod assembly past the ridge can be difficult, and damage to the piston ring lands could occur. If the ridge is not removed before installing a new piston or not removed at all, piston ring breakage and piston damage may occur.

➡ **It is always recommended that you remove any cylinder ridges before removing the piston and connecting rod assemblies. If you know that new pistons are going to be installed and the engine block will be bored oversize, you may be able to forego this step. However, some ridges may actually prevent the assemblies from being removed, necessitating its removal.**

There are several different types of ridge reamers on the market, none of which are inexpensive. Unless a great deal of engine rebuilding is anticipated, borrow or rent a reamer.

1. Turn the crankshaft until the piston is at the bottom of its travel.
2. Cover the head of the piston with a rag.
3. Follow the tool manufacturers instructions and cut away the ridge, exercising extreme care to avoid cutting too deeply.
4. Remove the ridge reamer, the rag and as many of the cuttings as possible. Continue until all of the cylinder ridges have been removed.

DISASSEMBLY

▶ **See Figures 155 and 156**

The engine disassembly instructions following assume that you have the engine mounted on an engine stand. If not, it is easiest to disassemble the engine on a bench or the floor with it resting on the bell housing or transmission mounting surface. You must be able to access the connecting rod fasteners and turn the crankshaft during disassembly. Also, all engine covers (timing, front, side, oil pan, whatever) should have already been removed. Engines which are seized or locked up may not be able to be completely disassembled, and a core (salvage yard) engine should be purchased.

If not done during the cylinder head removal, remove the pushrods and lifters, keeping them in order for assembly. Remove

TCCS3803

Fig. 155 Place rubber hose over the connecting rod studs to protect the crankshaft and cylinder bores from damage

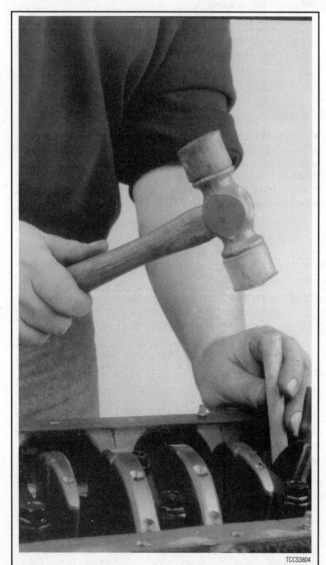

TCCS3804

Fig. 156 Carefully tap the piston out of the bore using a wooden dowel

the timing gears and/or timing chain assembly, then remove the oil pump drive assembly and withdraw the camshaft from the engine block. Remove the oil pick-up and pump assembly. If equipped, remove any balance or auxiliary shafts. If necessary, remove the cylinder ridge from the top of the bore. See the cylinder ridge removal procedure earlier in this section.

Rotate the engine over so that the crankshaft is exposed. Use a number punch or scribe and mark each connecting rod with its respective cylinder number. The cylinder closest to the front of the engine is always number 1. However, depending on the engine placement, the front of the engine could either be the flywheel or damper/pulley end. Generally the front of the engine faces the front of the boat. Use a number punch or scribe and also mark the main bearing caps from front to rear with the front most cap being number 1 (if there are five caps, mark them 1 through 5, front to rear).

✳✳ WARNING

Take special care when pushing the connecting rod up from the crankshaft because the sharp threads of the rod bolts/studs will score the crankshaft journal. Insure that special plastic caps are installed over them, or cut two pieces of rubber hose to do the same.

Again, rotate the engine, this time to position the number one cylinder bore (head surface) up. Turn the crankshaft until the number one piston is at the bottom of its travel, this should allow the maximum access to its connecting rod. Remove the number one connecting rods fasteners and cap and place two lengths of rubber hose over the rod bolts/studs to protect the crankshaft from damage. Using a sturdy wooden dowel and a hammer, push the connecting rod up about 1 in. (25mm) from the crankshaft and remove the upper bearing insert. Continue pushing or tapping the connecting rod up until the piston rings are out of the cylinder bore. Remove the piston and rod by hand, put the upper half of the bearing insert back into the rod, install the cap with its bearing insert installed, and hand-tighten the cap fasteners. If the parts are kept in order in this manner, they will not get lost and you will be able to tell which bearings came form what cylinder if any problems are discovered and diagnosis is necessary. Remove all the other piston assemblies in the same manner. On V-style engines, remove all of the pistons from one bank, then reposition the engine with the other cylinder bank head surface up, and remove that banks piston assemblies.

The only remaining component in the engine block should now be the crankshaft. Loosen the main bearing caps evenly until the fasteners can be turned by hand, then remove them and the caps. Remove the crankshaft from the engine block. Thoroughly clean all of the components.

INSPECTION

Now that the engine block and all of its components are clean, it's time to inspect them for wear and/or damage. To accurately inspect them, you will need some specialized tools:
• Two or three separate micrometers to measure the pistons and crankshaft journals
• A dial indicator
• Telescoping gauges for the cylinder bores
• A rod alignment fixture to check for bent connecting rods
If you do not have access to the proper tools, you may want to bring the components to a shop that does.

Generally, you shouldn't expect cracks in the engine block or its components unless it was known to leak, consume or mix engine fluids, it was severely overheated, or there was evidence of bad bearings and/or crankshaft damage. A visual inspection should be performed on all of the components, but just because you don't see a crack does not mean it is not there. Some more reliable methods for inspecting for cracks include Magnaflux®, a magnetic process or Zyglo®, a dye penetrant. Magnaflux® is used only on ferrous metal (cast iron). Zyglo® uses a spray on fluorescent mixture along with a black light to reveal the cracks. It is strongly recommended to have your engine block checked professionally for cracks, especially if the engine was known to have overheated and/or leaked or consumed coolant. Contact a local shop for availability and pricing of these services.

Engine Block

ENGINE BLOCK BEARING ALIGNMENT

Remove the main bearing caps and, if still installed, the main bearing inserts. Inspect all of the main bearing saddles and caps for damage, burrs or high spots. If damage is found, and it is caused from a spun main bearing, the block will need to be align-bored or, if severe enough, replacement. Any burrs or high spots should be carefully removed with a metal file.

Place a straightedge on the bearing saddles, in the engine block, along the centerline of the crankshaft. If any clearance exists between the straightedge and the saddles, the block must be align-bored.

Align-boring consists of machining the main bearing saddles and caps by means of a flycutter that runs through the bearing saddles.

DECK FLATNESS

The top of the engine block where the cylinder head mounts is called the deck. Insure that the deck surface is clean of dirt, carbon deposits and old gasket material. Place a straightedge across the surface of the deck along its centerline and, using feeler gauges, check the clearance along several points. Repeat the checking procedure with the straightedge placed along both diagonals of the deck surface. If the reading exceeds 0.003 in. (0.076mm) within a 6.0 in. (15.2cm) span, or 0.006 in. (0.152mm) over the total length of the deck, it must be machined.

CYLINDER BORES

▶ See Figures 157 thru 158

The cylinder bores house the pistons and are slightly larger than the pistons themselves. A common piston-to-bore clearance is 0.0015–0.0025 in. (0.0381mm–0.0635mm). Inspect and measure the cylinder bores. The bore should be checked for out-of-roundness, taper and size. The results of this inspection will determine whether the cylinder can be used in its existing size and condition, or a rebore to the next oversize is required (or in the case of removable sleeves, have replacements installed).

The amount of cylinder wall wear is always greater at the top of the cylinder than at the bottom. This wear is known as taper. Any cylinder that has a taper of 0.0012 in. (0.305mm) or more, must be rebored. Measurements are taken at a number of positions in each cylinder: at the top, middle and bottom and at two points at each position; that is, at a point 90 degrees from the crankshaft centerline, as well as a point parallel to the crankshaft centerline. The measurements are made with either a special dial indicator or a telescopic gauge and micrometer. If the necessary precision tools to

Fig. 157 Examples of a wet cylinder liner (left) and a dry liner (right) Note the corrosion on the wet liner from exposure to salt water

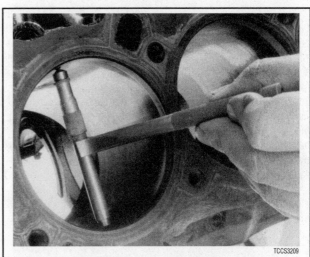

Fig. 158 Use a telescoping gauge to measure the cylinder bore diameter—take several readings within the same bore

check the bore are not available, take the block to a machine shop and have them mike it. Also if you don't have the tools to check the cylinder bores, chances are you will not have the necessary devices to check the pistons, connecting rods and crankshaft. Take these components with you and save yourself an extra trip.

For our procedures, we will use a telescopic gauge and a micrometer. You will need one of each, with a measuring range which covers your cylinder bore size.

1. Position the telescopic gauge in the cylinder bore, loosen the gauges lock and allow it to expand.

➡Your first two readings will be at the top of the cylinder bore, then proceed to the middle and finally the bottom, making a total of six measurements.

2. Hold the gauge square in the bore, 90 degrees from the crankshaft centerline, and gently tighten the lock. Tilt the gauge back to remove it from the bore.

3. Measure the gauge with the micrometer and record the reading.

4. Again, hold the gauge square in the bore, this time parallel to

the crankshaft centerline, and gently tighten the lock. Again, you will tilt the gauge back to remove it from the bore.

5. Measure the gauge with the micrometer and record this reading. The difference between these two readings is the out-of-round measurement of the cylinder.

6. Repeat steps 1 through 5, each time going to the next lower position, until you reach the bottom of the cylinder. Then go to the next cylinder, and continue until all of the cylinders have been measured.

The difference between these measurements will tell you all about the wear in your cylinders. The measurements which were taken 90 degrees from the crankshaft centerline will always reflect the most wear. That is because at this position is where the engine power presses the piston against the cylinder bore the hardest. This is known as thrust wear. Take your top, 90 degree measurement and compare it to your bottom, 90 degree measurement. The difference between them is the taper. When you measure your pistons, you will compare these readings to your piston sizes and determine piston-to-wall clearance.

Crankshaft

Inspect the crankshaft for visible signs of wear or damage. All of the journals should be perfectly round and smooth. Slight scores are normal for a used crankshaft, but you should hardly feel them with your fingernail. When measuring the crankshaft with a micrometer, you will take readings at the front and rear of each journal, then turn the micrometer 90 degrees and take two more readings, front and rear. The difference between the front-to-rear readings is the journal taper and the first-to-90 degree reading is the out-of-round measurement. Generally, there should be no taper or out-of-roundness found, however, up to 0.0005 in. (0.0127mm) for either can be overlooked. Also, the readings should fall within the factory specifications for journal diameters.

If the crankshaft journals fall within specifications, it is recommended that it be polished before being returned to service. Polishing the crankshaft insures that any minor burrs or high spots are smoothed, thereby reducing the chance of scoring the new bearings.

Pistons and Connecting Rods

PISTONS

▶ See Figures 159 thru 162

The piston should be visually inspected for any signs of cracking or burning (caused by hot spots or detonation), and scuffing or excessive wear on the skirts. The wrist pin attaches the piston to the connecting rod. The piston should move freely on the wrist pin, both sliding and pivoting. Grasp the connecting rod securely, or mount it in a vise, and try to rock the piston back and forth along the centerline of the wrist pin. There should not be any excessive play evident between the piston and the pin. If there are C-clips retaining the pin in the piston then you have wrist pin bushings in the rods. There should not be any excessive play between the wrist pin and the rod bushing. Normal clearance for the wrist pin is approx. 0.001–0.002 in. (0.025mm–0.051mm).

Use a micrometer and measure the diameter of the piston, perpendicular to the wrist pin, on the skirt. Compare the reading to its original cylinder measurement obtained earlier. The difference between the two readings is the piston-to-wall clearance. If the clearance is within specifications, the piston may be used as is. If the piston is out of specification, but the bore is not, you will need a

Fig. 159 An example of a badly damaged piston

Fig. 160 Notice the excessive carbon buildup on the top of the piston from excessive engine idling (the scoring of the skirt is from overheating)

Fig. 161 Another example of a damaged piston. The missing ring lands were caused from water entering the cylinder

Fig. 162 Measure the piston's outer diameter, perpendicular to the wrist pin, with a micrometer

new piston. If both are out of specification, you will need the cylinder rebored and oversize pistons installed. Generally if two or more pistons/bores are out of specification, it is best to rebore the entire block and purchase a complete set of oversize pistons.

CONNECTING ROD

♦ See Figure 163

You should have the connecting rod checked for straightness at a machine shop. If the connecting rod is bent, it will unevenly wear the bearing and piston, as well as place greater stress on these components. Any bent or twisted connecting rods must be replaced. If the rods are straight and the wrist pin clearance is within specifications, then only the bearing end of the rod need be checked. Place the connecting rod into a vice, with the bearing inserts in place, install the cap to the rod and torque the fasteners to specifications. Use a telescoping gauge and carefully measure the inside diameter of the bearings. Compare this reading to the rods original crankshaft journal diameter measurement. The difference is the oil clearance. If the oil clearance is not within specifications, install new bearings in the rod and take another measurement. If the clear-

Fig. 163 This bent connecting rod was caused by water entering through the exhaust valves from excessive cranking

ance is still out of specifications, and the crankshaft is not, the rod will need to be reconditioned by a machine shop.

➡You can also use Plastigage® to check the bearing clearances. The assembling section has complete instructions on its use.

Camshaft

Inspect the camshaft and lifters/followers as described earlier in this section.

Bearings

♦ See Figure 164

All of the engine bearings should be visually inspected for wear and/or damage. The bearing should look evenly worn all around with no deep scores or pits. If the bearing is severely worn, scored, pitted or heat blued, then the bearing, and the components that use it, should be brought to a machine shop for inspection. Full-circle bearings (used on most camshafts, auxiliary shafts, balance shafts, etc.) require specialized tools for removal and installation, and should be brought to a machine shop for service.

Fig. 164 Examples of worn engine bearings caused by dirty oil, which lead to the accelerated wear and scoring

Oil Pump

➡The oil pump is responsible for providing constant lubrication to the whole engine and so it is recommended that a new oil pump be installed when rebuilding the engine.

Completely disassemble the oil pump and thoroughly clean all of the components. Inspect the oil pump gears and housing for wear and/or damage. Insure that the pressure relief valve operates properly and there is no binding or sticking due to varnish or debris. If all of the parts are in proper working condition, lubricate the gears and relief valve, and assemble the pump.

REFINISHING

♦ See Figure 165

Almost all engine block refinishing must be performed by a machine shop. If the cylinders are not to be rebored, then the cylinder glaze can be removed with a ball hone. When removing cylinder glaze with a ball hone, use a light or penetrating type oil to lubricate the hone. Do not allow the hone to run dry as this may cause exces-

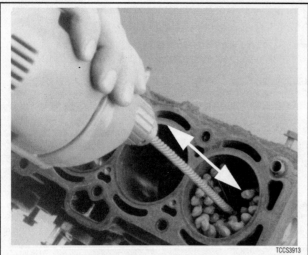

Fig. 165 Use a ball type cylinder hone to remove any glaze and provide a new surface for seating the piston rings

sive scoring of the cylinder bores and wear on the hone. If new pistons are required, they will need to be installed to the connecting rods. This should be performed by a machine shop as the pistons must be installed in the correct relationship to the rod or engine damage can occur.

Pistons and Connecting Rods

▶ See Figure 166

Only pistons with the wrist pin retained by C-clips are serviceable by the home-mechanic. Press fit pistons require special presses and/or heaters to remove/install the connecting rod and should only be performed by a machine shop.

All pistons will have a mark indicating the direction to the front of the engine and the must be installed into the engine in that manner. Usually it is a notch or arrow on the top of the piston, or it may be the letter F cast or stamped into the piston.

C-CLIP TYPE PISTONS

1. Note the location of the forward mark on the piston and mark the connecting rod in relation.

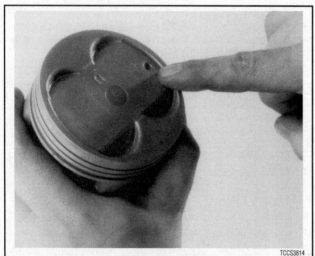

Fig. 166 Most pistons are marked to indicate positioning in the engine (usually a mark means the side facing the front)

2. Remove the C-clips from the piston and withdraw the wrist pin.

➡**Varnish build-up or C-clip groove burrs may increase the difficulty of removing the wrist pin. If necessary, use a punch or drift to carefully tap the wrist pin out.**

3. Insure that the wrist pin bushing in the connecting rod is usable, and lubricate it with assembly lube.

4. Remove the wrist pin from the new piston and lubricate the pin bores on the piston.

5. Align the forward marks on the piston and the connecting rod and install the wrist pin.

6. The new C-clips will have a flat and a rounded side to them. Install both C-clips with the flat side facing out.

7. Repeat all of the steps for each piston being replaced.

ASSEMBLY

Before you begin assembling the engine, first give yourself a clean, dirt free work area. Next, clean every engine component again. The key to a good assembly is cleanliness.

Mount the engine block into the engine stand and wash it one last time using water and detergent (dishwashing detergent works well). While washing it, scrub the cylinder bores with a soft bristle brush and thoroughly clean all of the oil passages. Completely dry the engine and spray the entire assembly down with an anti-rust solution such as WD-40® or similar product. Take a clean lint-free rag and wipe up any excess anti-rust solution from the bores, bearing saddles, etc. Repeat the final cleaning process on the crankshaft. Replace any freeze or oil galley plugs which were removed during disassembly.

Crankshaft

▶ See Figures 167, 168, 169 and 170

1. Remove the main bearing inserts from the block and bearing caps.

2. If the crankshaft main bearing journals have been refinished to a definite undersize, install the correct undersize bearing. Be sure that the bearing inserts and bearing bores are clean. Foreign material under inserts will distort bearing and cause failure.

3. Place the upper main bearing inserts in bores with tang in slot.

➡**The oil holes in the bearing inserts must be aligned with the oil holes in the cylinder block.**

4. Install the lower main bearing inserts in bearing caps.

5. Clean the mating surfaces of block and rear main bearing cap.

6. Carefully lower the crankshaft into place. Be careful not to damage bearing surfaces.

7. Check the clearance of each main bearing by using the following procedure:

   a. Place a piece of Plastigage® or its equivalent, on bearing surface across full width of bearing cap and about ¼ in. off center.

   b. Install cap and tighten bolts to specifications. Do not turn crankshaft while Plastigage® is in place.

   c. Remove the cap. Using the supplied Plastigage® scale, check width of Plastigage® at widest point to get maximum clearance. Difference between readings is taper of journal.

   d. If clearance exceeds specified limits, try a 0.001 in. or

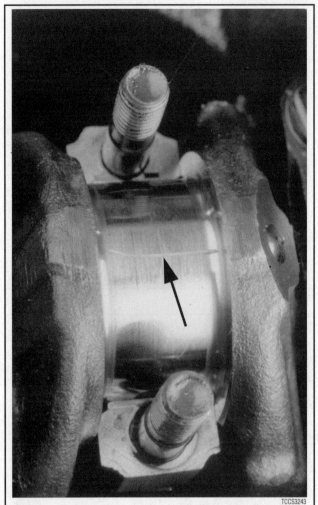

Fig. 167 Apply a strip of gauging material to the bearing journal, then install and torque the cap

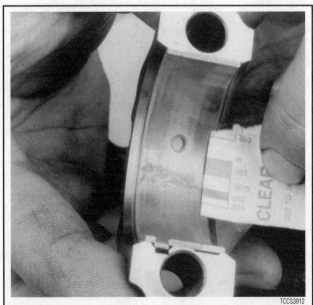

Fig. 168 After the cap is removed again, use the scale supplied with the gauging material to check the clearance

Fig. 169 A dial gauge may be used to check crankshaft end-play

0.002 in. undersize bearing in combination with the standard bearing. Bearing clearance must be within specified limits. If standard and 0.002 in. undersize bearing does not bring clearance within desired limits, refinish crankshaft journal, then install undersize bearings.

8. Install the rear main seal.

9. After the bearings have been fitted, apply a light coat of engine oil to the journals and bearings. Install the rear main bearing cap. Install all bearing caps except the thrust bearing cap. Be sure that main bearing caps are installed in original locations. Tighten the bearing cap bolts to specifications.

10. Install the thrust bearing cap with bolts finger-tight.

11. Pry the crankshaft forward against the thrust surface of upper half of bearing.

12. Hold the crankshaft forward and pry the thrust bearing cap to the rear. This aligns the thrust surfaces of both halves of the bearing.

13. Retain the forward pressure on the crankshaft. Tighten the cap bolts to specifications.

14. Measure the crankshaft end-play as follows:

Fig. 170 Carefully pry the crankshaft back and forth while reading the dial gauge for end-play

a. Mount a dial gauge to the engine block and position the tip of the gauge to read from the crankshaft end.

b. Carefully pry the crankshaft toward the rear of the engine and hold it there while you zero the gauge.

c. Carefully pry the crankshaft toward the front of the engine and read the gauge.

d. Confirm that the reading is within specifications. If not, install a new thrust bearing and repeat the procedure. If the reading is still out of specifications with a new bearing, have a machine shop inspect the thrust surfaces of the crankshaft, and if possible, repair it.

15. Rotate the crankshaft so as to position the first rod journal to the bottom of its stroke.

Pistons and Connecting Rods

▶ **See Figures 171 thru 178**

1. Before installing the piston/connecting rod assembly, oil the pistons, piston rings and the cylinder walls with light engine oil. Install connecting rod bolt protectors or rubber hose onto the connecting rod bolts/studs. Also perform the following:

a. Select the proper ring set for the size cylinder bore.

b. Position the ring in the bore in which it is going to be used.

c. Push the ring down into the bore area where normal ring wear is not encountered.

d. Use the head of the piston to position the ring in the bore so that the ring is square with the cylinder wall. Use caution to avoid damage to the ring or cylinder bore.

e. Measure the gap between the ends of the ring with a feeler gauge. Ring gap in a worn cylinder is normally greater than specification. If the ring gap is greater than the specified limits, try an oversize ring set.

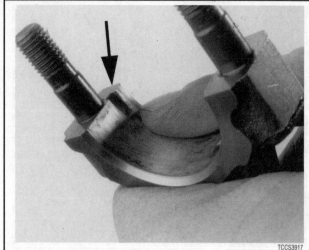

Fig. 172 The notch on the side of the bearing cap matches the tang on the bearing insert

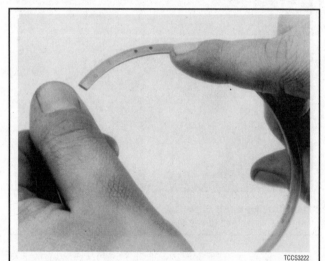

Fig. 173 Most rings are marked to show which side of the ring should face up when installed to the piston

Fig. 171 Checking the piston ring-to-ring groove side clearance using the ring and a feeler gauge

Fig. 174 Install the piston and rod assembly into the block using a ring compressor and the handle of a hammer

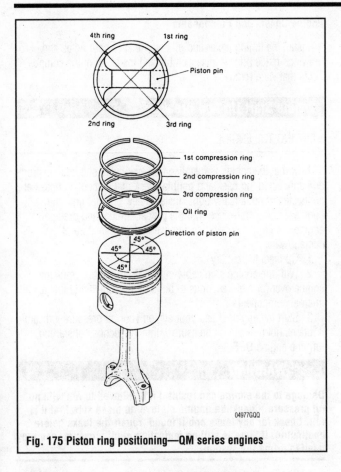

Fig. 175 Piston ring positioning—QM series engines

f. Check the ring side clearance of the compression rings with a feeler gauge inserted between the ring and its lower land according to specification. The gauge should slide freely around the entire ring circumference without binding. Any wear that occurs will form a step at the inner portion of the lower land. If the lower lands have high steps, the piston should be replaced.

2. Unless new pistons are installed, be sure to install the pistons in the cylinders from which they were removed. The numbers on the connecting rod and bearing cap must be on the same side when installed in the cylinder bore. If a connecting rod is ever transposed from one engine or cylinder to another, new bearings should be fitted and the connecting rod should be numbered to correspond with the new cylinder number. The notch on the piston head goes toward the front of the engine.

3. Install all of the rod bearing inserts into the rods and caps.

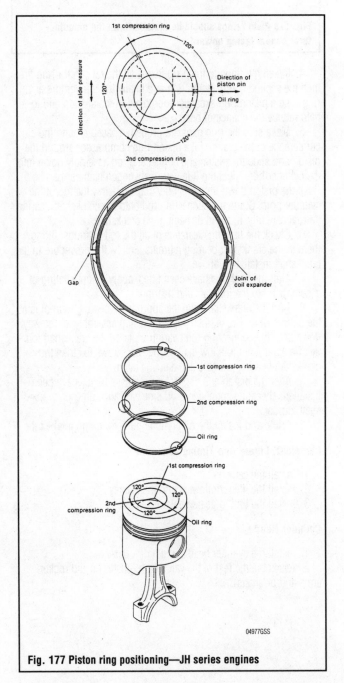

Fig. 177 Piston ring positioning—JH series engines

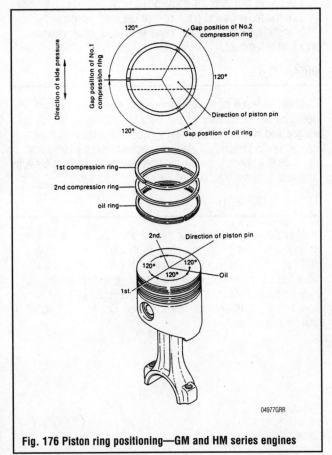

Fig. 176 Piston ring positioning—GM and HM series engines

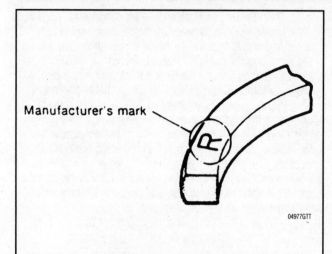

Fig. 178 Piston rings should be installed with the manufacturer's mark facing upward

4. Install the rings to the pistons. Install the oil control ring first, then the second compression ring and finally the top compression ring. Use a piston ring expander tool to aid in installation and to help reduce the chance of breakage.

5. Make sure the ring gaps are properly spaced around the circumference of the piston. Fit a piston ring compressor around the piston and slide the piston and connecting rod assembly down into the cylinder bore, pushing it in with the wooden hammer handle. Push the piston down until it is only slightly below the top of the cylinder bore. Guide the connecting rod onto the crankshaft bearing journal carefully, to avoid damaging the crankshaft.

6. Check the bearing clearance of all the rod bearings, fitting them to the crankshaft bearing journals. Follow the procedure in the crankshaft installation above.

7. After the bearings have been fitted, apply a light coating of assembly oil to the journals and bearings.

8. Turn the crankshaft until the appropriate bearing journal is at the bottom of its stroke, then push the piston assembly all the way down until the connecting rod bearing seats on the crankshaft journal. Be careful not to allow the bearing cap screws to strike the crankshaft bearing journals and damage them.

9. After the piston and connecting rod assemblies have been installed, check the connecting rod side clearance on each crankshaft journal.

10. Prime and install the oil pump and the oil pump intake tube.

Camshaft, Lifters And Timing Assembly

1. Install the camshaft.
2. Install the lifters/followers into their bores.
3. Install the timing gears/chain assembly.

Cylinder Head

1. Install the cylinder head using a new gasket.
2. Assemble the rest of the valve train (pushrods and rocker arms and/or shafts).

Engine Covers and Components

Install the timing cover and oil pan. Refer to your notes and drawings made prior to disassembly and install all of the components that were removed.

Engine Start-Up And Break-In

STARTING THE ENGINE

Now that the engine is installed and every wire and hose is properly connected, go back and double check that all coolant hoses are connected. If not already done, install a new oil filter onto the engine. Fill the crankcase with the proper amount and grade of engine oil. Fill the cooling system with a 50/50 mixture of coolant/water.

1. Connect the battery.
2. With the engine stop cable in the **OFF** position, crank the engine over for a few seconds to build oil pressure and lubricate all internal components.
3. Start the engine. Keep your eye on your oil pressure indicator; if it does not indicate oil pressure within 10 seconds of starting, turn the engine **OFF**.

❊❊ WARNING

Damage to the engine can result if it is allowed to run with no oil pressure. Check the engine oil level to make sure that it is full. Check for any leaks and if found, repair the leaks before continuing. If there is still no indication of oil pressure, you may need to prime the system.

4. Confirm that there are no fluid leaks (oil or other).
5. Allow the engine to reach normal operating temperature.
6. Install any remaining components such as the air cleaner which were removed.

BREAKING IT IN

Make the first hours on the new engine, easy ones. Vary the speed but do not accelerate hard. Most importantly, do not lug the engine, and avoid sustained high speeds. Check the engine oil and coolant levels frequently. Expect the engine to use a little oil until the rings seat. Change the oil and filter shortly after engine break in and then at normally scheduled intervals.

KEEP IT MAINTAINED

Now that you have just gone through all of that hard work, keep yourself from doing it all over again by thoroughly maintaining it. Not that you may not have maintained it before, heck you could have had a thousand hours on it before doing this. However, you may have bought the boat used, and the previous owner did not keep up on maintenance. Which is why you just went through all of that hard work. See?

Engine Torque Specifications

Component	Standard (ft. lbs.)	Metric (Nm)
Rocker Arm/Shafts		
QM	50.6-57.8	68.6-78.6
GM/HM	27	37
4JH	17.4-20.4	23.7-27.7
Cylinder Head Nuts/Bolts		
QM		
Studs (a)	65-69	88-94
Nuts		
Step 1	43	58
Step 2	87	118
Step 3	116-130	158-177
1GM		
Studs (a)	18-22	24-30
Nuts	54.2	73
1GM10		
Studs (a)	43	58
Nuts	54.2	73
2GM		
Studs (a)	29-33	39-45
Nuts	72.3	98.3
Bolts	18.1	26.6
2GM20		
Studs (a)	58	79
Nuts	86.8	118
Bolts 7-8	21.7	29.5
Bolts 2-4-6	86.8	118
3GM		
Studs (a)	29-33	39-45
Nuts	72.3	98.3
Bolts	18.1	26.6
3HM		
Studs (a)	29-33	39-45
Nuts	94	127.8
Bolts	21.7	29.5
3GM30		
Studs (a)	58	79
Nuts	86.8	118
Bolts 9-10-11	21.7	29.5
Bolts 1-2-3-4-6-8	86.8	118
3HM35		
Studs (a)	72	98
Nuts	94	127.8
Bolts 9-10-11	21.7	29.5
Bolts 1-2-3-4-6-8	94	127.8
4JH		
Bolts		
Step 1	25-32	34-43.5
Step 2	54-61.5	73.4-83.6

04977C01

Engine Torque Specifications

Component	Standard (ft. lbs.)	Metric (Nm)
Connecting Rod Bolts		
QM	39.9-43.4	54.3-59.1
GM	18.1	24.6
3HM	32.5	44.2
4JH	32.5-36.2	44.2-49.3
4JH2	36.2-39.8	49.3-54.2
Crankshaft Rear Main Bearing Housing		
QM	27.5	37.4
GM/HM	18.1	24.6
Crankshaft Main Bearing		
3QM (b)	47.2-50.6	64.2-68.9
2GM (b)	21.7-25.3	29.5-34.5
3GM (b)	21.7-25.3	29.5-34.5
3HM (b)	32.5-36.2	44.2-49.2
4JH	68.7-75.94	93.4-103.3
4JH2	75.94-83.17	103.3-113.1
Crankshaft Main Bearing Set Bolt		
3QM	57.8	78.6
2GM	32.5-36.2	44.2-49.2
3GM	32.5-36.2	44.2-49.2
3HM	50.6-54.2	68.9-73.7
Oil Pan		
QM	17-21	23-29
GM/HM	6.5	8.8
Timing Gear Cover		
QM	17.4-21	24-28.6
GM/HM		
M6 Bolts	6.5	8.8
M8 Bolts	18	24
4JH	17.4-21	24-28.6
Oil Pump		
GM/HM	6.5	8.8
Camshaft Gear Nut		
QM	43-58	58.5-78.9
GM/HM	50.6-57.9	68.8-78.7
Crankshaft Gear Nut	58-72.3	78.8-98.3
Crankshaft Pulley Nut/Bolt		
2QM	101-108.5	137.4-147.6
3QM	60-71.6	82-97.4
GM/HM	72.3	113
4JH	83-90.4	123
Flywheel		
2QM	290	394.4
3QM	68-76	92.5-103.4
GM/HM	47-50.6	63.9-68.8
4JH	50-58	68-78.9

(a) This specifications is to be used when tightening a cylinder head stud into the engine block that has worked loose, or is being installed. This torque specification is not to be confused with the specification for tightening the cylinder head to the block.

(b) This torque specification is for the bolts that attach the two halves of the main bearing carrier.

04977C02

Engine Rebuilding Specifications - 2QM20 and 3QM30

Component	Standard (in.)	Metric (mm)
Cylinder Block		
Stud Length		
2QM20		
a, b, c, d	4.25	108
e	1.97	50
f	4.76	121
3QM30		
a, b, c, d, e, f, g	4.25	108
h	4.76	121
Cylinder Bore		
Diameter		
Top	3.4646-3.4659	88-88.035
Bottom	3.3858-3.3872	86-86.035
Roundness	0.0008	0.02
Distortion	0.0039	0.1
Cylinder Liner Bore		
Cylinder Liner Diameter	3.4646-3.4659	88-88.035
Piston Outside Diameter	3.4646-3.4659	88-88.035
Piston to Bore Clearance	0.0055-0.0059	0.14-0.15
Circularity	0.0008-0.0039	0.02-0.1
Cylinder Liner Projection	0.0012-0.0039	0.03-0.10
Cylinder Head		
Distortion	0.00118-0.0039	0.03-0.1
Seat Width	0.0835-0.0984	2.12-2.5
Seat Angle	45 degrees	
Valves		
Valve Recess	0.0413-0.0571	1.05-1.45
Valve Diameter		
Intake	1.5157	38.5
Exhaust	1.2795	32.5
Valve Seat Width	0.0835	2.12
Valve Seat Angle	45 degrees	
Valve Stem Diameter	0.3149-0.3110	8.0-7.9
Valve Stem Bend	0.00118	0.03
Valve Guides		
Inside Diameter	0.315	8
Valve to Guide Clearance		
Intake	0.0016-0.0026	0.040-0.065
Exhaust	0.002-0.003	0.050-0.075
Valve Spring		
Free Length	1.50-1.48	38-37.7
Inclination	0.04	1.14
Pressure at Installed Height of 0.27 in. (6.75mm)	19.5-18.6 lbs	
Rocker Shaft Diameter	0.0006-0.0020	0.016-0.052
Valve Clearance	0.0059	0.15
Pistons		
Outside Diameter	3.4646-3.4567	88-87.3
Pin Bore Diameter	1.1811-1.1850	30-30.1
Pin Outside Diameter	1.1811-1.1791	30-29.95
Pin to Bore Clearance	0.0020-0.0007	0.05-0.019
Pin Bushing Inside Diameter	1.1811-1.18250	30-30.1
Piston Rings		
Ring to Groove Clearance		
First Compression Ring	0.0031-0.0045	0.080-0.115
Second and Third Compression Ring	0.0014-0.0028	0.035-0.070
Oil Ring Groove	0.00078-0.00216	0.020-0.055
Ring End Gap		
First, Second and Third Compression Ring	0.0118-0.0197	0.3-0.5
Oil Ring	0.0118-0.0197	0.3-0.5

04977C03

Engine Rebuilding Specifications - 2QM20 and 3QM30

Component	Standard (in.)	Metric (mm)
Connecting Rod		
Twist	0.00118	0.03
Parallelism	0.00315	0.08
Crank Pin Bushing		
Inside Diameter	2.1260-2.1299	54-54.1
Oil Clearance	0.0014-0.0037	0.036-0.095
Side Clearance	0.0059-0.0138	0.15-0.35
Crankshaft		
Main Journal Diameter		
Gear Case Side	2.7559-2.7520	70-69.9
Flywheel Side		
2QM20	2.7559-2.7520	70-69.9
3QM30	3.5433	90
Connecting Rod Journal Diameter	2.1260-2.1224	54-53.91
Out of Round	0.0004	0.01
Main Journal Oil Clearance		
Gear Case Side	0.0014-0.0039	0.036-0.099
Intermediate Bearing	0.0014-0.0037	0.036-0.095
Flywheel Side	0.0026-0.0052	0.066-0.132
Crankshaft Bend	0.006	0.015
End Clearance		
2QM20	0.0039-0.0079	0.10-0.20
3QM30	0.0035-0.0075	0.09-0.19
Crankshaft Gear Backlash	0.0031-0.0062	0.08-0.16
Camshaft		
Valve Head Clearance	0.0059	0.15
Valve Events		
Intake Open	20 deg. BTDC	
Intake Close	50 deg. ATDC	
Exhaust Open	50 deg. BTDC	
Exhaust Closed	20 deg. ATDC	
Lift	1.5209-1.5012	38.63-38.13
Side Clearance	0	
Camshaft Bearing		
Clearance	0.0020-0.0039	0.050-0.100
Backlash	0.0031-0.0062	0.08-0.16
Valve Tappet		
Diameter	0.4331	11
Tappet to Bore Clearance	0.0002-0.00138	0.006-0.035

04977C04

Engine Rebuilding Specifications - 1GM, 2GM, 3GM and 3HM

Component	Standard (in.)	Metric (mm)
Cylinder Block		
Stud Length		
1GM	3.1102	79
2GM, 3GM, 3HM		
a	3.7795	96
b	3.2283	82
Cylinder Bore		
1GM		
Diameter	2.8346-2.8358	72-72.03
Roundness	0.0004	0.01
Piston Outside Diameter	2.8346	72
Cylinder Liner Bore		
2GM, 3GM		
Cylinder Liner Diameter	2.8346	72
Piston Outside Diameter	2.8346	72
Piston to Bore Clearance	0.0055-0.0059	0.14-0.15
Circularity	0.0008	0.02
Liner Projection	0.0002-0.0030	0.005-0.075
3HM		
Cylinder Liner Diameter	2.9528	75
Piston Outside Diameter	2.9528	75
Piston to Bore Clearance	0.0015-0.0058	0.038-0.148
Circularity	0.0008	0.02
Liner Projection	0.0002-0.0030	0.005-0.075
Cylinder Head		
Distortion	0.0028	0.07
Seat Width	0.06969	1.77
Seat Angle	45 degrees	
Valves		
Valve Recess		
1GM, 2GM, 3GM	0.0374-0.0492	0.95-1.25
3HM	0.0492-0.0610	1.25-1.55
Valve Diameter		
1GM, 2GM, 3GM		
Intake	1.2598	32
Exhaust	1.0236	32
3HM		
Intake	1.2598	32
Exhaust	1.063	27
Valve Seat Width	0.0835	2.12
1GM, 2GM, 3GM	0.124	3.15
3HM	0.1197	3.04
Valve Seat Angle	45 degrees	
Valve Stem Diameter	0.2756-0.2717	7-6.9
Valve Stem Bend	0.0012	0.03
Valve Guides		
Inside Diameter		
Intake	0.0016-0.0026	0.040-0.065
Exhaust	0.002-0.003	0.050-0.075
Valve to Guide Clearance		
Intake	0.0016-0.0026	0.040-0.065
Exhaust	0.0018-0.0028	0.45-0.70
Valve Guide to Cylinder Head Clearance		
1GM	0.0002-0.0013	0.005-0.034
2GM, 3GM	0.0007-0.0019	0.018-0.047
3HM	0.0007-0.0019	0.018-0.047
Protrusion	0.2756	7
Valve Spring		
Free Length	1.5157	38.5
Installed Height		
1GM, 2GM, 3GM	1.1496	29.2
3HM	1.189	30.2
Pressure at Installed Height		
1GM, 2GM, 3GM	35.63 lbs	
3HM	31.81 lbs	

04977C05

Engine Rebuilding Specifications - 1GM, 2GM, 3GM and 3HM

Component	Standard (in.)	Metric (mm)
Rocker Shaft		
Diameter		
1GM	0.4724	12
2GM	0.5512	14
3GM, 3HM	0.5512	14
Clearance	0.0006-0.0020	0.016-0.052
Valve Clearance		
Intake and Exhaust	0.0079	0.2
Pistons		
1GM, 2GM, 3GM		
Outside Diameter	2.8312-2.8324	72.057-71.913
Pin Bore Diameter	0.7872-0.7877	19.995-20.008
Pin Outside Diameter	0.7870-0.7874	19.991-20
Pin to Bore Clearance	0.0010-0.0019	0.025-0.047
Pin Bushing Inside Diameter	0.7874	20
3HM		
Outside Diameter	2.9491-2.9503	74.907-75.063
Pin Bore Diameter	0.9053-0.9058	22.995-23.008
Pin Outside Diameter	0.9052-0.9055	22.991-23
Pin to Bore Clearance	0.0010-0.0019	0.025-0.047
Pin Bushing Inside Diameter	0.9055	23
Piston to Head Clearance		
1GM, 2GM, 3GM	0.0276	0.7
3HM	0.0315	0.8
Piston Rings		
Ring to Groove Clearance		
1GM, 2GM, 3GM		
First Compression Ring	0.0024-0.0039	0.06-0.10
Second Compression Ring	0.0014-0.0028	0.035-0.070
Oil Ring Groove	0.0008-0.0022	0.020-0.055
3HM		
First Compression Ring	0.0026-0.0039	0.065-0.10
Second Compression Ring	0.0014-0.0028	0.035-0.070
Oil Ring Groove	0.0008-0.0022	0.020-0.055
Ring End Gap	0.0079-0.0157	0.20-0.40
Connecting Rod		
Twist	0.00118	0.03
Parallelism	0.00315	0.08
Crank Pin Bushing		
Inside Diameter		
1GM, 2GM, 3GM	1.5748-1.5787	40-40.1
3HM	1.7323-1.7362	44-44.10
Oil Clearance		
1GM, 2GM, 3GM	0.0011-0.0034	0.028-0.086
3HM	0.0014-0.0037	0.036-0.095
Side Clearance	0.0079-0.0157	0.2-0.4
Crankshaft		
Main Journal Diameter		
Gear Case Side		
1GM, 2GM, 3GM	1.7303-1.7309	43.964-43.95
3HM	1.8484-1.8490	46.964-46.95
Intermediate Bearing		
1GM, 2GM, 3GM	1.7303-1.7309	43.964-43.95
3HM	1.8484-1.8490	46.964-46.95
Flywheel Side		
1GM, 2GM, 3GM	2.3602-2.3608	59.964-59.95
3HM	2.5571-2.5576	64.964-64.95
Connecting Rod Journal Diameter		
1GM, 2GM, 3GM	1.5728-1.5734	39.964-39.95
3HM	1.7303-1.7309	43.964-43.95
Out of Round	0.0004	0.01

04977C06

Engine Rebuilding Specifications - 1GM, 2GM, 3GM and 3HM

Component	Standard (in.)	Metric (mm)
Main Journal Oil Clearance		
Gear Case Side	0.0014-0.0037	0.036-0.095
Intermediate Bearing	0.0014-0.0037	0.036-0.095
Flywheel Side	0.0014-0.0037	0.036-0.095
Connecting Rod Journal Oil Clearance		
1GM, 2GM, 3GM	0.0011-0.0034	0.028-0.086
3HM	0.0014-0.0037	0.036-0.095
Crankshaft Bend	0.006	0.015
End Clearance		
1GM	0.0024-0.0075	0.06-0.19
2GM, 3GM	0.0035-0.0075	0.09-0.19
3HM	0.0035-0.0071	0.09-0.18
Camshaft		
Valve Head Clearance	0.0079	0.2
Valve Events		
Intake Open		20 deg. BTDC
Intake Close		50 deg. ATDC
Exhaust Open		50 deg. BTDC
Exhaust Closed		20 deg. ATDC
Lift		
1GM	1.1417	29
2GM, 3GM, 3HM	1.378	35
Camshaft Bearing		
Clearance	0.0020-0.0039	0.050-0.100
Backlash	0.0031-0.0062	0.08-0.16
Valve Tappet		
Diameter	0.3937	10
Tappet to Bore Clearance		
1GM	0.0010-0.0024	0.025-0.060
2GM, 3GM, 3HM	0.0004-0.0016	0.010-0.040
Pushrods		
Length		
1GM10	5.6299	143
2GM20, 3GM30	5.3543	136
3HM35	6.7323	171
Pushrod Bend	0.00118 max	0.03 max

04977C07

Engine Rebuilding Specifications - 1GM10, 2GM20, 3GM30 and 3HM35

Component	Standard (in.)	Metric (mm)
Cylinder Block		
Cylinder Bore		
1GM10, 2GM20, 3GM30		
Diameter	2.9528-2.9540	75-75.03
Roundness	0.0004	0.01
3HM35		
Diameter	3.1496-3.1508	80-80.03
Roundness	0.0004	0.01
Block Distortion	0.002	0.05
Cylinder Head		
Distortion	0.0028	0.07
Seat Width	0.06969	1.77
Seat Angle	45 degrees	
Valves		
Valve Recess		
1GM10, 2GM20, 3GM30	0.0374-0.0492	0.95-1.25
3HM35	0.0492-0.0610	1.25-1.55
Valve Diameter		
1GM10, 2GM20, 3GM30		
Intake	1.2598	32
Exhaust	1.0236	32
3HM35		
Intake	1.2598	32
Exhaust	1.063	27
Valve Stem Diameter	0.2756-0.2717	7-6.9
Valve Stem Bend	0.0012	0.03
Valve Guides		
Inside Diameter		
Intake	0.0016-0.0026	0.040-0.065
Exhaust	0.002-0.003	0.050-0.075
Valve to Guide Clearance		
Intake	0.0016-0.0026	0.040-0.065
Exhaust	0.0018-0.0028	0.45-0.70
Valve Guide to Cylinder Head Clearance		
1GM10	0.0002-0.0013	0.005-0.034
2GM20, 3GM30	0.0007-0.0019	0.018-0.047
3HM35	0.0007-0.0019	0.018-0.047
Protrusion	0.2756	7
Valve Spring		
Free Length	1.5157	38.5
Installed Height		
1GM10, 2GM20, 3GM30	1.1496	29.2
3HM35	1.189	30.2
Pressure at Installed Height		
1GM10, 2GM20, 3GM30	35.63 lbs	
3HM35	31.81 lbs	
Rocker Shaft		
Diameter		
1GM10	0.4724	12
2GM20	0.5512	14
3GM30, 3HM35	0.5512	14
Clearance	0.0006-0.0020	0.016-0.052
Valve Clearance		
Intake and Exhaust	0.0079	0.2

04977C08

Engine Rebuilding Specifications - 1GM10, 2GM20, 3GM30 and 3HM35

Component	Standard (in.)	Metric (mm)
Pistons		
1GM10, 2GM20, 3GM30		
Outside Diameter	2.9492-2.9504	74.91-74.94
Pin Bushing Inside Diameter	0.7872-0.7877	19.995-20.008
Pin to Bore Clearance	0.0002-0.0007	0.005-0.017
3HM35		
Outside Diameter	3.1457-3.1470	79.902-79.932
Pin Bushing Inside Diameter	0.9053-.9058	22.995-23.008
Pin to Bore Clearance	0.0002-0.0007	0.005-0.017
Piston to Head Clearance		
1GM10, 2GM20, 3GM30	0.0268-0.0346	0.68-0.88
3HM35	0.0260-0.0339	0.66-0.86
Piston Rings		
Ring to Groove Clearance		
First Compression Ring	0.0026-0.0039	0.065-0.10
Second Compression Ring	0.0014-0.0028	0.035-0.070
Oil Ring Groove	0.0008-0.0022	0.020-0.055
Ring End Gap	0.0079-0.0157	0.20-0.40
Connecting Rod		
Twist	0.00118	0.03
Parallelism	0.00315	0.08
Crank Pin Bushing		
Inside Diameter		
1GM10, 2GM20, 3GM30	1.5748-1.5787	40-40.1
3HM35	1.7323-1.7362	44-44.10
Oil Clearance		
1GM10, 2GM20, 3GM30	0.0011-0.0034	0.028-0.086
3HM35	0.0014-0.0037	0.036-0.095
Side Clearance	0.0079-0.0157	0.2-0.4
Crankshaft		
Main Journal Diameter		
Gear Case Side		
1GM10, 2GM20, 3GM30	1.7303-1.7309	43.964-43.95
3HM35	1.8484-1.8490	46.964-46.95
Intermediate Bearing		
1GM10, 2GM20, 3GM30	1.7303-1.7309	43.964-43.95
3HM35	1.8484-1.8490	46.964-46.95
Flywheel Side		
1GM10, 2GM20, 3GM30	2.3602-2.3608	59.964-59.95
3HM35	2.5571-2.5576	64.964-64.95
Connecting Rod Journal Diameter		
1GM10, 2GM20, 3GM30	1.5728-1.5734	39.964-39.95
3HM35	1.7303-1.7309	43.964-43.95
Out of Round	0.0004	0.01
Main Journal Oil Clearance		
Gear Case Side	0.0014-0.0037	0.036-0.095
Intermediate Bearing	0.0014-0.0037	0.036-0.095
Flywheel Side	0.0014-0.0037	0.036-0.095
Connecting Rod Journal Oil Clearance		
1GM10, 2GM20, 3GM30	0.0011-0.0034	0.028-0.086
3HM35	0.0014-0.0037	0.036-0.095
Crankshaft Bend	0.006	0.015
End Clearance		
1GM10	0.0024-0.0075	0.06-0.19
2GM20, 3GM30	0.0035-0.0075	0.09-0.19
3HM35	0.0035-0.0071	0.09-0.18

04977C09

Engine Rebuilding Specifications - 1GM10, 2GM20, 3GM30 and 3HM35

Component	Standard (in.)	Metric (mm)
Camshaft		
Valve Head Clearance	0.0079	0.2
Valve Events		
Intake Open		20 deg. BTDC
Intake Close		50 deg. ATDC
Exhaust Open		50 deg. BTDC
Exhaust Closed		20 deg. ATDC
Lift		
1GM10	1.1417	29
2GM20, 3GM30, 3HM35	1.378	35
Camshaft Bearing		
Clearance	0.0020-0.0039	0.050-0.100
Backlash	0.0031-0.0062	0.08-0.16
Valve Tappet		
Diameter	0.3937	10
Tappet to Bore Clearance		
1GM10	0.0010-0.0024	0.025-0.060
2GM20, 3GM30, 3HM35	0.0004-0.0016	0.010-0.040
Pushrods		
Length		
1GM10	5.6299	143
2GM20, 3GM30	5.3543	136
3HM35	6.7323	171
Pushrod Bend	0.00118 max	0.03 max

04977C10

Engine Rebuilding Specifications - 4JHE, 4JH-TE, 4JH-THE, 4JH-DTE

Component	Standard (in.)	Metric (mm)
Cylinder Block		
Cylinder Bore		
Diameter	3.2283-3.2295	82-82.03
Roundness	0-0.0004	0-0.01
Block Distorition	0.002	0.05
Cylinder Head		
Distortion	0.002	0.05
Seat Width	0.06969	1.77
Intake	0.0504	1.28
Exhaust	0.0697	1.77
Seat Angle		
Intake	30 degrees	
Exhaust	45 degrees	
Valves		
Valve Recess	0.0157-0.0236	0.4-0.6
Valve Stem Diameter		
Intake	0.3134-0.3140	7.960-7.975
Exhaust	0.3132-0.3138	7.955-7.970
Valve to Piston Clearance	0.0280-0.0350	0.71-0.89
Valve Guides		
Inside Diameter		
Intake	0.3154-0.3159	8.010-8.025
Exhaust	0.3156-0.3161	8.015-8.030
Valve Spring		
Free Length	1.748	44.4
Installed Height	1.5748	40
Pressure at Installed Height	26.46 lbs	
Valve Clearance		
Intake and Exhaust	0.0079	0.2
Pistons		
Outside Diameter	3.2252-3.2263	81.92-81.95
Pin Bushing Inside Diameter	1.1033-1.1039	28.025-28.038
Piston Pin Outside Diameter	1.1019-1.1024	27.99-28
Pin to Bore Clearance	0.0009-0.002	0.025-0.051
Piston Rings		
Ring to Groove Clearance		
First Compression Ring	0.0027-0.0039	0.070-0.10
Second Compression Ring	0.0013-0.0027	0.035-0.070
Oil Ring Groove	0.0011-0.0023	0.030-0.060
Ring End Gap		
First Ring	0.0098-0.0157	0.25-0.40
Second and Third Ring	0.0079-0.0157	0.20-0.40
	0.0079-0.0157	0.20-0.40
Connecting Rod		
Twist	0.0019	0.05
Side Clearance	0.0079-0.0157	0.2-0.4
Crankshaft		
Main Journal Diameter	1.9666-1.9670	49.952-49.962
Main Journal Oil Clearance	0.0014-0.0036	0.038-0.093
Out of Round	0.0004	0.01
Connecting Rod Journal Diameter	1.8878-1.8882	47.952-47.962
Connecting Rod Journal Oil Clearance	0.0014-0.0036	0.038-0.093
Crankshaft Bend	0.0012	0.03
End Clearance	0.0035-0.0106	0.090-0.271

04977C11

Engine Rebuilding Specifications - 4JHE, 4JH-TE, 4JH-THE, 4JH-DTE

Component	Standard (in.)	Metric (mm)
Camshaft		
Valve Events		
4JHE		
Intake Open	10-20 deg. BTDC	
Intake Close	48-58 deg. ATDC	
Exhaust Open	51-61 deg. BTDC	
Exhaust Closed	13-23 deg. ATDC	
4JH-TE, 4JH-HTE, 4JH-DTE		
Intake Open	26-36 deg. BTDC	
Intake Close	38-48 deg. ATDC	
Exhaust Open	49-59 deg. BTDC	
Exhaust Closed	29-39 deg. ATDC	
Lift		
Intake	1.5220-1.5251	38.66-38.74
Exhaust		
4JHE	1.5220-1.5251	38.66-38.74
4JH-TE, 4JH-HTE, 4JH-DTE	1.5299-1.5330	38.86-38.94
Camshaft Bearing		
Clearance		
Gear Case Side	0.0015-0.0049	0.050-0.100
Intermediate	0.0025-0.0045	0.065-0.115
Flywheel Side	0.0019-0.0039	0.050-0.100
End Gap	0.0019-0.0079	0.05-0.20
Deflection	0.0007	0.02
Valve Tappet		
Diameter	0.4714-0.4720	11.975-11.990
Tappet to Bore Clearance	0.0003-0.0016	0.010-0.043
Pushrods		
Length	7.0177-7.0374	178.25-178.75
Pushrod Bend	0.0011 max	0.03 max

04977C12

Engine Rebuilding Specifications - 4JH2E, 4JH2-TE, 4JH2-THE, 4JH2-DTE

Component	Standard (in.)	Metric (mm)
Cylinder Block		
Cylinder Bore		
Diameter	3.2283-3.2295	82-82.03
Roundness	0-0.0004	0-0.01
Block Distortion	0.002	0.05
Cylinder Head		
Distortion	0.002	0.05
Seat Width	0.06969	1.77
Intake	0.0504	1.28
Exhaust	0.0697	1.77
Seat Angle		
Intake	30 degrees	
Exhaust	45 degrees	
Valves		
Valve Recess	0.0157-0.0236	0.4-0.6
Valve Stem Diameter		
Intake	0.3134-0.3140	7.960-7.975
Exhaust	0.3132-0.3138	7.955-7.970
Valve to Piston Clearance	0.0280-0.0350	0.71-0.89
Valve Guides		
Inside Diameter		
Intake	0.3154-0.3159	8.010-8.025
Exhaust	0.3156-0.3161	8.015-8.030
Valve Spring		
Free Length	1.748	44.4
Installed Height	1.5748	40
Pressure at Installed Height	26.46 lbs	
Valve Clearance		
Intake and Exhaust	0.0079	0.2
Pistons		
Outside Diameter	3.2252-3.2263	81.92-81.95
Pin Bushing Inside Diameter	1.1033-1.1039	28.025-28.038
Piston Pin Outside Diameter	1.1019-1.1024	27.99-28
Pin to Bore Clearance	0.0009-0.002	0.025-0.051
Piston Rings		
Ring to Groove Clearance		
First Compression Ring	0.0027-0.0039	0.070-0.10
Second Compression Ring	0.0013-0.0027	0.035-0.070
Oil Ring Groove	0.0011-0.0023	0.030-0.060
Ring End Gap		
First Ring	0.0098-0.0157	0.25-0.40
Second and Third Ring	0.0079-0.0157	0.20-0.40
	0.0079-0.0157	0.20-0.40
Connecting Rod		
Twist	0.0019	0.05
Side Clearance	0.0079-0.0157	0.2-0.4
Crankshaft		
Main Journal Diameter	1.9666-1.9670	49.952-49.962
Main Journal Oil Clearance	0.0014-0.0036	0.038-0.093
Out of Round	0.0004	0.01
Connecting Rod Journal Diameter	1.8878-1.8882	47.952-47.962
Connecting Rod Journal Oil Clearance	0.0014-0.0036	0.038-0.093
Crankshaft Bend	0.0012	0.03
End Clearance	0.0035-0.0106	0.090-0.271

04977C13

Engine Rebuilding Specifications - 4JH2E, 4JH2-TE, 4JH2-THE, 4JH2-DTE

Component	Standard (in.)	Metric (mm)
Camshaft		
Valve Events		
4JH2E		
Intake Open	10-20 deg. BTDC	
Intake Close	48-58 deg. ATDC	
Exhaust Open	51-61 deg. BTDC	
Exhaust Closed	13-23 deg. ATDC	
4JH2-TE, 4JH2-HTE, 4JH2-DTE		
Intake Open	26-36 deg. BTDC	
Intake Close	38-48 deg. ATDC	
Exhaust Open	49-59 deg. BTDC	
Exhaust Closed	29-39 deg. ATDC	
Lift		
Intake	1.5220-1.5251	38.66-38.74
Exhaust		
4JH2E	1.5220-1.5251	38.66-38.74
4JH2-TE, 4JH2-HTE, 4JH2-DTE	1.5299-1.5330	38.86-38.94
Camshaft Bearing		
Clearance		
Gear Case Side	0.0015-0.0049	0.050-0.100
Intermediate	0.0025-0.0045	0.065-0.115
Flywheel Side	0.0019-0.0039	0.050-0.100
End Gap	0.0019-0.0079	0.05-0.20
Deflection	0.0007	0.02
Valve Tappet		
Diameter	0.4714-0.4720	11.975-11.990
Tappet to Bore Clearance	0.0003-0.0016	0.010-0.043
Pushrods		
Length	7.0177-7.0374	178.25-178.75
Pushrod Bend	0.0011 max	0.03 max

04977C14

8

TRANSMISSIONS

TRANSMISSIONS 8-2

TRANSMISSIONS

General Information

In to transfer the power of an inboard engine to the propeller and then propel the boat, some sort of transmission is required between the engine and the propeller shaft. With the exception of jet drive units, propulsion force is usually provided by a propeller located at or near the stern. Since the power of the engine may not be of the amount or type the propeller needs and in order to provide forward, neutral and reverse gears, a transmission is necessary.

In the conventional installation, the propeller is trying to force itself forward, together with the propeller shaft, into the boat. Similarly, when backing, the propeller and shaft are pull away from the boat. These forces, or thrust loads, must be absorbed by the transmission and hence thrust bearings are usually provided. If a thrust bearing is not a part of the transmission, it must be provided separately. While separate thrust bearings and housings were once fairly common, they are seldom used in most boats simply because thrust protection is built into most modern transmissions.

In addition to thrust protection and directional control the marine transmission may also serve to change the drive ratio between the engine and the propeller. While many boats can get by without varying the propeller shaft speed in relation to engine speed, other boats will require a reduction gear that will reduce the propeller shaft speed from that of the engine speed in order to multiply the torque available at the shaft. The ability of the transmission to provide this change in propeller shaft rpm is important in order to use the most efficiently sized propeller.

Generally, the smaller, lighter and faster a boat is, the smaller the reduction gear ratio. Small runabouts and cruisers often will not require a reduction gear since high rpm will be desirable at the propeller. Consequently, engines aboard such boats will have direct drives turning at a 1:1 ratio. Many boats require even less than a direct drive gear, in which case an overdrive or step-up gear ratio is required. In these cases, the propeller shaft will be driven at a higher speed than that provided by the engine. Such boats, however, are usually of the high speed sport or competition craft with very fast turning propellers.

Transmission Types

♦ **See Figures 1 and 2**

➥GM series engines are usually mated to Yanmar transmissions, while transmissions on JH series engines are at the discretion of the boat builder.

The Yanmar diesel engines covered in this manual come equipped with three types of transmissions: straight drive, angle drive, and V-drive. Although they are different in design, their purpose is universal—to provide power from the engine to the propeller shaft at the proper speed and direction.

Yanmar includes transmissions of all three types with their engines, and in some cases, the transmissions can be interchanged from one model to another. However, the transmission in your vessel was most likely specified by the boat builder, and custom fitted to the boat.

In most cases, retrofitting a different type of transmission would require extensive modifications to the engine mounting. Even fitting another transmission of the same type can cause problems, since

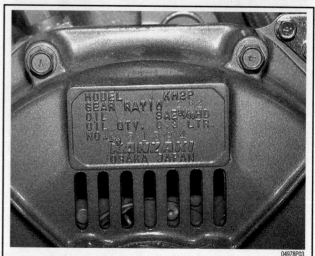

Fig. 1 On the case of each transmission, important information is included on the nameplate

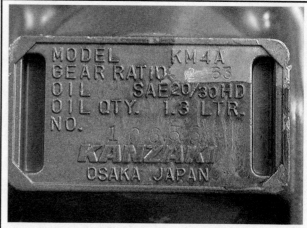

Fig. 2 A closer look at the nameplate. Notice the oil type and quantity are included, along with the model number and gear ratio

the gear ratios are tailored for the engine and propeller for maximum efficiency.

As with any installation, below deck quarters and the size and shape of the hull will dictate the location of the engine. In most cases, the engineer who designed the boat will have specified engine and transmission type and location in the hull before the boat is even built.

STRAIGHT DRIVE

The most common transmission found on Yanmar engines is a basic straight drive transmission. The output shaft on a straight drive transmission is inline with the crankshaft of the engine, hence the name.

The main drawback of a straight drive transmission is the angle in which the engine must be mounted to align with the propeller shaft.

When the engine is mounted on an angle, it occupies more space than if it were mounted level with the hull. The angle (unless extreme) has little effect on the ability of the engine to operate properly.

Another potential drawback of a straight drive transmission is that it requires that the engine be located further forward in the hull to compensate for the angle of the propeller shaft. Of course this angle also depends on the shape of the hull.

Outweighing the drawbacks, straight drive transmissions are simple in construction, contain few moving parts, and are very reliable and trouble-free. Other than changing the oil and adjusting the shifter cable(s), straight line transmissions are virtually indestructible.

Yanmar supplied straight drive transmissions include:
- KBW10A, KBW10D, KBW10E
- KBW20
- KBW21
- KH-18
- KM2-A, KM2-C, KM2-P
- KM3-A, KM2-C, KM3-P
- YP-7M
- YP-10M

ANGLE DRIVE

Angle drive transmissions are also quite common with most Yanmar engines. Angle drive transmissions are not easily distinguished between straight drive transmissions, until a closer look at the relationship between the output shaft and the engine reveal a slight difference in angle.

In many cases, the slight tilting of the output shaft on an angle drive transmission is enough to allow the engine to be mounted level with the hull. If the engine is not mounted level with the hull, the mounting angle of the engine is at least reduced from the angle of the transmission.

The construction of an angle drive transmission are very similar to a straight drive unit. The tilted output shaft of the transmission is achieved by using beveled gears on the output shaft. Angle drive transmissions typically do not contain any more moving parts than a straight drive transmission, and are equally reliable and trouble free.

Yanmar supplied angle drive transmissions include the KM4A.

V-DRIVE

With this type of transmission, the engine is essentially installed backwards (the flywheel faces the bow) and the propeller shaft runs underneath the engine, and attaches to the transmission. A V-drive transmission allows the engine to be located further astern, allowing for more usable space below deck.

Mechanically, V-drive transmissions are more complex than their straight drive and angle drive counterparts. The construction of a V-drive transmission combines the design of a straight drive unit combined with the beveled gear design on an angle drive.

As with any engine installation, there are both drawbacks and advantages to a V-drive transmission. Although the shape of the transmission allows the engine to be located further astern, the engine also must be mounted higher in the hull to allow for the propeller shaft underneath the engine.

Yanmar supplied angle drive transmissions include the KM3V.

Transmission Assembly

REMOVAL & INSTALLATION

♦ **See Figures 3 thru 9**

Removal of the transmission can be a simple procedure, or a lengthy, difficult task depending on the design of the boat, installation of the engine and propeller shaft, and the available space in the engine bay.

The procedure listed here is generic, and should only be used as a guideline for transmission removal. Since every boat is constructed differently, and engine installation unique, specific procedures cannot be given. Some removal procedures may require the use of hydraulic jacks, or ratchet pulleys. Use the steps listed as a "checklist" for removing the transmission from your engine.

1. Assess the amount of room available for transmission removal and installation. Before removing any fasteners, you will

Fig. 3 Some transmissions have oil coolers that must be disconnected for removal of the transmission

Fig. 4 In some cases, it may be necessary to remove cable or wire brackets for removal of the transmission

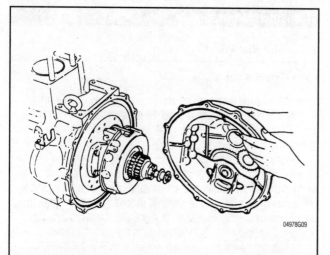

Fig. 5 YP7M and YP10M transmissions used on some QM series engines use a clutch mechanism that attaches to the flywheel

Fig. 8 Lightly grease the splines on the input shaft before installation

Fig. 6 The transmission can be removed from the engine by removing the bolts around the flywheel housing

Fig. 9 When installing the transmission, it may be necessary to rotate the input shaft (or flywheel) slightly for the splines to align

Fig. 7 Once the transmission is removed, the input shaft seal can be replaced if necessary

need to confirm that the transmission can be removed from the engine and lifted from the engine bay. In most cases, the boat builder engineered reasonable access to the engine and transmission for servicing. However, it may be necessary to remove some fixtures and other components to allow for the transmission to be removed from the engine bay of the boat.

2. Remove any components that would hinder with removing the transmission from the engine bay.

Once you've assessed the feasibility of removing the transmission, remove the necessary components in and around the engine bay to allow enough clearance for removal.

3. Remove the oil from the transmission. Although not necessary, a small amount of weight can be saved by evacuating the oil from the transmission before lifting it from the engine bay.

4. Disconnect the control cable(s) and other connections from the transmission. Before getting started on removing the transmission from the engine, disconnect the shifter cable(s) and/or linkage from the levers. Also, on some transmissions, an oil cooler may be used. Make sure to drain the coolant or water and disconnect the hoses from the transmission.

Troubleshooting Marine Transmissions

Condition	Cause	Correction
Output coupling does not turn when transmission is in gear	Cable length incorrect	Adjust cable
Propeller does does not turn when transmission coupling turns	Internal transmisison malfunction	Service transmission
	Coupling bolts sheared	Determine cause of shearing, then replace coupling bolts
	Coupling slipping on propeller shaft	Tighten or replace set screws, keys, pins or coupling bolts
Remote control lever does not move operating lever on transmission	Cable broken or disconnected	Repair or replace cable
Remote control lever does not move operating lever on transmission through full range of motion	Cable slipping or kinked	Repair or replace cable
Remote control lever is not moving freely	Cable corroded or binding	Lubricate cable
Tranmsision buzzing	Transmission binding	Service transmission
	Transmisison oil low	Correct oil level and run transmisison in neutral to evacuate air
Transmission gear range does not correspond to range selected by remote control lever	Cable length incorrect	Adjust cable

0497BC01

5. Disconnect the output coupling from the propeller shaft. This procedure can vary in difficulty depending on what components the boat builder used to connect the propeller shaft to the output shaft on the transmission. In most cases, simple loosening or removal of fasteners is all that is required to disconnect the propeller shaft from the transmission. Rigid couplings are typically bolted together; simply removing the bolts will separate the coupling. Other couplings, like rubber flex and two-piece collar types may be more difficult to disconnect. Consult the boat manufacturer for more information if necessary.

6. Loosen or disconnect the engine/transmission mounts. Disconnecting or removing the engine and transmission mounts depends on the mounting of the engine in the hull, and the need for sufficient clearance between the disconnected propeller shaft and the transmission. Most transmissions need about 3 inches of clearance to allow the input shaft of the transmission to be drawn from the flex plate on the flywheel. Some older QM series engines equipped with YP-7M and YP-10M transmissions may require up to 12 inches (or more) of clearance between the propeller shaft and output shaft of the transmission in order to clear the clutch assembly that is bolted to the flywheel. If your propeller shaft can be pushed toward the stern after it is disconnected from the transmission, tilting the engine may not be necessary. (Make sure the stuffing box does not leak if the boat is in the water.) If the propeller shaft cannot be moved, the engine must be tilted or moved forward to allow the transmission to be removed from the engine. It may be necessary to completely disconnect the engine mounts in order to avoid bending or distorting them. Make sure to support the engine safely.

✳✳ WARNING

Use EXTREME caution when lifting the engine or transmission. If you are unsure of the method you have chosen to support or suspend the engine or transmission, do NOT continue until an experienced marine mechanic has confirmed your method is safe. The weight of an engine or transmission can cause serious bodily injury if dropped.

7. Remove the bolts that secure the transmission to the engine. Once sufficient clearance has been obtained to remove the transmission from the engine, remove the bolts that attach the transmission to the engine. Make sure to support the transmission when removing the bolts. Start with the lower bolts first, and leave the top two or three in place until the transmission is ready to be removed.

8. Remove the transmission from the engine. Before removing the upper bolts, make sure the transmission is secure (having

helpers hold the transmission steady is recommended) while the remaining bolts are removed. Once all the bolts are removed, carefully pull the transmission straight back, allowing the input shaft of the transmission to disengage from the damper disc on the flywheel. Once the transmission is free, lift it from the engine bay. Of the three types of transmissions Yanmar supplies with its engines, a V-drive is the heaviest. One strong individual will likely be able to lift the transmission from the engine bay unassisted. But for safety's sake, always use two or more people to lift the transmission.

To install a transmission, the steps can simply be reversed. However, there are some important steps that must be performed when installing a new transmission.

9. Apply a light coating of grease to the input shaft on the transmission to allow the splines to align with the damper disc on the flywheel. It may be necessary to slightly rotate the engine or input shaft to allow the splines to come into alignment.

10. The output shaft on the transmission MUST be aligned with the propeller shaft. For more information on this procedure, refer to the Powertrain Installation & Alignment section of this manual.

11. The control cables or linkage may require adjustment after installation. Be sure to adjust them accordingly.

12. Make sure oil is added to the transmission before the boat is operated.

ADJUSTMENTS

In most cases, the cables or linkage that control the transmission are the culprit for a "bad" transmission. Failure of the transmission to operate properly may be something as simple as a broken or stretched cable. Always check that the linkage from the remote control is operating and adjusted properly. Refer to the "Remote Control" section of this manual for more information.

OVERHAUL

▶ See Figures 10 thru 17

Due to the amount of special tools and equipment necessary for proper overhaul, procedures for transmission disassembly are not provided in this manual. A transmission repair or overhaul is better left to a qualified repair facility that is equipped with the special tools and other necessary equipment.

Most often, exploded views are used to understand the function of the components of a transmission. By viewing the inner workings of a transmission, a better understanding of the flow of power through the gears can be achieved.

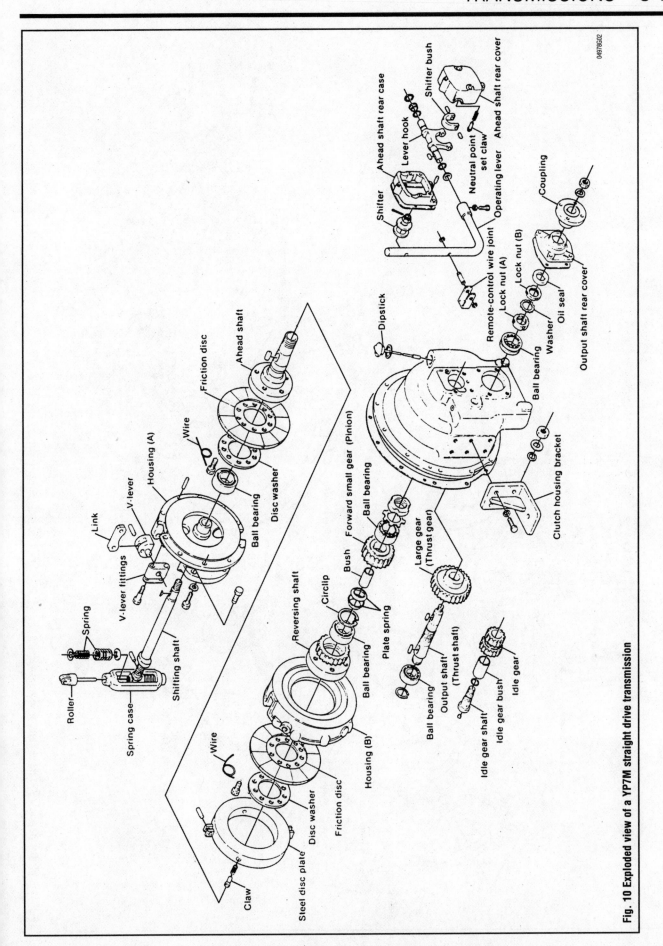

04976G02

Fig. 10 Exploded view of a YP7M straight drive transmission

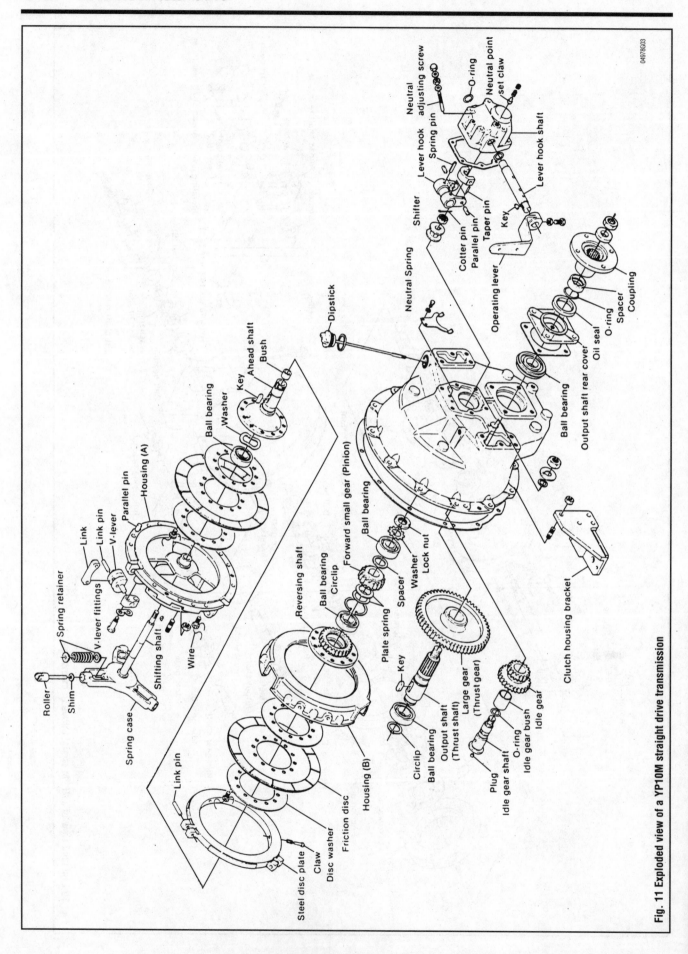

Fig. 11 Exploded view of a YP10M straight drive transmission

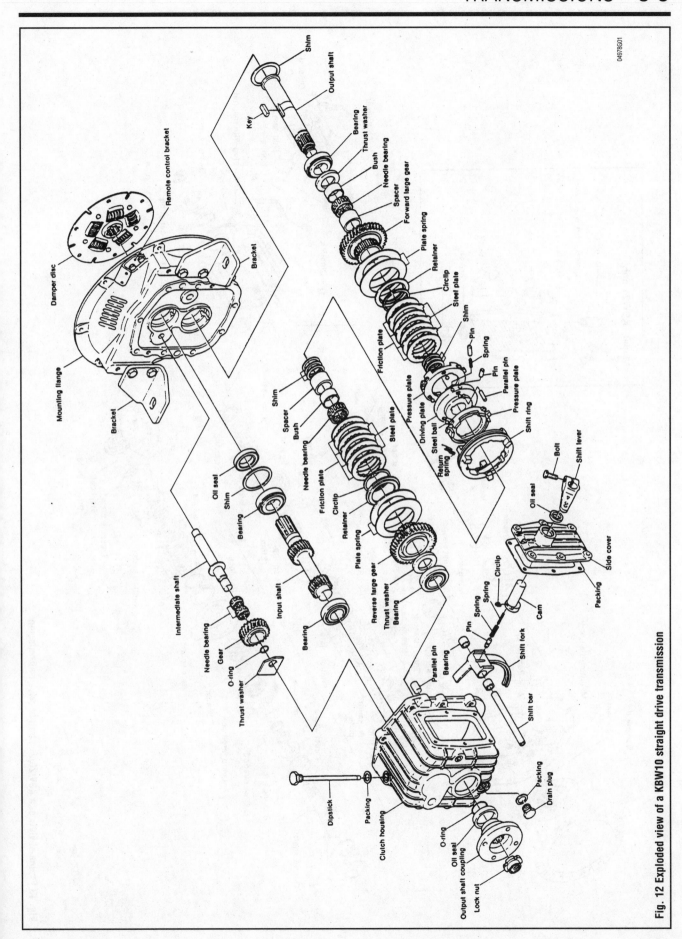

Fig. 12 Exploded view of a KBW10 straight drive transmission

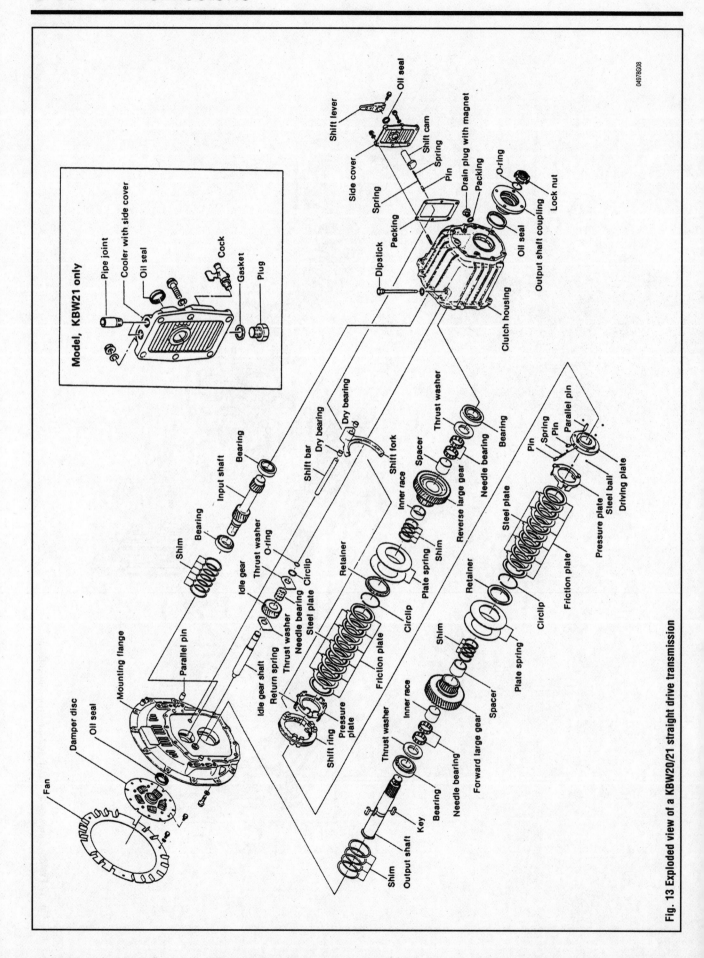

Fig. 13 Exploded view of a KBW20/21 straight drive transmission

Model, KBW21 only

Pipe joint
Cooler with side cover
Oil seal
Cock
Gasket
Plug

Shift lever
Oil seal
Shift cam
Spring
Side cover
Spring
Pin
Drain plug with magnet
Packing
O-ring
Packing
Oil seal
Dipstick
Output shaft coupling
Lock nut
Clutch housing

Input shaft
Bearing
Shift bar
Dry bearing
Dry bearing
Shift fork
Thrust washer
Bearing
Spacer
Bearing
Shim
Idle gear
Thrust washer
O-ring
Inner race
Reverse large gear
Needle bearing
Parallel pin
Mounting flange
Idle gear shaft
Return spring
Thrust washer
Needle bearing
Steel plate
Circlip
Retainer
Plate spring
Shim
Retainer
Spring
Pin
Spring
Pin
Parallel pin
Pin
Steel plate
Steel ball
Driving plate
Pressure plate
Damper disc
Oil seal
Shift ring
Pressure plate
Friction plate
Circlip
Thrust washer
Inner race
Shim
Circlip
Friction plate
Fan
Thrust washer
Needle bearing
Forward large gear
Spacer
Plate spring
Shim
Output shaft
Key
Bearing

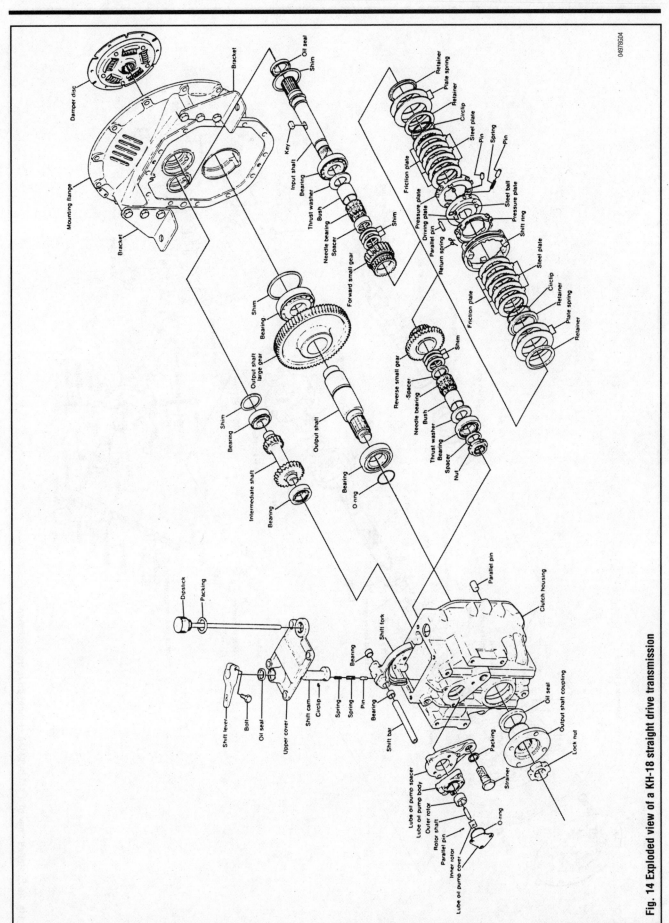

04978G04

Fig. 14 Exploded view of a KH-18 straight drive transmission

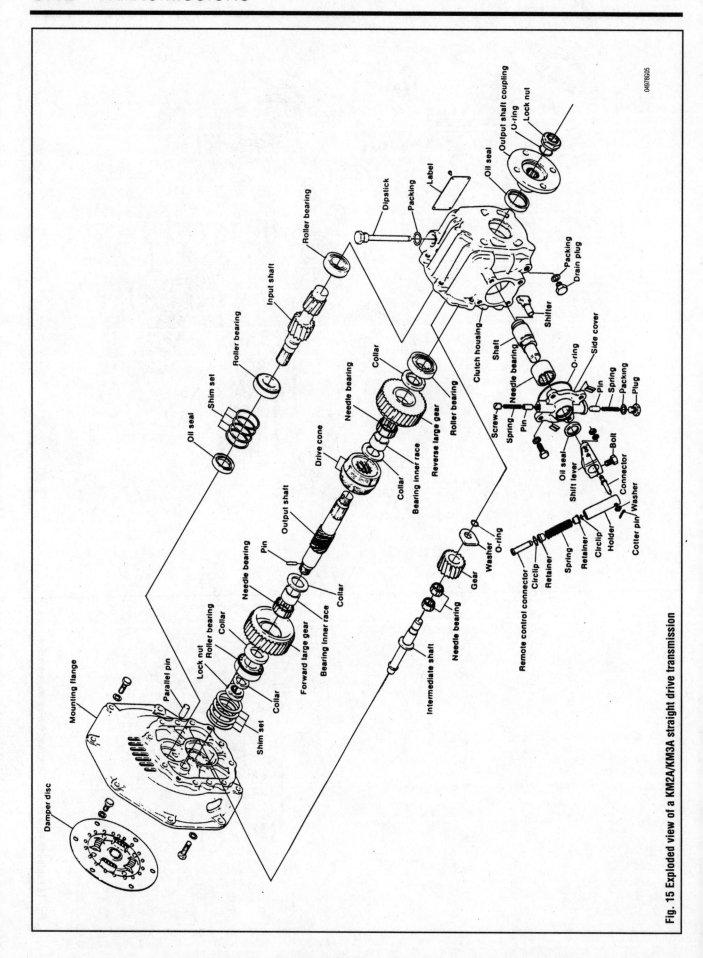

Fig. 15 Exploded view of a KM2A/KM3A straight drive transmission

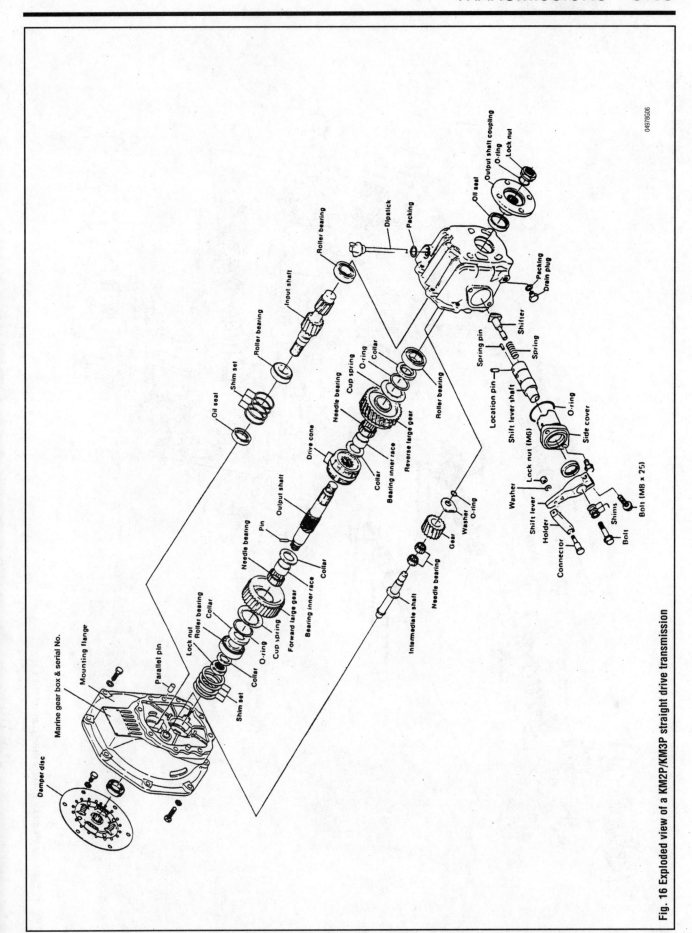

04978G06

Fig. 16 Exploded view of a KM2P/KM3P straight drive transmission

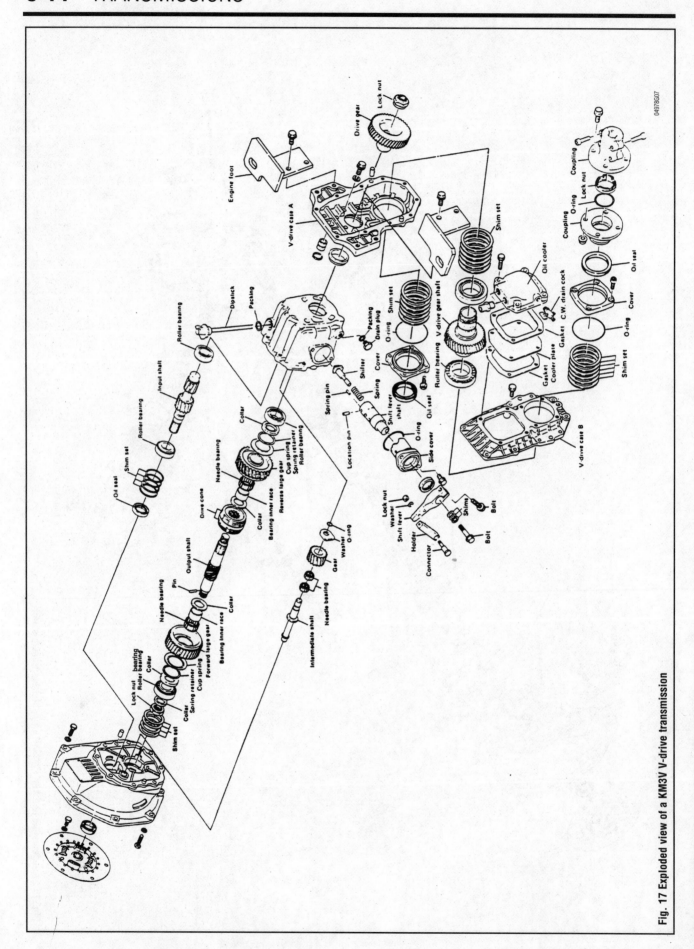

Fig. 17 Exploded view of a KM3V V-drive transmission

9

POWERTRAIN INSTALLATION AND ALIGNMENT

POWERTRAIN INSTALLATION

Sizing the Engine

Diesels are susceptible to damage from over and underloading. When overloaded, overheating can lead to severe engine damage. The damage from underloading is in many ways more harmful. Running a diesel at high speeds with a small load due to a mismatched propeller or idling while charging the battery can do more harm than most boaters think. The latter problem is particularly common on sailboaters when charging a battery or running a refrigerator at anchor.

An underloaded engine takes time to reach proper operating temperatures and at low speeds also tends to run unevenly. These two factors encourage the formation of sulfuric acid in the lubricating oil and carbon deposits throughout the engine. The cylinder walls are likely to become glazed, piston rings will get gummed in their grooves and all this will resulting in blow-by and a loss of compression. Valves may stick in their guides, while carbon will plug up the exhaust system. A carbon sludge will form in the oil if oil change procedures are neglected. Eventually, the sludge will plug oil passages and lead to total bearing failure.

➡**Repeated running of a diesel engine at low loads is a destructive practice, which greatly increases maintenance costs and reduces engine life.**

So how much horsepower do you need? Well, first you have to think about how a displacement hull (most often used in sailboats) works.

A displacement hull is one that remains immersed at all times. It has a pre-determined top speed (hull speed) regardless of available power. This top speed is governed by physical properties of the waves the boat makes as it passes through the water. A clean displacement hull can be driven at close to its hull speed in smooth water by a relatively small engine but as hull speed is approached resistance increases rapidly and any additional speed can only be gained by a great increase in power.

A planing hull, on the other hand, breaks free of the constraints imposed by the waves it generates. A certain minimum amount of power is required to come up to a plane. Thereafter, the boat's top speed is at least in part related to available power.

Various formulas have been derived for determining the horsepower requirements of displacement. The formulas will enable you to take into account the effects of a foul bottom, wave action, head winds and other factors.

➡**Only in exceptional circumstances will you find that a displacement hull requires more than .002 horsepower per pound of full loaded boat.**

Horsepower in most engine manufacturers' specifications are measured before adding the transmission, any reduction gears or the propeller shaft. This is known as the engine's Brake Horse Power (BHP). However, the true measure of an engines muscle is the power available at the propeller, otherwise known as Shaft Horse Power (SHP).

As boats get bigger and more luxurious, the effect of belt-driven auxiliary equipment is often of more concern. The DC loads on boats are steadily increasing as boat owners add more and more gadgets. In order to keep up with this increasing load, the tendency is to install optional alternators-130 amp and 160 amp models are becoming quite common. At full load, these alternators rob more

than 2 horsepower from an engine. Engine-driven refrigeration compressors can make similar demands.

On large engines, the impact of such loads is small, but on a sailboat with a small auxiliary engine, the load will be great and must be taken into account. Every horsepower absorbed by auxiliary devices is lost at the propeller shaft.

In extreme cases, the engine size may need to be increased to compensate for these losses. When comparing manufacturers' specifications, you must differentiate between an engine's horsepower rating between continuous duty and intermittent duty. An intermittently rated engine is designed to be operated at full power only for limited periods of time.

➡**A consultation with the engine manufacturer should always be a part of sizing an engine for a particular boat.**

Engine Room

LAYOUT

The overall layout of the engine room is planned for easy inspection, servicing and handling of the engine, front power take-off and auxiliary machinery. Allow space for the fuel tank, battery and seacock and their related piping, wiring and remote control cables. Thoroughly study all equipment and devices to be installed, consult with the boat builder draw up a plan to provide optimum engine room space. The following conditions should be met.

VENTILATION

When the engine room temperature rises after long operation, the air intake volume decreases (as air is heated, its volume expands oxygen gets thinner), adversely affecting combustion, raising exhaust gas temperatures, increasing exhaust gas density and lowering horsepower. Accordingly, it is important to feed fresh air to the engine room through proper ventilation.

➡**Temperature of the engine intake air at the inlet of the intake air cleaner should not exceed 113°F (45°C).**

Air Ducts

▶ **See Figure 1**

The simplest ventilation is made by the intake and exhaust air ducts. An air intake cover is attached at the aperture end of the intake air duct to receive the air while the boat is cruising. The intake air from the duct is released near the engine's air inlet. Proper arrangements must be made to prevent seawater from entering the inlet of the duct.

For ventilation using ducts only, install both inlet and outlet ducts. However, if the engine room is not completely closed, it is sufficient only to install an inlet duct and to use the engine room door or window as the air outlet.

Attach the exhaust duct at the lowest possible position opposite the intake duct. Place the duct end in the open air outboard. The duct pipe should be as short as possible. Avoid bending as much as possible.

➡**If the temperature of the engine room cannot be lowered, use a larger diameter duct pipe or increase the number of pipes.**

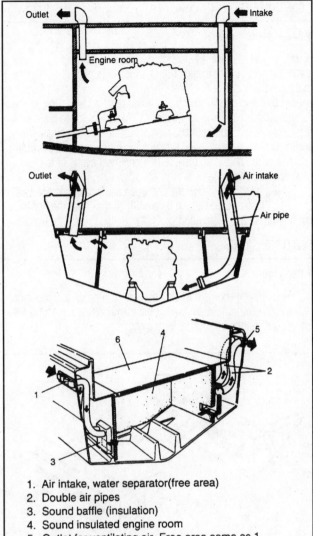

1. Air intake, water separator(free area)
2. Double air pipes
3. Sound baffle (insulation)
4. Sound insulated engine room
5. Outlet for ventilating air. Free area same as 1
6. Sound insulated hatches

04979G01

Fig. 1 Typical layout of air ventilation ducts for marine engines

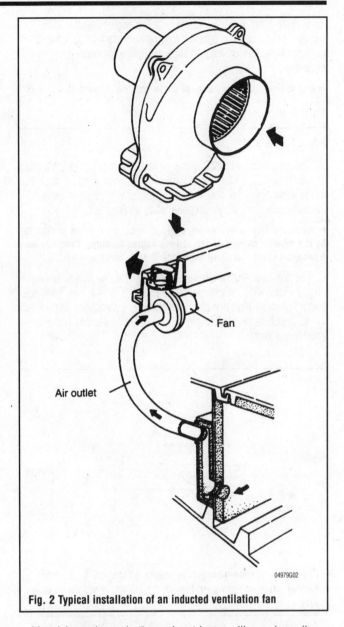

04979G02

Fig. 2 Typical installation of an inducted ventilation fan

Electrical Draft Fans

▶ **See Figure 2**

There are two ventilation methods of this type; the forced draft fan (attached to the intake duct side) system the induced draft fan (attached to the exhaust duct side) system. In both cases, use a fan within the rated capacity of the electrical equipment.

When ventilation is by draft fans only two draft fans with the same capacity are used for the inlet and outlet. When ventilation is by draft fan and duct the draft fan should be used for the outlet.

Engine Bed (Stringers)

The typical inboard installation uses engine stringers to take the load of the engine and distribute it to the hull. In most boats, the engine stringers are part of the structure and are already installed. If a boat is to be repowered and a change is required in the engine stringers, the previous installation can be used as a guide. If building a new boat, the engine stringers are usually detailed on the plans.

Materials used to make the engine stringers will vary depending on the hull construction material and other factors. Wood is used for engine stringers on most wood and plywood boats, while wood may also be used in fiberglass boats, either in the same way as they are used in wood boats bonded and fiberglassed into the hull. In some cases, special foam and fiberglass engine stringers a molded-in grid, may be provided for mounting engines in a fiberglass hull. With metal boats, the engine stringers (or foundations) are usually of the same material as the hull and designed as an integral structure.

In all boats, the engine stringers must be rigidly and strongly installed. Rolling, pitching, slamming, thrust torque forces can cause tremendous loads to be exerted onto these members as well as onto the hull structure itself. Steel plates or angles are often used on top of wood and fiberglass stringers for additional strength.

The spacing of the engine stringers must match that of the engine used. If the engine will be mounted directly to the stringers, they must have the same center-to-center spacing as the engine mounts. If the engine will be mounted to an engine bed which will, in turn, be mounted to the stringers, the stringers must be spaced far enough apart to allow clearance for the engine mounts to fit between them. In some cases, it may be necessary to cut away a portion of a engine stringer to allow clearance for some part of the engine.

➡ **Any stringer that has been cut away must be reinforced.**

ENGINE INCLINATION

▶ **See Figure 3**

The maximum permissible engine installation angle is the angle of the crankshaft center line to the water line when the hull is at rest in the water. Propulsion efficiency decreases with larger engine installation angles (larger propeller shaft angles).

➡ **When the engine inclination angle is large, air may be sucked in by the lube oil pump causing serious engine damage. This may be a problem when installing angle and V-drive transmissions.**

The propeller shaft angle varies according to the hull shape or engine installation arrangement. The propeller center line, however, should be more than one propeller diameter beneath the water line. When the propeller shaft angle is large, there is a large output and boat speed loss.

CONSTRUCTION

▶ **See Figures 4 and 5**

Use a level to place the boat on the horizontal (in both the bow-stern and port-starboard directions). When determining the engine bed's dimensions, full consideration must be given to the stress caused by the propeller's propulsion and reaction force, the engine's torque and boat cruising requirements. To distribute the load, make the contact area of the bed and mounts as large as possible. The engine bed must be attached firmly to the hull. Reinforce the front and rear of the bed. Also provide reinforcement every 20 in. (500mm) in between.

It is not advisable to bolt the L-shaped angle to the engine bed. Use it as illustrated and attach it carefully to maximize the horizontal equilibrium of the engine bed.

BED DIMENSIONS

▶ **See Figure 6**

Determine the engine bed dimensions (including the dimensions of the engine mount) by looking at the manufacturer's specification. Ensure sufficient contact area for the engine mount.

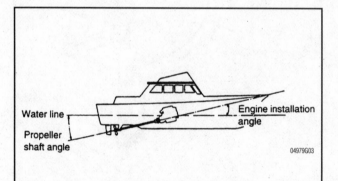

Fig. 3 The maximum amount of engine inclination is 15° with the hull static in the water. Propeller shaft angle should be 8°or less

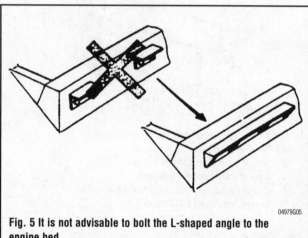

Fig. 5 It is not advisable to bolt the L-shaped angle to the engine bed

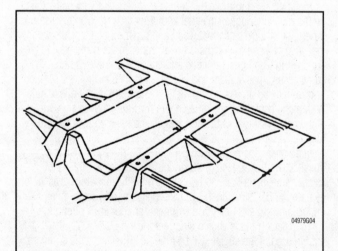

Fig. 4 The engine bed must be attached firmly to the hull and supported every 20 inches

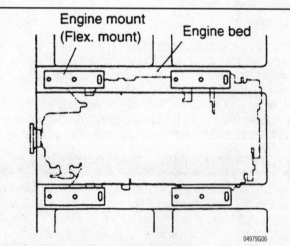

Fig. 6 For the dimension of the engine outline and engine mount specifications, refer to the engine manufacturer's specifications for your particular engine

INDENT OF ENGINE BED

▶ **See Figure 7**

When fabricating the engine bed, be sure to leave a 1 in. (25mm) or larger clearance between the engine bed and the powertrain. In addition, be sure to leave a 1 in. (25mm) or larger clearance between the hull bottom and the powertrain. Measure these values with the adjusting nuts of the flexible mount brought down to their lowest point, where they make contact with the fixing nuts of the stud bolts.

➡**The use of flexible engine mounts for too many hours makes the rubber lose its tension. This reduces the clearance and may cause interference between the engine and the hull bottom.**

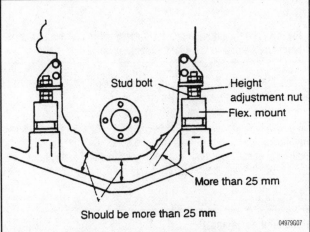

Fig. 7 A clearance of 1 in. (25mm) or more should be left between hull and powertrain. This clearance should also be observed between the engine bed and the powertrain

Engine Mounts

▶ **See Figures 8 thru 13**

Be sure to use a flexible engine mount for the installation of every Yanmar engine model. Never install the engine directly on the engine bed. The use of a flexible engine mount reduces vibration and noise by absorbing the vibrations at the couplings between the engine and the engine bed. When using flexible engine mounts, don't forget to make allowance for the exhaust, fuel oil and cooling water piping.

➡**The dimensions for both front and rear-side flexible engine mounts are identical. However the rubber elastic modulus is different for front and rear, so be sure remember their numbers.**

The illustration shows the vibration transmission rate and the engine amplitude rate in relation to the engine speed. Lowering the spring constant lowers the resonance point. In the engine's operational range, the vibration transmission rate (engine amplitude rate) is kept at under "t0".

The target value for hull vibration with the use of Yanmar flexible engine mounts is 80 dB in the engine's operational range. This target value has already been established by sea trial tests. 80 dB vibration is the same level of vibration as in a normal passenger car. The decrease of each 6 dB of vibration reduces the previous vibrations by half.

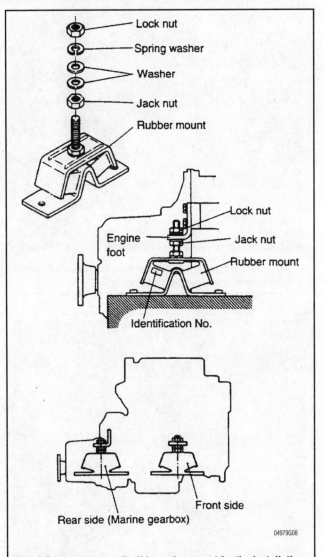

Fig. 8 Be sure to use a flexible engine mount for the installation of every Yanmar engine model. Never install the engine directly on the engine bed

Fig. 9 On Yanmar diesel engine, the front flexible engine mounts . . .

Fig. 10 . . . appear the same as the rear, but they are different. Take care to note part numbers when installing

Fig. 11 A properly mounted engine sitting in an engine bed

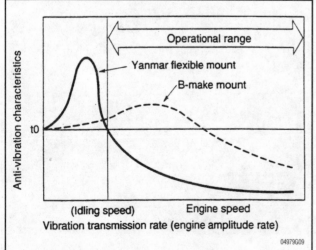

Fig. 12 Vibration transmission rate and engine amplitude rate in relation to the engine speed

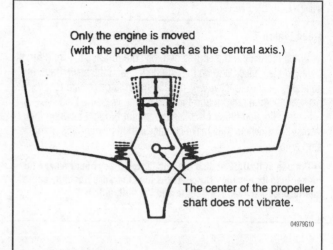

Fig. 13 Flexible engine mounts are designed to make vibration amplitudes small in the engine's operational range

As shown in the illustration, the Yanmar flexible engine mount is designed to make the amplitudes small in the engine's operational range. At the resonance point, which is encountered at both engine starting and stopping, the amplitude may be large, but the mode of vibration at such times is rolling, centered around the propeller shaft. The upper part of the engine may be vibrated to a certain degree, but there is almost no vibration in the propeller.

STATIC DISTORTION

▶ See Figures 14 and 15

Flexible engine mounts are compressed by the engine load. The compression value varies according to the engine model and the positioning of the flexible engine mount. It is important to know this compression value in order to make the propeller shaft hole on the hull at the correct height above the engine bed.

To determine if the engine mounts are still serviceable, inspect the distance between the metal portions at the center of the mount. When the distance is ³⁄₁₆ inches or less, the mount should be replaced.

DURABILITY

The rubber tension of the flexible engine mounts is lost after many hours use. This leads to a drop in vibration absorption performance also causes centering misalignment of the propeller shaft. Be sure to replace the flexible engine mount after two years.

Engine Drip Pans

In many boats is it desirable to install an engine drip pan under the engine. Such a drip pan is more common in powerboats than in sailboats because a deep enough pan is often not possible in sailboats to overcome the effects of heeling which can cause any oil in the pan to overflow and enter the bilge, thereby negating the purpose of the pan in the first place. Drip pans can be made out of a variety of metals such as copper or galvanized steel or can be molded to suit from fiberglass. A drain cock is often placed in the drip pan at the low point so that the contents can be drained into a container or pumped out. Insulation should be provided under the drip pan since the vibration of the engine can cause drumming between it and the hull structure. If a drip pan is desired, it should be installed prior to the engine should be larger in area than the engine itself and of ample capacity.

Engine Couplings

▶ See Figures 16 and 17

For flexibly mounted engines, a flexible coupling is a must since the engine must be allowed to float free (i.e. vibrate). However, a proper flex mounted setup needs another consideration. The propeller shaft should not be rigidly mounted in such a situation, as this will cause the rigid bearing to wear out since there will be no way to absorb any of the vibration transferred to this point along the

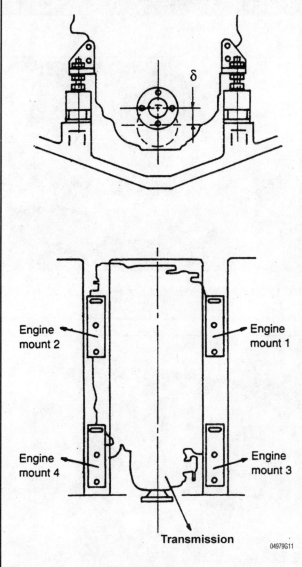

Fig. 14 Flexible engine mounts, which are compressed by the engine load, should be replaced every two years

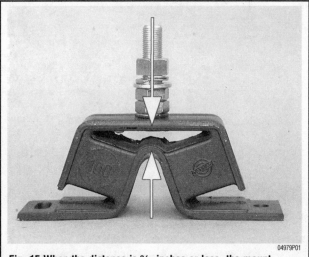

Fig. 15 When the distance is ³⁄₁₆ inches or less, the mount should be replaced.

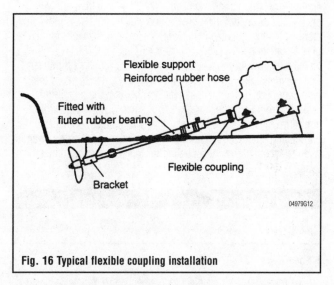

Fig. 16 Typical flexible coupling installation

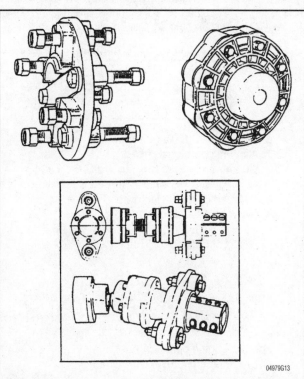

Fig. 17 Examples of several flexible couplings commonly used today

shaft. To overcome this situation, a rubber water lubricated bearing should be installed at the end of the shaft a packing gland with rubber hose (as is used with the conventional self-aligning shaft log or stuffing box) used where the shaft passes through the hull. Then the movement along the shaft will be absorbed throughout the shaft. Of course, there must be plenty of clearance in the shaft bore or shaft tube to allow the shaft to move.

If inner and outer rigid stern bearings (two rigid shaft supports) are used and the engine is mounted on flexible mounts, the problem can be overcome by using two flexible couplings joined between with a separate shaft section. While it is desirable to align the engine carefully, such an installation can accept several degrees of misalignment. One of the advantages of a flexible coupling is, that even though the engine may be very carefully aligned during the engine installation and/or construction of the boat, the hull will often settle or distort after some period which could cause misalignment with consequent vibration and noise problems with rigid coupled installations.

Installation with most flexible shaft couplings is not difficult. The engine is aligned in the normal way using the standard flange-type couplings. Once the alignment is checked, the coupling is removed from the propeller shaft and replaced with the flexible coupling. Since there is no metal-to-metal contact between the engine and shaft with most flexible couplings, a grounding strap is often provided across the coupling for boats with bonded electrical systems.

Propeller Shaft

DIAMETER

▶ See Figure 18

The propeller shaft diameter recommended by Yanmar is calculated as follows. Be sure to install the propeller shaft bearing within 3.9 feet (1.2 m) from the propeller shaft coupling.

$$d = 154 \times \frac{1}{\sqrt[3]{T}} \times \sqrt[3]{\frac{(1+Y)H}{R}}$$

d: Propeller shaft dia.(mm)
T: Allowable torsional stress,
    5.0 kgf/mm = Stainless steel, JIS. SUS304
    10.5 kgf/mm = Stainless steel, JIS. SUS630
Y: Torque fluctuation factor,
    (4-cycle diesel engine)

No. of Cylinder	Factor
1	0.6
2	0.4
3	0.3
4	0.2
Over 6	0.1

H: Max. engine output (hp)
R: Propeller shaft revolutions (rpm)

04979G14

Fig. 18 Propeller shaft diameter calculation formula

Propeller

▶ See Figure 19, 20 and 21

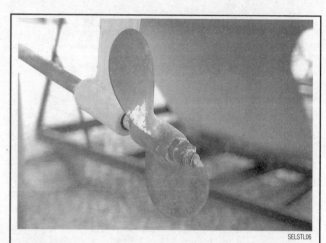

SELSTL06

Fig. 19 Three types of propeller are commonly available, the two blade propeller . . .

SELSTL28

Fig. 20 . . . the folding propeller . . .

SELSTL21

Fig. 21 . . . and the high speed propeller. This high speed version also has adjustable pitch

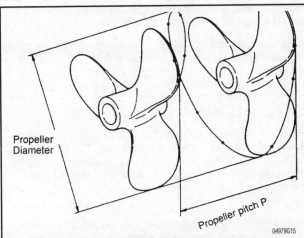

Fig. 22 Select a propeller with a suitable diameter and pitch for the displacement of the boat, as well as for its intended usage

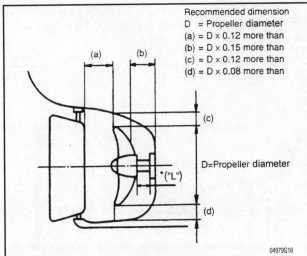

Recommended dimension
D = Propeller diameter
(a) = D × 0.12 more than
(b) = D × 0.15 more than
(c) = D × 0.12 more than
(d) = D × 0.08 more than

Fig. 23 The position of the propeller section shaft center must be at least one propeller diameter beneath the surface of the water with the boat fully loaded

DIMENSIONS

▶ **See Figure 22**

Select a propeller that is suitable for the size, shape and displacement of the boat, as well as for its intended usage. An extremely small or large propeller will reduce the speed of the boat and overload the engine, which may lead to engine damage. It is recommended that a propeller be selected in consultation with a reputable dealer. However, the best way to make sure the propeller fits the boat is to conduct a test run after installation.

POSITION (APERTURE)

▶ **See Figure 23**

The boat's full speed will not be obtained if the spacing between the propeller and hull is not equal to greater than, the value given in the figure. The position of the propeller section shaft center must be at least one propeller diameter beneath the surface of the water with the boat fully loaded.

➡ On the engines with flexible engine mounts, the "L" clearance should be more than .787 in. (20mm). Shorter clearance may cause the propeller to interfere with the boat hull on astern operations due to the flexion of the flexible engine mounts.

MATCHING

▶ **See Figure 24**

The type of boat, hull size, hull weight and loading weight decides the selection of a main engine for a boat. The selection of an appropriate propeller that matches the output of the main engine is very important for the economical operation and durability of the engine.

When the propeller is small for the engine output, there will be a loss in both engine output efficiency and boat speed. When the propeller is large for the engine output, the engine will be overloaded at full throttle. Prolonged engine use at full throttle will cause engine trouble.

The size of the propeller is determined at the design stage of the boat in consideration of the boat's speed and size, the shape and weight of the hull, main engine output, the reduction ratio of the transmission and the maximum number of the crew (maximum loading

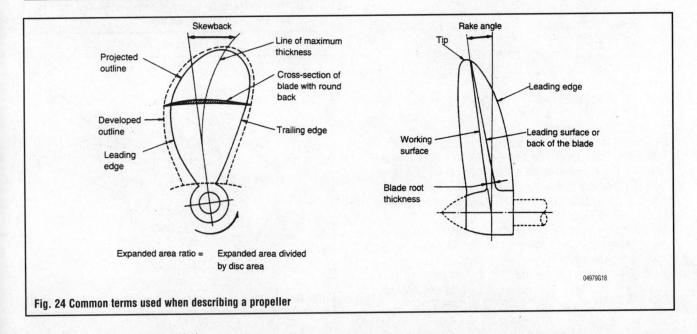

Fig. 24 Common terms used when describing a propeller

weight). However, it is extremely difficult for a boat to be built strictly according to the design drawings. Accordingly, the final checks of the boat speed, propeller matching and other aspects of boat performance must be made by sea trials. If the boat's performance is unsatisfactory, it can be improved by changing the propeller specifications (number of blades, diameter, pitch, extended area ratio of blades, etc.).

It is easier to change the propeller than the engine or the shape of the hull.

Various ways exist to calculate the proper propeller size for pleasure boats, but there is no definitive formula. We recommend that you consult the propeller makers.

Sea trials for estimating the load horsepower are made by measuring the engine's exhaust temperature underload. Special measuring equipment is used for the exhaust temperature. Consult with your Yanmar dealer if planning to conduct such measurements.

Ensure appropriate propeller matching in consideration of the following factors, keep some output in reserve:
- Increase of resistance due to contamination of hull.
- Increase of displacement due to increase of load and crew.
- Horsepower required to drive auxiliary equipment such as a generator or pump powered by the main engine.
- Engine de-rating factors (factors that reduce engine output)such as lack of air intake volume, excessive temperature, excessive exhaust back pressure or high fuel temperature.

PROPELLER MATCHING PROCEDURE

It is possible to find if the propeller is too large for the engine output (the engine is overloaded) or too small (a loss of engine output) by the following simple procedure during the sea trial:
1. Check the engine's maximum speed (high idle) at no-load
   a. Move the shift lever of the transmission to neutral.
   b. Raise the engine speed gradually with the engine speed control handle (remote control handle), moving the shift lever to the full throttle position.
   c. Check whether the engine speed at the full throttle position is within the no-load maximum speed range.
2. Refer to the no-load maximum engine speed for each model. If the engine speed is below the no-load maximum engine speed range, the problem is due to the absence of remote control stroke or faulty contact of the remote control cable.
3. Check the engine's maximum speed at load
   a. Move the transmission shift lever to forward at the idling speed of the engine.
   b. Raise the engine speed gradually with the engine speed control handle (remote control handle) to move the shift lever to full throttle.
   c. Check the loaded maximum speed at full throttle.
   d. Compare the engine speeds obtained with the engine speed at maximum rating output.
4. Refer to the no-load maximum engine speed for each model. If the engine speeds are within the range between maximum rating output and no-load maximum engine speed, the propeller size is normal.
   a. If the no-load and loaded maximum engine speeds are identical, the propeller is too small. There is unused engine output in reserve.
   b. If the engine speed is lower than the engine speed at maximum rating output, the propeller is too large. The engine will be overloaded at full throttle.
   c. Correct the situation by changing to a smaller propeller or keeping the maximum service speed at approximately 100 rpm below the engine speed at full throttle at all times.

➡When checking maximum engine speed underload, also load the auxiliary equipment driven by the main engine.

Powertrain Assembly

INSTALLATION CHECKLIST

◆ **See Figures 25, 26 and 27**

➡If you are building a new boat, it is best to install and mount the engine prior to closing in the hull with the deck. This statement applies also to the installation of tanks, exhaust systems any other components that will be easier to install without the various structural units causing interference.

The typical inboard engine installation, together with all the ancillary systems, takes quite a bit of planning and head scratching if the installation is to progress in an orderly sequence, so plan ahead. Here are some basic steps to guide you in the installation of your engine:
1. Prior to purchasing an engine, consult with your local dealer or the engine manufacturer about matching an engine to your boat. Nothing is worse than spending good money for a new engine and having it not.be able to do what you expect.

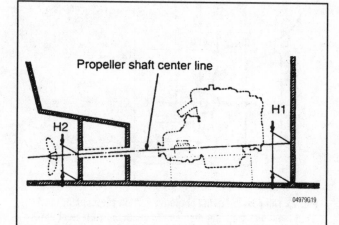

Fig. 25 When installing an engine in a new application, the propeller shaft centerline must be set first

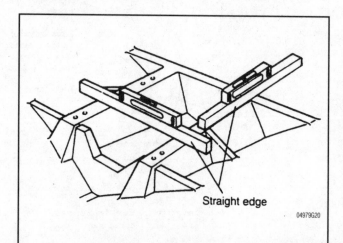

Fig. 26 Use a straightedge to check that the engine bed is level

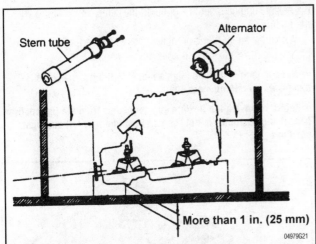

Fig. 27 Ensure there is enough clearance for the stern tube and any accessories. Also ensure the bottom of the engine/transmission is at least 1 in. (25mm) from the hull

Stern tube

Alternator

More than 1 in. (25 mm)

04979G21

2. Make sure all engine parts and standard accessories are included. Make yourself a checklist and check it twice. Any parts that must be ordered during installation will delay the process and possibly cost you money and aggravation.

3. If the engine is being installed in a new boat, the propeller shaft angle must be found. A taut string or wire can be lead through the prop shaft hole to locate the centerline. This string line will be used later to reference the location of the engine stringers and/or engine beds.

4. Install the engine beds using the propeller shaft as a center-line.

5. If the engine is being installed in a new boat, install all necessary accessories such as the seacock and fuel tank.

6. If the engine has not been test run at the factory, now is the time to run it. Connect water and electrical lines to the engine. A small fuel tank will provide enough fuel for a short test. Start the engine and check for proper operation and leaks.

7. Install the engine in the boat and roughly align it with the propeller shaft. The engine should be dropped into the correct location and temporarily held in position with clamps holding the engine beds to the engine stringers.

8. Loosen the clamps and shift the engine as required to achieve accurate alignment with the engine coupling. When everything is in alignment, install bolts the engine beds and engine stringers. Carriage bolts are commonly used, however, other bolts may be used as long as the heads are accessible. Use at least four ³⁄₈ in. bolts per member.

➡It may also be necessary to align the engine crossways. This is usually done by slotting the engine mounting holes slightly or using bolts slightly smaller than in the engine mounts.

9. Connect all the necessary fuel, water and exhaust pipes to the engine. Connect all the necessary electrical harnesses. Connect all the necessary control cables.

10. Perform an overall visual inspection of the powertrain and hull. Did you miss anything? Fix it now!

11. Go around the engine compartment and check for loose nuts and bolts. Don't just feel them, put a wrench on each and every one. Pay special attention to all electrical harnesses and control cable fittings. Visually inspect every one for proper installation and functionality.

12. Ensure the fresh-water cooled engines are filled with coolant, the fuel tank is full and the seacock is open.

13. Perform a bulb check on the control panel lighting and a functional check of all control panel switches and gauges.

14. Launch the boat and perform a no load engine test. Visually inspect the cooling water scupper for proper flow. Check the engine gauges to ensure all is functioning properly.

15. Visually inspect the powertrain for leakage. Correct these conditions immediately.

16. Visually inspect the color and amount of exhaust gas.

17. Listen for abnormal sounds and shut the engine down if any exist.

18. Perform a fine adjustment of the propeller shaft-to-engine alignment. This is necessary due to the settling of the hull with the boat in the water.

19. Take the boat out for a sea trial. This should be a short run around the local area. Do not be tempted to take the boat on a major trip just yet. Remember, a tow back to the marina will cost you $125 an hour from the time the tow boat leaves his port till the time he returns.

20. During the sea trial, slowly bring the boat up to maximum speed. Listen for any abnormal noise and keep checking for leaks.

21. If an exhaust gas temperature gauge is available, use it to measure the exhaust gas temperature for engine output. This is important to maximizing your engine output and propeller. Report your findings to the engine and propeller manufacturer.

22. Check the engine idle speed and full engine load and boat speed.

23. Check the hull for an abnormal vibration. Note the speed and operating conditions where the vibration occurs. This will be important in diagnosing what is causing the vibration. Consult the "Vibration Diagnosis" section for more information.

24. After all leaks are repaired, adjustments made and inspections completed, go out and have some fun!

25. Finally, read the beginning of the "Maintenance" section, especially where it talks about how maintenance is cheaper than repair. Keep a log and maintain your new engine so it will provide you with years of faithful service.

Engine Alignment

Regardless of the mounting method, the engine should be carefully aligned in position. The shaft should be blocked into position either at the shaft hole or at the shaft log position. With the self-aligning shaft log, there will no doubt be considerable spring in the shaft from the stern bearing or strut. The position of the shaft should be adjusted in such a way that it will turn freely and not bind at the bearing, checking carefully to assure that it is located on the centerline of the boat. A temporary brace can be used across the stringers to hold the shaft in place for alignment.

When the engine coupling is aligned to the shaft coupling, great care should be taken. Bring the engine into alignment to the shaft coupling flange. Don't force the shaft to line up with the engine. With the engine in neutral, the shaft should turn easily and smoothly without binding. If a spot binds, check the alignment again. In aligning the couplings, a .003 in. (0.075 mm) or thinner feeler gage should be used. Check around the entire flange to assure correct alignment. If the alignment is not correct, shim, shift adjust the mount until it is. If a feeler gage is not available, use four strips of paper located around the flange equally spaced between the two flange faces. Any variation in alignment will be indicated by the looseness of one or more of the strips. Extra time spent in careful alignment will be worthwhile in terms of smoothness and added reliability. While somewhat tedious, the process is not really difficult. The alignment procedures are often specified by the engine manufacturer.

Many types of boats are somewhat flexible and will shape or settle somewhat after being launched. An additional check should be made of the alignment between the connecting flanges after the boat has been launched in the case of certain boats, at protracted intervals thereafter. In larger boats, it is advisable to leave the flanges loose and do the final adjusting when the boat is in the water. Additional checking should be made until no more change is noticed.

ALIGNMENT PROCEDURE

♦ **See Figures 28 thru 32**

Prepare the alignment device as illustrated attach it to the flange of the propeller shaft. Turn the propeller shaft to determine its center line.

Attach the centering string in the center of the propeller shaft coupling flange. Turn the propeller shaft and move the string up and down at the other end until the string and the tip of the aligning device line up. When they are completely aligned, set the string at the other end.

➡ **The length of the centering string must be adjusted at the other end to allow the propeller shaft coupling flange to shift back and forth.**

It will be easier to determine the installation dimensions of the flexible engine mount on the engine bed if you fabricate a jig like this:

After extending the centering string, determine the engine installation position by using the jig as illustrated. Align the propeller shaft coupling flange guide on the jig adjust the angle of the jig and the flexible engine mount height until the center of the string and the jig line up.

When the center of the propeller shaft is aligned, mark the flexible engine mount fixing bolt holes on the engine bed.

➡ **When fixing the flexible engine mount to the jig, make the H-dimension an intermediate one.**

➡ **Before removing the flexible engine mounts from the jig, confirm each fixing position. Do not confuse front and rear left and right positions.**

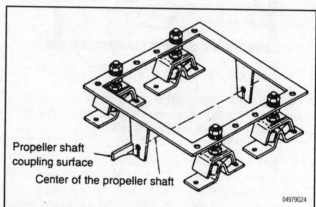

Propeller shaft
coupling surface
Center of the propeller shaft

04979G24

Fig. 30 This jig can be fabricated to determine the engine installation dimensions

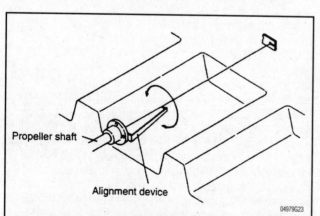

Propeller shaft coupling
bolt hole pitch circle dia.

100 mm

Center line

Alignment device

Adjustable according to each engine

04979G22

Fig. 28 An alignment tool can be fabricated to assist in the critical task of aligning the propeller shaft

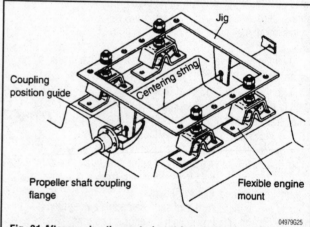

Jig

Coupling
position guide

Centering string

Propeller shaft coupling
flange

Flexible engine
mount

04979G25

Fig. 31 After running the centering string, determine the engine installation position by using the jig as illustrated

Propeller shaft

Alignment device

04979G23

Fig. 29 When the centering string and the tip of the aligning device line up the propeller shaft is centered

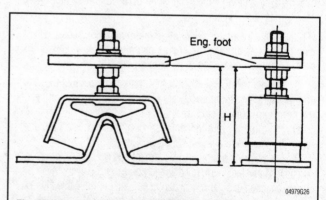

Eng. foot

H

04979G26

Fig. 32 When the center of the propeller shaft is aligned, mark the flexible engine mount bolts holes on the engine bed

CENTERING THE ENGINE

▶ See Figures 33, 34 and 35

Before connecting the transmission drive shaft to the propeller shaft, make sure that the flange surfaces of both parts are parallel to each other that their centers are aligned. Then adjust the centering of the engine as follows. When the propeller shaft is long, place it in the center by moving it up and down then center.

Install a clearance gauge on the propeller shaft coupling and measure the circumference against the output shaft flange face run-out (at four equally spaced points around the circumference).

Then lock the output shaft flange, turn the propeller shaft and dial gauge measure the outside periphery of the output shaft flange and adjust it to the value previously measured.

➡**Even when using the flexible stern tube and flexible coupling, it is necessary to ensure the same propeller shaft centering accuracy.**

ADJUSTING METHOD

▶ See Figures 36, 37 and 38

When the values deviate from acceptable specifications, make the following adjustments:

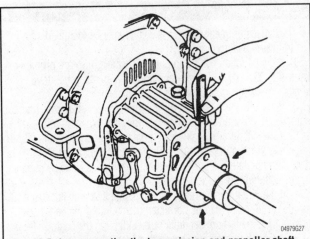

Fig. 33 Before connecting the transmission and propeller shaft, make sure that the flange surfaces. of both parts are parallel

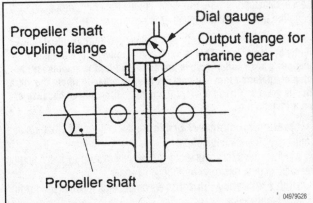

Fig. 34 Install a clearance gauge on the propeller shaft coupling and measure the circumference against the output shaft flange face run-out

1. For a deviation of the coupling face, adjust in both the vertical and lateral directions.
2. For a centering adjustment in the vertical direction, change the "H" dimension by adjusting the jack-nut position.

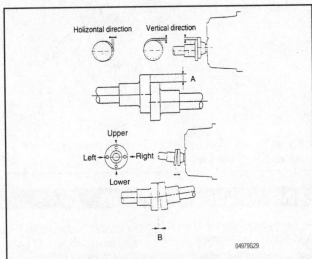

Fig. 35 Allowable centering is 0.004–0.012 in. (0.1–0.3 mm) off center and 0–0.008 in. (0–0.2 mm) out of round

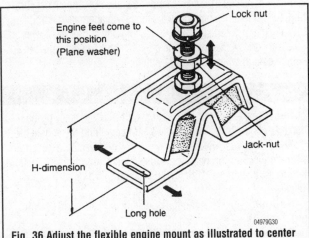

Fig. 36 Adjust the flexible engine mount as illustrated to center the engine

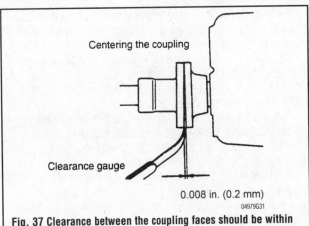

Fig. 37 Clearance between the coupling faces should be within 0.008 in. (0.2 mm)

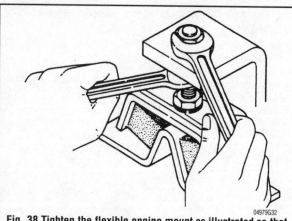

Fig. 38 Tighten the flexible engine mount as illustrated so that the mount is firmly fixed

3. For a centering adjustment in the lateral direction, move the long hole side of the flexible engine mount laterally.

➡The rubber flexible engine mounts are compressed with use, so raise the jack-nut by one turn to install the engine a little bit higher.

4. After ensuring the dimensions of both A and B, insert the faucet of the propeller shaft coupling about halfway confirm that the clearance between the coupling faces is within 0.008 in. (0.2 mm). If the clearance exceeds 0.008 in. (0.2 mm), readjust.

5. After finishing centering, tighten the flexible engine mount and the engine tightening bolts and nuts.

6. After tightening the bolts, make sure that the shaft coupling face centering has not slipped. If slippage is found, readjust.

7. Re-check shaft centering after launching when the adhering agent is completely hardened. If there is any change in the centering, readjust.

NOISE AND VIBRATION DIAGNOSIS

Engine noise and vibration are usually indicators of a problem. Although they are difficult to quantify, most owners know their boats well enough to hear or feel a change in the engine the second it occurs. Even if your experience with engines is limited, trust your senses and investigate the problem further. Letting it go will almost always lead to bigger problem in the end.

If you're having trouble locating the source of a noise or vibration, try using a mechanic's stethoscope. An alternate method is to use a metal or wooden dowel. By placing one end of the dowel on different parts of the engine and the other end against your ear, you will be able to tell which part is vibrating.

Noise

Knocking—a hard mechanical sound with frequency proportional to engine rpm. Sounds like a hammer hitting the engine block.

Knocking is usually caused by a defective injector, excess fuel, worn connecting rod bearings, loose connecting rod bolts, piston hitting a valve, injection timing too far advanced or a loose flywheel.

Rattling—one or a handful of nuts being shaken in an empty metal can.

Rattling is usually caused by excessive valve clearances or loose accessories.

Rumbling—a slow-speed, dull sound, like assorted rocks being turned in a large drum.

Rumbling is usually caused by a propeller shaft out of balance, worn shaft bearing, propeller out of balance, worn gearbox bearings or worn or loose drive plate.

Squealing—sounds like a car who's tires spin causing a distinctive sound during a racing start.

Squealing is usually caused by a belt slipping, lack of piston lubrication or gasoline in fuel tank.

Hissing—sounds like escaping gas. Occurs intermittently as a piston approaches the top of its compression stroke.

Hissing is caused by a leaking injector washer, a leaking cylinder head gasket or a leaking intake valve seat.

Clicking—a light metallic sound that can occur once or continuously.

Clicking is usually caused by a starter solenoid engaging or excessive valve clearances.

High-Pitched Whir—a very high speed zing that rapidly increases in frequency.

High-Pitched Whir is usually caused by a starter still energized or engaged.

Vibration

The causes of linear vibrations can usually be identified using the following assumptions:

1. If the vibration cycles amplify with speed, they are probably caused by centrifugal forces bending components of the drive shafts. Check for imbalance and misalignment.

2. If the vibrations occur within a narrow speed range, inspect equipment attached to the engine. Pipes, air cleaners, etc.

3. When vibrations peak out at a narrow speed range, the vibrating component is in resonance. These vibrations can be stopped by changing the natural frequency of the part. This is done by stiffening or softening its mounting.

4. If the vibrations increase as load is applied, this is caused by torque reaction and can be corrected by mounting the engine or driven equipment more securely or by stiffening the base. Defective or worn couplings can also cause this problem.

Low Frequency Vibration—components moving and turning at slower speeds.

Low frequency vibrations are usually caused by natural resonance, damaged or dirty propeller, engine misfiring, propeller shaft coupling loose, bent propshaft or a loose flywheel.

➡Smaller engines with 1, 2, or 3 cylinders will often vibrate violently at slower rpm. This is particularly noticeable with lighter, engines whose softer mounts are optimized for the smoothest running at operating rpm. All engines have rpm bands where the vibration is greater. Often the vibration is more pronounced because of poor installation.

Medium Frequency Vibration—the majority of vibration at a frequency close to engine rpm.

Medium frequency vibrations are usually caused by loose engine mounts, engine misfiring or a loose flywheel.

High Frequency Vibration—components moving and turning at very high speeds.

High frequency vibrations are usually caused by starter stuck/engaged or an alternator fan out of balance.

10

REMOTE CONTROLS

REMOTE CONTROLS

▶ **See Figure 1**

To allow the engine and transmission to be operated from outside the engine compartment, a remote control system is used. A typical remote control system consists of a control head and cables. The cables are routed from the handle (usually on deck) to the engine compartment, and connected to the operating levers on the engine. Cable operated remote control mechanisms are simple and reliable, if maintained properly.

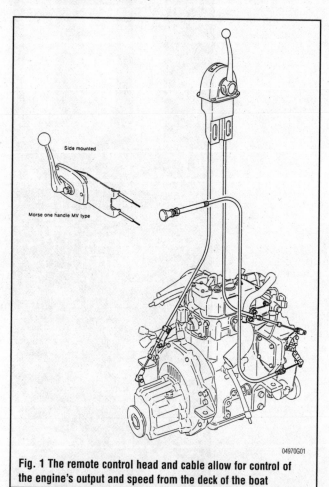

Fig. 1 The remote control head and cable allow for control of the engine's output and speed from the deck of the boat

Remote Control Head

▶ **See Figures 2, 3, 4 and 5**

Remote control heads can vary from on manufacturer to another, but they all perform the same function: moving the cables attached to the engine for proper speed and direction control.

The remote control head consists of lever(s) that push and pull cables with cams and other linkage inside the head. Most control heads are entirely mechanical and rarely fail. If a control head does break, the reason for breakage is usually from applying excessive force to the head. If the lever(s) on the control head require excessive force to move, the likely cause is not within the control head mechanism, but caused by improperly routed, corroded, or damaged cables.

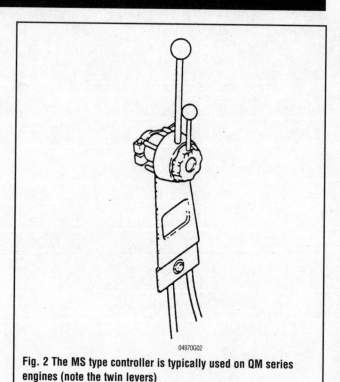

Fig. 2 The MS type controller is typically used on QM series engines (note the twin levers)

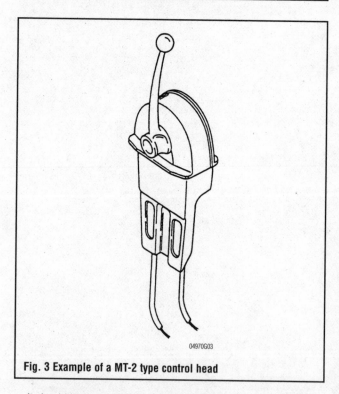

Fig. 3 Example of a MT-2 type control head

It should be noted that some remote control heads may contain a neutral safety switch that prevents the engine from being started unless the transmission in neutral. If the switch breaks, the engine cannot be started unless the switch is bypassed. The control head may require partial disassembly to test and repair the problem. Refer to the Electrical section of this book for more information.

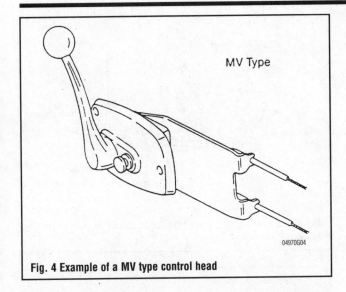

Fig. 4 Example of a MV type control head

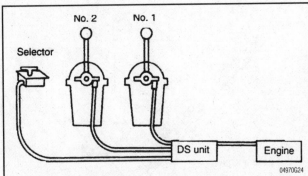

Fig. 5 Some larger boats may be equipped with a more complex twin remote control system

ADJUSTMENT

▶ See Figure 6

The most important aspect of maintaining the control head is making sure that the movement of the lever(s) on the handle coincide with the movement of the cable, and the lever that is being operated. For instance, when the engine speed lever is placed to the full ahead position, the engine speed lever (or governor lever) on the injector pump should be in the maximum speed position. Respectively, the engine speed lever on the injector pump should be in the idle position when the lever on the control head is in the idle position. Transmission control should operate in the same manner.

Typically, cable adjustments are not made at the control head, but at the end of the cable in the engine compartment where the cable attaches to the engine or transmission. If the threads on the end of the cable cannot compensate for wear, the cable should be replaced.

LUBRICATION

When the control head movement becomes difficult (and the cables are found to be in good condition), the control head internals may need to be lubricated. Because of the many different brands and types of control heads, detailed procedures cannot be given for

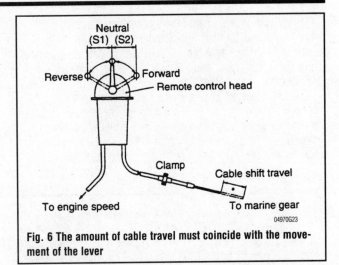

Fig. 6 The amount of cable travel must coincide with the movement of the lever

disassembly. However, most control heads usually can be lubricated by simply removing the back cover (after it is unbolted from its mounting) and applying grease to the pivoting and sliding surfaces. Make sure to use a grease that is compatible with the existing grease.

Cables

Typically, most of the control cables used on boats are of the "Morse" type, and use a solid wire core surrounded by a flexible sheath. These type of cables are referred to a "push-pull" cables. In most applications, cables not only move a lever when "pulled" but also move the lever in the opposite direction when "pushed" hence the name.

INSPECTION

Control cables should be inspected regularly for proper operation.

To check a cable, simply remove the clevis or cable end, and operate the appropriate lever on the control head. The cable should not bind or take a great deal of effort to move with the lever on the control head. If the cable feels tight, it should be thoroughly lubricated.

LUBRICATION

Cables that are lubricated on a regular basis will provide years of service.

The best method for lubricating cables is to completely remove the cable from its sheath. Most engine speed cables can be separated. To separate the cable, simply remove the pinch bolt on the lever that retains the cable to the lever. The cable should also be removed from the control head. Once the cable is completely drawn from its sheath, it can be inspected, greased, and installed. Before greasing the cable, check for rust spots on the cable; remove them with light grade sandpaper if necessary. Then the cable should be greased with a Teflon-based waterproof grease, and reinstalled into the sheath. It may be necessary to round off the end of the cable with sandpaper to keep it from catching inside the sheath. Once the cable is placed back into the sheath, attach the ends of the cable, and adjust it as necessary.

There may be cases where a cable cannot be separated, like some cables used for transmission control. If the cable cannot be separated, the only way to properly lubricate the cable is to remove it from the control head, and use oil to drip down inside the sheath until it comes out the other side.

Once all cables are lubricated, the pivot points of the levers that the cables actuate should also be lubricated. The best way to lubricate the cable clevis and lever pivot points is by disassembling them and applying grease as necessary. Also, grease the exposed ends of the cable to prevent water from entering the sheath.

REPLACEMENT

If a cable is badly corroded, kinked, or broken, it will require complete replacement. When replacing cables, it is highly recommended that the cable be replaced with one of the same type. This way, when the cable is routed, there will not be any need for additional mounting. The most common cause of stiff cable operation is improper routing. The minimum radius of any bend should be 8 inches.

Another reason for replacing the cables with one of the same type is the available stroke of the cable. The stroke of the cable is important because the cable needs to slide back and forth enough to properly move the lever its controlling. In the case of an engine speed cable, the engine may not be able to achieve full speed because the cable does not have enough range of motion.

If you spend any length of time away from your home port, it is recommended that an extra set of cables be kept aboard at all times.

ADJUSTMENT

Engine Speed Cable

▶ See Figures 7 thru 12

➡On GM/HM and JH series engines, a spring loaded mechanism is used on the governor lever. It is very important for the mechanism to be mounted in the proper direction, or the engine speed governor will not function properly. Keep the mechanism lubricated for proper function.

1. Place the lever on the control head to the ahead (full speed) position, with the engine **OFF**.

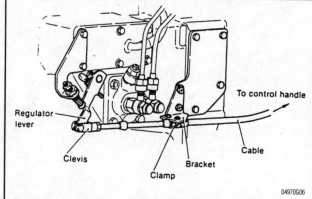

Fig. 8 The clevis on the end of the cable can be used to adjust the cable

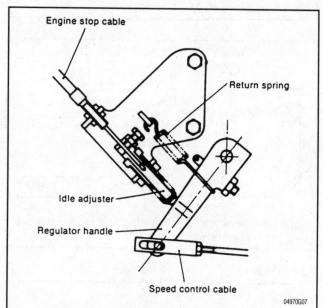

Fig. 9 QM series engine speed control and engine stop control lever detail

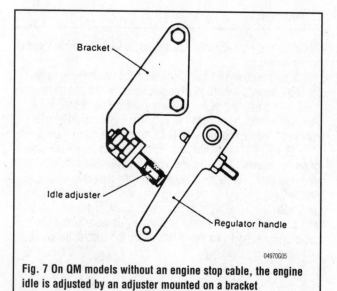

Fig. 7 On QM models without an engine stop cable, the engine idle is adjusted by an adjuster mounted on a bracket

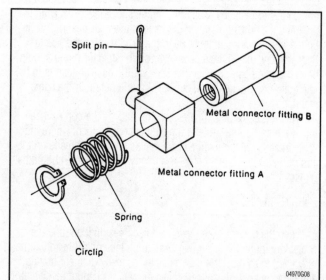

Fig. 10 GM/HM engine speed cable connecting detail (note the location of the spring)

2. Check that the cable moves the governor lever on the injector pump so it rests against the maximum speed set screw.

3. If the governor lever is not against the maximum speed set screw, loosen the locknut on the clevis. On QM series engines,

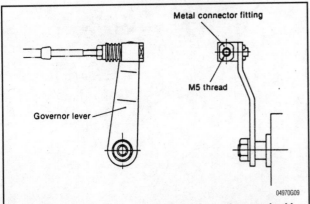

Fig. 11 A closer look at the connection of the engine speed cable hardware. Note that the cable end threads into the connector

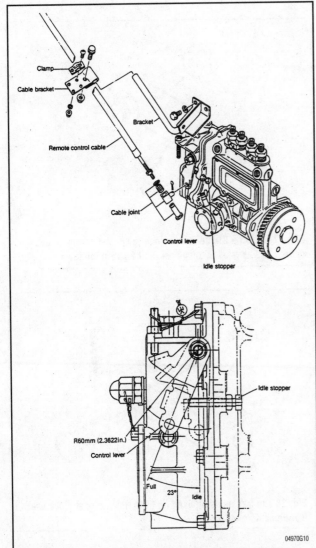

Fig. 12 JH series engine speed cable linkage detail. Note the connector for the cable is identical to a GM/HM series

the clevis should be removed from the lever, adjusted, then reinstalled.

➡ While adjusting the cable, take the opportunity to apply a fresh coating of grease on the exposed portion of the cable, and the pivoting points on the clevis assembly.

4. To adjust the cable length, turn the clevis on the cable end threads after the locknut has been loosened.

5. The cable should be adjusted so the governor lever rests against the maximum speed set screw when the control head lever is set to the full ahead (full speed) position.

6. Once the cable is adjusted, move the control head lever to the idle position and check that the governor lever rests against the idle speed set screw.

7. Once the cable has been adjusted properly, tighten the clevis locknut.

8. ON QM series engines, install a new cotter pin.

❊❊ WARNING

DO NOT attempt to adjust the governor lever set screws if they are bound with wire; the factory has properly adjusted the governor lever at the time of manufacture, and adjustment is not necessary.

Transmission Cable

▶ See Figures 13 thru 19

1. Place the lever on the control head to the ahead position, with the engine **OFF**.

2. Check that the cable is pulling/pushing the control lever on the transmission control lever its maximum stroke. If you are unsure whether or not the lever is at its full stroke, disconnect the clevis from the lever, and try to push/pull the lever further through its stroke. If the lever moves further, the cable may need to be adjusted.

➡ While adjusting the cable, take the opportunity to apply a fresh coating of grease on the exposed portion of the cable, and the pivoting points on the clevis assembly.

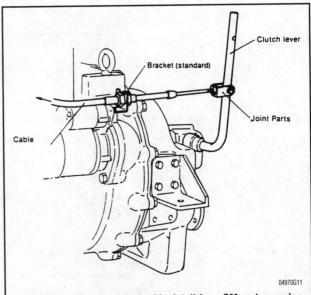

Fig. 13 Transmission control cable detail for a QM series engine with a YP-7M transmission

3. To adjust the cable length, turn the clevis on the cable end threads after the locknut has been loosened.

4. With the locknut still loose, temporarily install the clevis and check that the cable has been adjusted properly by checking the relationship between the control head lever and the transmission control lever. Also check that the cable pushes/pulls the transmission control lever to the reverse position when the control head lever is placed in the full astern position.

5. Once the cable has been adjusted properly, tighten the clevis locknut, and install the clevis on the transmission control lever. Use a new cotter pin.

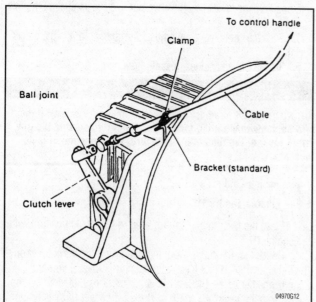

Fig. 14 Transmission control cable mounting detail for a QM series engine with a KBW10 transmission

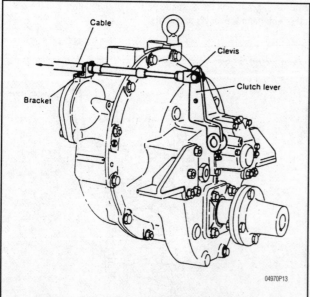

Fig. 15 Transmission control cable mounting detail for a 3QM series engine with a YP-10M transmission

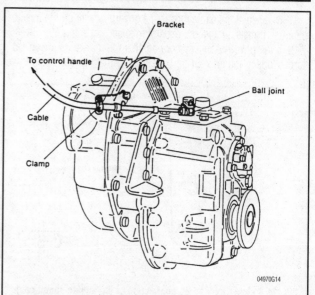

Fig. 16 Transmission control cable mounting detail for a 3QM series engine with a KH-18 transmission

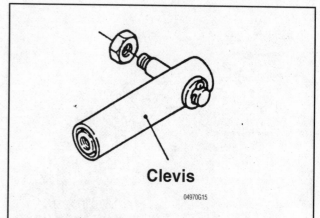

Fig. 17 On some GM series engines, a special spring loaded cable end link is used to control the transmission lever

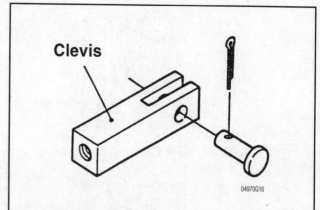

Fig. 18 The transmission used on 3GM/HM series engines uses a solid cable end link

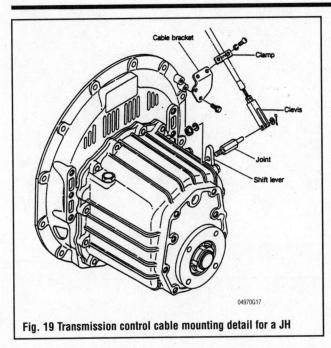

Fig. 19 Transmission control cable mounting detail for a JH

Engine Stop Cable

▶ See Figures 20, 21, 22 and 23

1. Place the engine stop cable in the RUN position, with the engine **OFF**.

2. Check that the cable is pulling/pushing the lever on the injector pump through its full stroke. If you are unsure whether or not the lever is at its full stroke, loosen the set bolt on the cable connector, and push the lever through the rest of its stroke.

3. If the lever moves further, hold the lever in position and tighten the set bolt on the cable connector Do not overtighten the bolt. Once the bolt is tightened, hold it in position while tightening the locknut.

➡ **While adjusting the cable, take the opportunity to apply a fresh coating of grease on the exposed portion of the cable, and the pivoting points on the clevis assembly.**

4. Once the bolt is tightened, place the engine stop cable in the STOP position with the engine **OFF**.

5. Check that the stop lever on the injector pump is at the end of its stroke.

6. If the cable sheath hinders the movement of the lever, use the two locknuts to reposition the sheath, and readjust the cable as necessary.

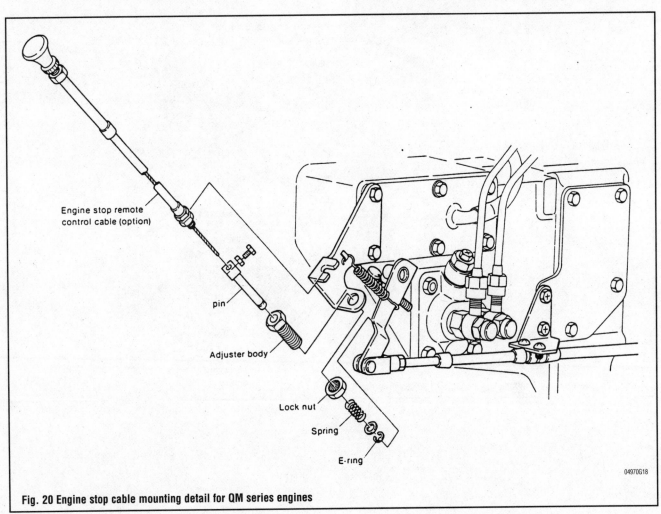

Fig. 20 Engine stop cable mounting detail for QM series engines

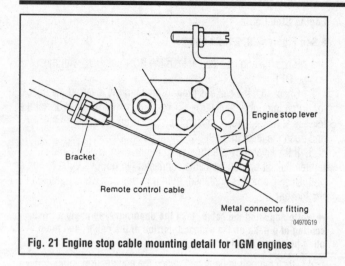

Fig. 21 Engine stop cable mounting detail for 1GM engines

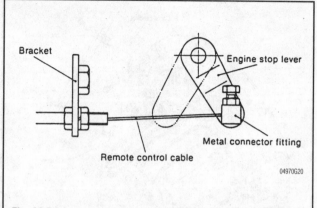

Fig. 22 Engine stop cable mounting detail for 2GM and 3GM/HM series engine

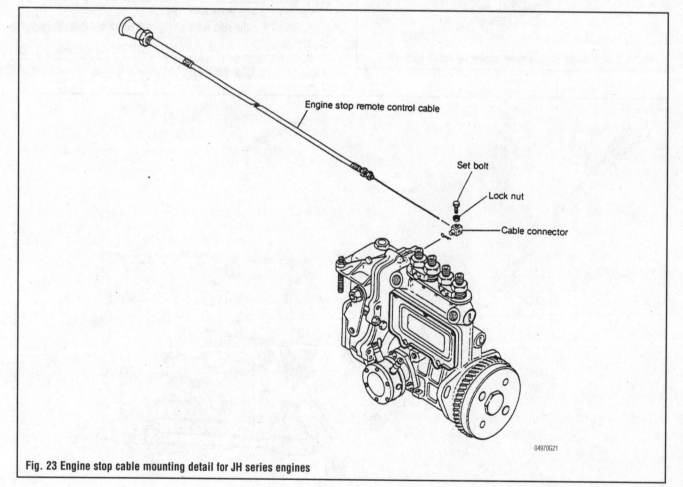

Fig. 23 Engine stop cable mounting detail for JH series engines

GLOSSARY

Understanding your mechanic is as important as understanding your marine engine. Most boaters know about their boats, but many boaters have difficulty understanding marine terminology. Talking the language of boats makes it easier to effectively communicate with professional mechanics. It isn't necessary (or recommended) that you diagnose the problem for him, but it will save him time, and you money, if you can accurately describe what is happening. It will also help you to know why your boat does what it is doing, and what repairs were made.

AFTER TOP DEAD CENTER (ATDC): The point after the piston reaches the top of its travel on the compression stroke.

AIR CLEANER: An assembly consisting of a housing, filter and any connecting ductwork. The filter element is made up of a porous paper or a wire mesh screening, and is designed to prevent airborne particles from entering the engine. Also see Intake Silencer.

The air cleaner assembly consists of a housing, filter and any connecting ductwork

AIR/FUEL RATIO: The ratio of air-to-fuel, by weight, drawn into the engine.

ALTERNATING CURRENT (AC): Electric current that flows first in one direction, then in the opposite direction, continually reversing flow.

ALTERNATOR: A device which produces AC (alternating current) which is converted to DC (direct current) to charge the battery.

AMMETER: An instrument, calibrated in amperes, used to measure the flow of an electrical current in a circuit. Ammeters are always connected in series with the circuit being tested.

AMP/HR. RATING (BATTERY): Measurement of the ability of a battery to deliver a stated amount of current for a stated period of time. The higher the amp/hr. rating, the better the battery.

AMPERE: The rate of flow of electrical current present when one volt of electrical pressure is applied against one ohm of electrical resistance.

ANTIFREEZE: A substance (ethylene or propylene glycol) added to the coolant to prevent freezing in cold weather.

Marine alternators may look like their automotive cousins, but have special spark arresting features to prevent fires onboard

ARMATURE: A laminated, soft iron core wrapped by a wire that converts electrical energy to mechanical energy as in a motor or relay. When rotated in a magnetic field, it changes mechanical energy into electrical energy as in a generator.

ATDC: After Top Dead Center.

ATMOSPHERIC PRESSURE: The pressure on the Earth's surface caused by the weight of the air in the atmosphere. At sea level, this pressure is 14.7 psi at 32°F (101 kPa at 0°C).

ATOMIZATION: The breaking down of a liquid into a fine mist that can be suspended in air.

AXIAL PLAY: Movement parallel to a shaft or bearing bore.

BACKFIRE: The sudden combustion of gases in the intake or exhaust system that results in a loud explosion.

BACKLASH: The clearance or play between two parts, such as meshed gears.

BALL BEARING: A bearing made up of hardened inner and outer races between which hardened steel balls roll.

BATTERY: A direct current electrical storage unit, consisting of the basic active materials of lead and sulphuric acid, which converts chemical energy into electrical energy. Used to provide current for the operation of the starter as well as other equipment, such as the radio, lighting, etc.

BEARING: A friction reducing, supportive device usually located between a stationary part and a moving part.

BEFORE TOP DEAD CENTER (BTDC): The point just before the piston reaches the top of its travel on the compression stroke.

BLOCK: See Engine Block.

Bearings are located between the crankshaft and engine block, and also between the connecting rods and crankshaft

The combustion chamber is the part of the cylinder head where combustion takes place

BLOW-BY: Combustion gases, composed of water vapor and unburned fuel, that leak past the piston rings into the crankcase during normal engine operation. These gases are removed by the evacuation system to prevent the buildup of harmful acids in the crankcase.

BORE: Diameter of a cylinder.

BTDC: Before Top Dead Center.

BUSHING: A liner, usually removable, for a bearing; an anti-friction liner used in place of a bearing.

CAMSHAFT: A shaft in the engine on which are the lobes (cams) which operate the valves. The camshaft is driven by the crankshaft, via a belt, chain or gears, at one half the crankshaft speed.

CARBON MONOXIDE (CO): A colorless, odorless gas given off as a normal byproduct of combustion. It is poisonous and extremely dangerous in confined areas, building up slowly to toxic levels without warning if adequate ventilation is not available.

CETANE RATING: A measure of the ignition value of diesel fuel. The higher the cetane rating, the better the fuel. Diesel fuel cetane rating is roughly comparable to gasoline octane rating.

CHECK VALVE: Any one-way valve installed to permit the flow of air, fuel or vacuum in one direction only.

CIRCLIP: A split steel snapring that fits into a groove to hold various parts in place.

CIRCUIT BREAKER: A switch which protects an electrical circuit from overload by opening the circuit when the current flow exceeds a pre-determined level. Some circuit breakers must be reset manually, while most reset automatically.

CIRCUIT: Any unbroken path through which an electrical current can flow. Also used to describe fuel flow in some instances.

COMBUSTION CHAMBER: The part of the engine in the cylinder head where combustion takes place.

COMPRESSION CHECK: A test involving cranking the engine with a special high pressure gauge connected to an individual cylinder. Individual cylinder pressure as well as pressure variance across cylinders is used to determine general operating condition of the engine.

COMPRESSION RATIO: The ratio of the volume between the piston and cylinder head when the piston is at the bottom of its stroke (bottom dead center) and when the piston is at the top of its stroke (top dead center).

CONDUCTOR: Any material through which an electrical current can be transmitted easily.

CONNECTING ROD: The connecting link between the crankshaft and piston.

CONTINUITY: Continuous or complete circuit. Can be checked with an ohmmeter.

COOLANT: Mixture of water and anti-freeze circulated through the engine to carry off heat produced by the engine.

CRANKCASE: The lower part of an engine in which the crankshaft and related parts operate.

CRANKSHAFT: Engine component (connected to pistons by connecting rods) which converts the reciprocating (up and down) motion of pistons to rotary motion used to turn the driveshaft.

CYLINDER BLOCK: See engine block.

CYLINDER HEAD: The detachable portion of the engine, usually fastened to the top of the cylinder block and containing all or most of the combustion chambers. On overhead valve engines, it contains the valves and their operating parts.

CYLINDER: In an engine, the round hole in the engine block in which the piston(s) ride.

DETONATION: An unwanted explosion of the air/fuel mixture in the combustion chamber caused by excess heat and compression, advanced timing, or an overly lean mixture. Also referred to as "ping".

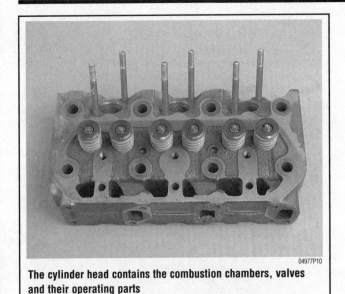

The cylinder head contains the combustion chambers, valves and their operating parts

DIAPHRAGM: A thin, flexible wall separating two cavities, such as in a vacuum advance unit.

DIESELING: The engine continues to run after the it is shut off; caused by fuel continuing to be burned in the combustion chamber.

DIGITAL VOLT OHMMETER: An electronic diagnostic tool used to measure voltage, ohms and amps as well as several other functions, with the readings displayed on a digital screen in tenths, hundredths and thousandths.

DIODE: An electrical device that will allow current to flow in one direction only.

DIRECT CURRENT (DC): Electrical current that flows in one direction only.

DISPLACEMENT: The total volume of air that is displaced by all pistons as the engine turns through one complete revolution.

DOHC: Double overhead camshaft.

DOUBLE OVERHEAD CAMSHAFT: The engine utilizes two camshafts mounted in one cylinder head. One camshaft operates the exhaust valves, while the other operates the intake valves.

DVOM: Digital volt ohmmeter

ELECTROLYTE: A solution of water and sulfuric acid used to activate the battery. Electrolyte is extremely corrosive.

END-PLAY: The measured amount of axial movement in a shaft.

ENGINE: The primary motor or power apparatus of a vessel, which converts fuel into mechanical energy.

ENGINE BLOCK: The basic engine casting containing the cylinders, the crankshaft main bearings, as well as machined surfaces for the mounting of other components such as the cylinder head, oil pan, transmission, etc.

ETHYLENE GLYCOL: The base substance of antifreeze.

EXHAUST MANIFOLD: A set of cast passages or pipes which conduct exhaust gases from the engine.

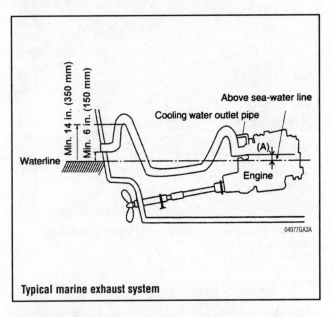

Typical marine exhaust system

FEELER GAUGE: A blade, usually metal, of precisely predetermined thickness, used to measure the clearance between two parts.

FIRING ORDER: The order in which combustion occurs in the cylinders of an engine.

FLAME FRONT: The term used to describe certain aspects of the fuel explosion in the cylinders. The flame front should move in a controlled pattern across the cylinder, rather than simply exploding immediately.

FLAT SPOT: A point during acceleration when the engine seems to lose power for an instant.

FLYWHEEL: A heavy disc of metal attached to the rear of the crankshaft. It smoothes the firing impulses of the engine and keeps the crankshaft turning during periods when no firing takes place. The starter also engages the flywheel to start the engine.

FOOT POUND (ft. lbs. or sometimes, ft. lb.): The amount of energy or work needed to raise an item weighing one pound, a distance of one foot.

FREEZE PLUG: A plug in the engine block which will be pushed out if the coolant freezes. Sometimes called expansion plugs, they protect the block from cracking should the coolant freeze.

FUEL FILTER: A component of the fuel system containing a porous paper element used to prevent any impurities from entering the engine through the fuel system. It usually takes the form of a canister-like housing, mounted in-line with the fuel hose, located anywhere on a vessel between the fuel tank and engine.

FUEL INJECTION: A system that sprays fuel into the cylinder through nozzles. The amount of fuel can be more precisely controlled with fuel injection.

FUSE: A protective device in a circuit which prevents circuit overload by breaking the circuit when a specific amperage is present. The device is constructed around a strip or wire of a lower amperage rating than the circuit it

The fuel filter prevents impurities from entering the engine through the fuel system

04973P50

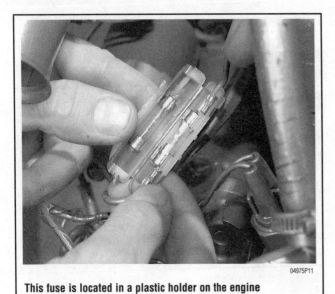

This fuse is located in a plastic holder on the engine

04975P11

is designed to protect. When an amperage higher than that stamped on the fuse is present in the circuit, the strip or wire melts, opening the circuit.

FUSIBLE LINK: A piece of wire in a wiring harness that performs the same job as a fuse. If overloaded, the fusible link will melt and interrupt the circuit.

HORSEPOWER: A measurement of the amount of work; one horsepower is the amount of work necessary to lift 33,000 lbs. one foot in one minute. Brake horsepower (bhp) is the horsepower delivered by an engine on a dynamometer. Net horsepower is the power remaining (measured at the flywheel of the engine) that can be used to power the vessel after power is consumed through friction and running the engine accessories (water pump, alternator, fan etc.)

HYDROCARBON (HC): Any chemical compound made up of hydrogen and carbon. A major pollutant formed by the engine as a by-product of combustion.

HYDROMETER: An instrument used to measure the specific gravity of a solution.

IMPELLER: The portion of the water pump which provides the propulsion for the coolant to circulate it through the system.

INCH POUND (inch lbs.; sometimes in. lb. or in. lbs.): One twelfth of a foot pound.

INJECTOR: A device which receives metered fuel under relatively low pressure and is activated to inject the fuel into the engine under relatively high pressure at a predetermined time.

A typical diesel fuel injector

04974P61

INTAKE MANIFOLD: A casting of passages or pipes used to conduct air or a fuel/air mixture to the cylinders.

INTAKE SILENCER: An assembly consisting of a housing, and sometimes a filter. The filter element is made up of a porous paper or a wire mesh screening, and is designed to prevent airborne particles from entering the engine. Also see Air Cleaner.

JOURNAL: The bearing surface within which a shaft operates.

JUMPER CABLES: Two heavy duty wires with large alligator clips used to provide power from a charged battery to a discharged battery.

JUMPSTART: Utilizing one sufficiently charged battery to start the engine of another vessel with a discharged battery by the use of jumper cables.

KNOCK: Noise which results from the spontaneous ignition of a portion of the air-fuel mixture in the engine cylinder.

LITHIUM-BASE GREASE: Bearing grease using lithium as a base. Not compatible with sodium-base grease.

LOCK RING: See Circlip or Snapring

MANIFOLD VACUUM: Low pressure in an engine intake manifold formed just below the throttle plates. Manifold vacuum is highest at idle and drops under acceleration.

MANIFOLD: A casting of passages or set of pipes which connect the cylinders to an inlet or outlet source.

MISFIRE: Condition occurring when the fuel mixture in a cylinder fails to ignite, causing the engine to run roughly.

MULTI-WEIGHT: Type of oil that provides adequate lubrication at both high and low temperatures.

NEEDLE BEARING: A bearing which consists of a number (usually a large number) of long, thin rollers.

NITROGEN OXIDE (NOx): One of the three basic pollutants found in the exhaust emission of an internal combustion engine. The amount of NOx usually varies in an inverse proportion to the amount of HC and CO.

OEM: Original Equipment Manufactured. OEM equipment is that furnished standard by the manufacturer.

OHM: The unit used to measure the resistance of conductor-to-electrical flow. One ohm is the amount of resistance that limits current flow to one ampere in a circuit with one volt of pressure.

OHMMETER: An instrument used for measuring the resistance, in ohms, in an electrical circuit.

OVERHEAD CAMSHAFT (OHC): An engine configuration in which the camshaft is mounted on top of the cylinder head and operates the valve either directly or by means of rocker arms.

OVERHEAD VALVE (OHV): An engine configuration in which all of the valves are located in the cylinder head and the camshaft is located in the cylinder block. The camshaft operates the valves via lifters and pushrods.

OXIDES OF NITROGEN: See nitrogen oxide (NOx).

PING: A metallic rattling sound produced by the engine during acceleration. It is usually due to incorrect timing or a poor grade of fuel.

PISTON RING: An open-ended ring which fits into a groove on the outer diameter of the piston. Its chief function is to form a seal between the piston and cylinder wall. Most pistons have three rings: two for compression sealing; one for oil sealing.

POLARITY: Indication (positive or negative) of the two poles of a battery.

POWERTRAIN: See Drivetrain.

PPM: Parts per million; unit used to measure exhaust emissions.

PREIGNITION: Early ignition of fuel in the cylinder, sometimes due to glowing carbon deposits in the combustion chamber.

PRELOAD: A predetermined load placed on a bearing during assembly or by adjustment.

PRESS FIT: The mating of two parts under pressure, due to the inner diameter of one being smaller than the outer diameter of the other, or vice versa; an interference fit.

PSI: Pounds per square inch; a measurement of pressure.

PUSHROD: A steel rod between the hydraulic valve lifter and the valve rocker arm in overhead valve (OHV) engines.

RACE: The surface on the inner or outer ring of a bearing on which the balls, needles or rollers move.

RADIATOR: Part of the cooling system for some water-cooled engines. Through the radiator, excess combustion heat is dissipated into the atmosphere through forced convection using a water and glycol based mixture that circulates through, and cools, the engine.

REAR MAIN OIL SEAL: A synthetic or rope-type seal that prevents oil from leaking out of the engine past the rear main crankshaft bearing.

RECTIFIER: A device (used primarily in alternators) that permits electrical current to flow in one direction only.

REGULATOR: A device which maintains the amperage and/or voltage levels of a circuit at predetermined values.

RELAY: A switch which automatically opens and/or closes a circuit.

RESISTANCE: The opposition to the flow of current through a circuit or electrical device, and is measured in ohms. Resistance is equal to the voltage divided by the amperage.

RESISTOR: A device, usually made of wire, which offers a preset amount of resistance in an electrical circuit.

ROCKER ARM: A lever which rotates around a shaft pushing down (opening) the valve with an end when the other end is pushed up by the pushrod. Spring pressure will later close the valve.

ROLLER BEARING: A bearing made up of hardened inner and outer races between which hardened steel rollers move.

RPM: Revolutions per minute (usually indicates engine speed).

RUN-ON: Condition when the engine continues to run, even when the key is turned off. See dieseling.

SENDING UNIT: A mechanical, electrical, hydraulic or electromagnetic device which transmits information to a gauge.

SENSOR: Any device designed to measure engine operating conditions or ambient pressures and temperatures. Usually electronic in nature and designed to send a voltage signal to an on-board computer, some sensors may operate as a simple on/off switch or they may provide a variable voltage signal (like a potentiometer) as conditions or measured parameters change.

SHIM: Spacers of precise, predetermined thickness used between parts to establish a proper working relationship.

SHORT CIRCUIT: An electrical malfunction where current takes the path of least resistance to ground (usually through damaged insulation). Current flow is excessive from low resistance resulting in a blown fuse.

SINGLE OVERHEAD CAMSHAFT: See overhead camshaft.

SLUDGE: Thick, black deposits in engine formed from dirt, oil, water, etc. It is usually formed in engines when oil changes are neglected.

SNAP RING: A circular retaining clip used inside or outside a shaft or part to secure a shaft, such as a floating wrist pin.

SOHC: Single overhead camshaft.

SOLENOID: An electrically operated, magnetic switching device.

SPECIFIC GRAVITY (BATTERY): The relative weight of liquid (battery electrolyte) as compared to the weight of an equal volume of water.

SPLINES: Ridges machined or cast onto the outer diameter of a shaft or inner diameter of a bore to enable parts to mate without rotation.

STARTER: A high-torque electric motor used for the purpose of starting the engine, typically through a high ratio geared drive connected to the flywheel ring gear.

STROKE: The distance the piston travels from bottom dead center to top dead center.

TACHOMETER: A device used to measure the rotary speed of an engine, shaft, gear, etc., usually in rotations per minute.

TDC: Top dead center. The exact top of the piston's stroke.

THERMOSTAT: A valve, located in the cooling system of an engine, which is closed when cold and opens gradually in response to engine heating, controlling the temperature of the coolant and rate of coolant flow.

A typical marine thermostat

TOP DEAD CENTER (TDC): The point at which the piston reaches the top of its travel on the compression stroke.

TORQUE: Measurement of turning or twisting force, expressed as foot-pounds or inch-pounds.

TUNE-UP: A regular maintenance function, usually associated with the replacement and adjustment of parts and components in the electrical and fuel systems of a engine for the purpose of attaining optimum performance.

TURBOCHARGER: An exhaust driven pump which compresses intake air and forces it into the combustion chambers at higher than atmospheric pressures. The increased air pressure allows more fuel to be burned and results in increased horsepower being produced.

VALVE CLEARANCE: The measured gap between the end of the valve stem and the rocker arm, cam lobe or follower that activates the valve.

VALVE GUIDES: The guide through which the stem of the valve passes. The guide is designed to keep the valve in proper alignment.

VALVE LASH (clearance): The operating clearance in the valve train.

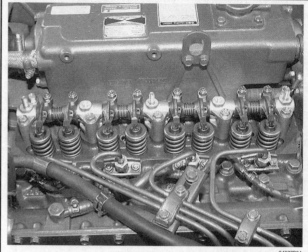

The valve train on this diesel engine is clearly visible

VALVE TRAIN: The system that operates intake and exhaust valves, consisting of camshaft, valves and springs, lifters, pushrods and rocker arms.

VALVE: A device which control the pressure, direction of flow or rate of flow of a liquid or gas.

VISCOSITY: The ability of a fluid to flow. The lower the viscosity rating, the easier the fluid will flow. 10 weight motor oil will flow much easier than 40 weight motor oil.

VOLT: Unit used to measure the force or pressure of electricity. It is defined as the pressure

VOLTAGE REGULATOR: A device that controls the current output of the alternator or generator.

VOLTMETER: An instrument used for measuring electrical force in units called volts. Voltmeters are always connected parallel with the circuit being tested.

WATER PUMP: A belt driven component of the cooling system that mounts on the engine, circulating the coolant under pressure.

A marine water pump mounted to the front of a diesel engine

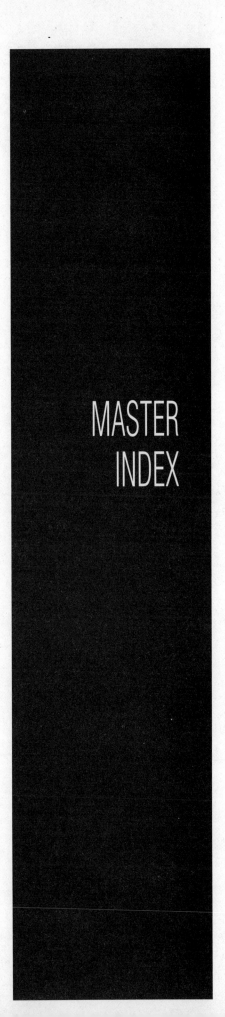

MASTER

INDEX

2P1VerA

Total Car Care, continued

Pick-Ups and Montero 1983-95
PART NO. 8666/50500

NISSAN
Datsun 210/1200 1973-81
PART NO. 52300
Datsun 200SX/510/610/710/
810/Maxima 1973-84
PART NO. 52302
Nissan Maxima 1985-92
PART NO. 8261/52450
Maxima 1993-98
PART NO. 52452
Pick-Ups and Pathfinder 1970-88
PART NO. 8585/52500
Pick-Ups and Pathfinder 1989-95
PART NO. 8145/52502
Sentra/Pulsar/NX 1982-96
PART NO. 8263/52700
Stanza/200SX/240SX 1982-92
PART NO. 8262/52750
240SX/Altima 1993-98
PART NO. 52752
Datsun/Nissan Z and ZX 1970-88

PART NO. 8846/52800
RENAULT
Coupes/Sedans/Wagons 1975-85
PART NO. 58300
SATURN
Coupes/Sedans/Wagons 1991-98
PART NO. 8419/62300
SUBARU
Coupes/Sedan/Wagons 1970-84
PART NO. 8790/64300
Coupes/Sedans/Wagons 1985-96
PART NO. 8259/64302
SUZUKI
Samurai/Sidekick/Tracker 1986-98
PART NO. 66500
TOYOTA
Camry 1983-96
PART NO. 8265/68200
Celica/Supra 1971-85
PART NO. 68250
Celica 1986-93
PART NO. 8413/68252

Celica 1994-98
PART NO. 68254
Corolla 1970-87
PART NO. 8586/68300
Corolla 1988-97
PART NO. 8414/68302
Cressida/Corona/Crown/MkII 1970-82
PART NO. 68350
Cressida/Van 1983-90
PART NO. 68352
Pick-ups/Land Cruiser/4Runner 1970-88
PART NO. 8578/68600
Pick-ups/Land Cruiser/4Runner 1989-98
PART NO. 8163/68602
Previa 1991-97
PART NO. 68640
Tercel 1984-94
PART NO. 8595/68700
VOLKSWAGEN
Air-Cooled 1949-69
PART NO. 70200

Air-Cooled 1970-81
PART NO. 70202
Front Wheel Drive 1974-89
PART NO. 8663/70400
Golf/Jetta/Cabriolet 1990-93
PART NO. 8429/70402
VOLVO
Coupes/Sedans/Wagons 1970-89
PART NO. 8786/72300
Coupes/Sedans/Wagons 1990-98
PART NO. 8428/72302

Total Service Series

ATV Handbook
PART NO. 9123
Auto Detailing
PART NO. 8394
Auto Body Repair
PART NO. 7898
Automatic Transmissions/Transaxles
Diagnosis and Repair
PART NO. 8944
Brake System Diagnosis and Repair
PART NO. 8945
Chevrolet Engine Overhaul Manual
PART NO. 8794
Easy Car Care
PART NO. 8042
Engine Code Manual
PART NO. 8851
Ford Engine Overhaul Manual
PART NO. 8793
Fuel Injection Diagnosis and Repair
PART NO. 8946
Motorcycle Handbook
PART NO. 9099
Small Engine Repair
(Up to 20 Hp)
PART NO. 8325
Snowmobile Handbook
PART NO. 9124

Collector's Hard-Cover Manuals

Auto Repair Manual 1993-97
PART NO. 7919
Auto Repair Manual 1988-92
PART NO. 7906
Auto Repair Manual 1980-87
PART NO. 7670
Auto Repair Manual 1972-79
PART NO. 6914
Auto Repair Manual 1964-71
PART NO. 5974
Auto Repair Manual 1954-63
PART NO. 5652

Auto Repair Manual 1940-53
PART NO. 5631
Import Car Repair Manual 1993-97
PART NO. 7920
Import Car Repair Manual 1988-92
PART NO.7907
Import Car Repair Manual 1980-87
PART NO. 7672
Truck and Van Repair Manual 1993-97
PART NO. 7921
Truck and Van Repair Manual 1991-95
PART NO. 7911

Truck and Van Repair Manual 1986-90
PART NO. 7902
Truck and Van Repair Manual 1979-86
PART NO. 7655
Truck and Van Repair Manual 1971-78
PART NO. 7012
Truck Repair Manual 1961-71
PART NO. 6198

System-Specific Manuals

Guide to Air Conditioning Repair and
Service 1982-85
PART NO. 7580
Guide to Automatic Transmission
Repair 1984-89
PART NO. 8054
Guide to Automatic Transmission
Repair 1984-89
Domestic cars and trucks
PART NO. 8053

Guide to Automatic Transmission
Repair 1980-84
Domestic cars and trucks
PART NO. 7891
Guide to Automatic Transmission
Repair 1974-80
Import cars and trucks
PART NO. 7645
Guide to Brakes, Steering, and
Suspension 1980-87
PART NO. 7819

Guide to Fuel Injection and Electronic
Engine Controls 1984-88
Domestic cars and trucks
PART NO.7766
Guide to Electronic Engine Controls
1978-85
PART NO. 7535
Guide to Engine Repair and Rebuilding
PART NO. 7643
Guide to Vacuum Diagrams 1980-86
Domestic cars and trucks
PART NO. 7821

Multi-Vehicle Spanish Repair Manuals

Auto Repair Manual 1992-96
PART NO. 8947
Import Repair Manual 1992-96
PART NO. 8948
Truck and Van Repair Manual
1992-96
PART NO. 8949
Auto Repair Manual 1987-91
PART NO. 8138
Auto Repair Manual
1980-87
PART NO. 7795
Auto Repair Manual
1976-83
PART NO. 7476

2P2VerA